REVOLUTION IN TIME

Clocks and the Making of the
Modern World

DAVID S. LANDES

THE BELKNAP PRESS OF
HARVARD UNIVERSITY PRESS
Cambridge, Massachusetts, and London, England

Library of Congress Cataloging in Publication Data

Landes, David S.
Revolution in time.

Includes index.
1. Clocks and watches—History. 2. Horology—History.
I. Title.
TS542.L24 1983   681.1′13′09   83-8489
ISBN 0-674-76800-0 (cloth)
ISBN 0-674-76802-7 (paper)

In memory of teachers and mentors:
Arthur Harrison Cole
Donald Cope McKay
Sir Michael Moyse Postan
Abbott Payson Usher

Contents

III MAKING TIME

Illustrations

Preface

THIS BOOK goes back to a ritual wine tasting in Berkeley, California, almost twenty-five years ago. It was Tuesday evening, and we economic historians had just finished one of our weekly team lectures. There was Carlo Cipolla, "medieval and early modern," bon vivant, *bec fin,* oenologist, he of the custom-tailored suits with the *real* sleeve buttons. Carlo would show us how the buttons opened—for what reason, no one could say. A vestigial reminder of some earlier stage in the history of costume. And there was Henry Rosovsky, an early rider on the wave of the Japanese economic miracle. He would one day be dean of the Faculty of Arts and Sciences at Harvard—the biggest academic job in the United States; but then he was a young scholar and a genius at teaching by disagreement. And Albert Fishlow, a specialist in United States history, younger than all of us and, appropriately, a representative of the new, quantitative, econometric history. We urged him to learn the economic history of Latin America, and he got snared in the toils of Brazilian development. The world was no doubt better off, but not economic history. We all argued so much in class that the students could not understand how we remained such good friends.

That evening Carlo pulled out—without warning—a beautiful gold pocket watch and made it chime. "What's that?" we asked. "A striker watch." "What's a striker watch?" "A watch that chimes the time"—to the last minute (or sometimes the last five minutes or quarter, but none of us knew that) on demand: one ring for each hour, double rings for each quarter, and single, lighter rings for each minute. I had never seen such a mechanical marvel and vowed that one day I would have one.

The years passed. I had one or two opportunities to buy what I had now learned to call by its technical name, a repeater watch. But I never had the ready cash. Then in 1972 I was teaching in Paris. And I remembered that watch. If I don't get one now, I told my wife, I'll never get one; or if I do, I won't have time to enjoy it. So I found a shop that specialized in antique watches, R. Laforêt's *Au Vieux Cadran* on the rue Bonaparte. It's gone now and so alas is he, but in its day it was a Mecca for collectors. There I spent a lot of money for a large, complicated timepiece that could do all kinds of interesting things. It wasn't signed, and I later learned that signed watches (signed anything) are worth more. But it looked beautiful to me, and I called it Superwatch.

I didn't sleep that night. Not for pleasure at my acquisition but for guilt at spending so much. On myself, no less. Was this reasonable behavior for the father of a family? The next morning I did what I should have done before I bought the watch: I went comparison shopping.

That proved reassuring. The important thing, though, is that in the course of this expedition, I chanced to run into a French *collector* of antique watches. Adolphe Chapiro was (and is) a chemist, *directeur de recherches* with the French Conseil National de la Recherche Scientifique; but he was also a top-flight horologist. He told me I was just plain lucky. "You must never do that, you know—buy an expensive watch without knowing anything about watches. You can make terrible mistakes that way."

Know about watches? The idea had never occurred to me. I'd always owned a watch, yet never thought to find out why and how it worked. Dereliction of duty: I'd spent years studying and writing about technological and industrial development, visiting museums and walking forge and factory sites. I'd listened to the din of weaving sheds and admired the fireworks of a Bessemer converter. I'd even stood inside the blast furnace where in 1709 Abraham Darby first smelted iron ore with coal—the brick cradle of the Industrial Revolution. Yet I had never paid any attention to clocks and watches. Few historians had.

"How do you learn about watches?" I asked. "There are books." "Books? What books?" So Chapiro gave me some titles, and I bought one or two straightway. That was an experience: I couldn't even understand half the technical words. But I learned to read the diagrams.

Shortly thereafter we were in London, then and now the capital of the trade in antique timepieces. There are lots of little shops in Mayfair and environs that offer really interesting watches two hundred and more years old. And there's no better way to learn about these things than to handle them, open them, look at them under the loupe. "You must be here for tomorrow's sale," ventured one merchant. "Tomorrow's sale?" "Yes, at Sotheby's."

At Sotheby's, I watched dealers working through trays of watches at dizzying speed. They knew just what to do, how to open the tricky catches, just what to look for; and they took little cryptic notes to guide them in their bidding. To me they were denizens of a strange, closed world. They all seemed to know one another. (They did.) One look at me, and they knew I was a rank outsider. I tried to hear what they said, but I've since learned that it's better not to listen: you can't believe a word.

Watch auctions in London are largely for the trade. Some of the dealers work in a ring: they do not compete with one another at the sale, but have a private knock-out auction later on. It's hard on the owners of the watches offered, but kind to the purses of the dealers. That day the dealers let me buy a small lot—an American watch, inscribed in memory of the San Francisco earthquake and fire. Again, an unsigned piece; but I liked the association to a place where I had lived.

That did it: I was smitten—caught by the combination of mechanical ingenuity, craftsmanship, artistry, and elegance. Coming from a utilitarian culture, I could not get over the talent and time that must have gone into the making of these objects—the devotion that expressed itself in painstaking finish, even in normally invisible details. And I was equally taken by the people: the dealers, some of whom could have served as curators in the best museums, others who did not know much about horology but had an unerring nose for price; the habitués of the salerooms, with their favorite seats and distinctive mannerisms—they might have stepped out of the pages of Balzac or the prints of Daumier; the artists and craftsmen, holdovers from an earlier age, men of golden fingers, mechanical geniuses some of them, and they know it; and the collectors, whose passion and devotion underlie the entire structure. To enter a new world such as this is to drink at the fountain of youth. How could I resist?

In the years that followed I looked at and handled some thousands of watches and read some thousands of pages. Instead of weekly magazines and paperback fiction, I took watch books and catalogues to bed with me. I soon learned that there is an abundant literature on watches and their history, by collectors for collectors; also much good technical work by and for watchmakers and horologists; but very little by and for general historians. The material on the cultural, social, and economic aspects of time measurement is there, more than any one person can ever absorb. But it is scattered in archives, monographs, and specialized journals and has to be pulled together. Little by little, what had started as a hobby became a subject of ever more systematic investigation.

In 1978 I was invited to teach at the University of Zurich and the Eidgenössische Technische Hochschule of that city and decided to venture a seminar in the history of time measurement. That may have been the first time such a course was offered in a Swiss institution of higher education, and it took some temerity for an American to bring such coals to Newcastle. But I learned a lot from the experience, not only because teaching is the best way to learn anything, but because the students in the seminar did excellent work.

Zurich was followed that fall by three talks at the University of Virginia—the Richard lectures. These helped crystallize my interest and gave me the idea for this book.

I finished a first version of this manuscript in the spring of 1982 and used it as the basis of lectures at the University of Geneva and a seminar at the Ecole des Hautes Etudes en Sciences Sociales in Paris in the fall of that year. Once again, the teaching made all the difference: new material, new contacts, suggestive responses. I substantially rewrote parts of the manuscript and felt elated at what I perceived as marked improvement; but also concerned that so much improvement had apparently been possible. The inference was inescapable: if I could work on the text longer, it would be even better.

Still, like it or not, there comes a time for parturition. The book represents a first attempt at a general history of time measurement and its contribution, for better or worse, to what we call modern civilization. It is a triptych: a study in cultural history (why was the mechanical clock invented in Europe?); in the history of science and technology (how did we get from crude, approximate timekeepers to instruments of high precision?); and in social and economic history (who made these instruments? how? who used them? why?). The inquiry took me into new and unexpected paths and places—to such topics as religion and folklore, mathematics and mechanics, astronomy and navigation; and to such scenes as the courts of the Great Khan and the Holy Roman Emperor, the pretelescopic observatories of Renaissance Europe, the learned societies of the Ancien Régime, the deadly route of the Manila galleon, the fiercely silent chronometric combats at the observatories in Kew, Geneva, and Neuchâtel; from the cluttered workroom of the *cabinotier* of Geneva and the mountain shop of the craftsman of Neuchâtel, to the many-windowed factory buildings of Waltham, Massachusetts, and Elgin, Illinois, to the horological sweatshops of Southeast Asia.

One will understand, I think, that such an odyssey has been intellectually exhilarating, almost to the point of inebriation. Most exciting has been the sense that much of this is new, that I have been exploring a subject whose importance has not saved it from neglect. This, it is fair to say, is the rarest of coincidences. History is an old and diligent discipline, and most topics have been studied to a fare-thee-well. To come upon a major aspect of the development of modern society, economy, and civilization that is still largely uncharted territory is a rare strike. People have asked me how it is that so important a subject has been so little studied. I'm not sure that I know the answer; but this I hope: that time measurement will never again be so ignored. Once historians are sensitized to the problem, they should find far more—in archives of course, but also in well-known secondary works—than we can imagine. And if they don't, they should ask, like Sherlock Holmes, why not.

This book, then, is only prologue.

THE LIST of people who have helped me learn these divers subjects and write this book is very long, and with the best of will, I am not going to be able to remember them all. But I shall do my best and ask those I inadvertently overlook to forgive me. I hope the book itself will be an expression of their knowledge and generosity. Insofar as it falls short of their rightful expectations, the fault is obviously mine. They did their best to teach me.

First, the horologists, curators, dealers, collectors:

—in France: Adolphe Chapiro, Jean-Claude Sabrier, Anthony Turner, Catherine Cardinal, Jean-G. Laviolette, Charles Jacob, Henri Vidal, and the members of the Tuesday lunch club that gathers *chez* Michel Journe;

—in England: John Combridge; Andrew Crisford, Simon Bull, and Sebastian Whitestone of Bobinet Ltd.; George Daniels; David Penney; Tony Mercer; Richard Good (British Museum); Beresford Hutchinson (National Maritime Museum in Greenwich); Vaudrey Mercer; Cecil Clutton; Terence Camerer Cuss; Donald R. Hill; Charles Allix; John D. North (also of Groningen); Chris Ellmers (Museum of London); Geoffrey Greetham; Peter K. Weiss; Michael Denton;

—in Switzerland: François Jequier (University of Lausanne); André Curtit, Jean-Pierre Chollet, and the staff of the Musée International d'Horlogerie, La Chaux-de-Fonds; Margarida Archinard (Musée d'Histoire des Sciences, Geneva); Fabienne-X. Sturm, Anne Winter-Jensen, and staff of the Musée de l'Horlogerie et de l'Emaillerie, Geneva; Gérard Bauer (Centre Electronique Horloger); René Retornaz (Fédération Horlogère); Robert Chapuis; Henri-Daniel Piguet-Aubert; Simone Oppliger; Jacques Ketterer (Vacheron & Constantin); Jean Lebet (Le Coultre et Cie); Erwin Bernheim (Mondaine Watch Ltd); M. Panicali fils; André Gabus (Institut Battelle, Geneva); Theodor Beyer; Edgar Mannheimer; Peter Ineichen; and not least my students in Zurich and Geneva;

—in Germany: Klaus Maurice (Bayerisches Nationalmuseum, Munich); Gerhard Dohrn–van Rossum (University of Bielefeld); Horand M. Vogel;

—in Austria: Hans von Bertele;

—in Italy: Giuseppe Brusa;

—in Israel: Gabriel Moriah and Ohannes Markarian (L. A. Mayer Memorial Institute); Joseph Goodman;

—in the United States: Derek de Solla Price; William Andrewes and Seth Atwood (Time Museum, Rockford, Illinois); Henry Fried; Efroim Greenberg; William Scolnik; Joseph Conway; Jan Skala; Stephen Bogoff; Bruce Chandler; Winthrop Edcy; Otto Mayr and Deborah Warner (Smithsonian Institution); Dana J. Blackwell; Clare Vincent (Metropolitan Museum of Art); Michael Edidin; Donald Hoke; Richard B. Gardner; Joseph O'Sullivan; George C. Kenney and the other members of the American Section of the Antiquarian Horological Society; Wilbur and Kathy Pritchard and other members of the National Association of Watch and Clock Collectors.

I also received all kinds of help from nonhorologists, especially in fields of history and science where I was an outsider: Rudolf Braun, Hansjürg Siegenthaler, and other colleagues at the University of Zurich; Jean-François Bergier of the Eidgenössische Technische Hochschule; Paul Bairoch, Anne-Marie Piuz, Jean Starobinski, and other colleagues at the University of Geneva; Jacques Le Goff, François Furet, and others at the Ecole des Hautes Etudes in Paris; Glen Bowersock, Mason Hammond, and Zeph Stewart on points of Latin and ancient history; Giles Constable, Ihor Sevcenko, David Herlihy, and Richard A. Landes in medieval history; John K. Fairbank, Benjamin Schwartz, Beatrice S. Bartlett, Merle Goldman, and Robin Yates in Chinese history; William Bouwsma and Natalie Zemon Davis in Renaissance and early modern; Norman F. Ramsey, Thomas Kuhn, and William F. Hughes in matters of science; Patrice Higonnet, Prudence Steiner, Melvin Lasky, Eric Robinson, Merritt Roe Smith, Françoise Crouzet, Pierre Lérys, Robert Fogel, Abram Bergson, Zvi Griliches, John H. Weiss, Peter Hertner, Guido Goldman, Morton Keller, Eric Robinson, Norma Farber, Rabbi Ben-Zion Gold, and others who kept an eye out for and sent me clippings and quotations in matters horological; Walker Cowen of the University of Virginia; Ebenezer Gay of Historical Scientific Instruments at Harvard; and in more ways than they know, the members of the Economic History Workshop in Harvard University.

I have been exceptionally fortunate in my collaborators: Elly

Solmitz, Jan Bedau, Rebecca Ramsay, Carol Murphy—readers and typists who have so identified with the work as to amend and enrich it; the staffs in the Department of Economics and the Committee on Social Studies at Harvard; staff in Zurich, Geneva, and Paris, including unknown bringers of books and records; Armand Dionne and Richard McDonald of the Cruft Photo Laboratory at Harvard; William Minty, draftsman-cartographer; above all, my editors, especially Arthur Rosenthal and Aida Donald, who believed in the book and did everything to further it, and Maria Ascher, keen of ear and a hawk for mistakes, who saw the manuscript through, amended it, and saved me from all manner of pitfalls.

Finally, I owe an immeasurable debt of thanks to family—to Pierrette Coadou; to ever-interested parents and children; to grandchildren who thought when they were small that a watch was not a watch unless it chimed; above all to my wife and best friend Sonia, who was ready to hear about clepsydras and escapements in the small hours of the morning. She has always encouraged and shared my interest in time and its measurement, partly no doubt because the same museums, shops, and salerooms that show antique watches hold other beautiful objects. Synergy.

The clock is not merely a means of keeping track of the hours, but of synchronizing the actions of men.

The clock, not the steam-engine, is the key-machine of the modern industrial age ... In its relationship to determinable quantities of energy, to standardization, to automatic action, and finally to its own special product, accurate timing, the clock has been the foremost machine in modern technics; and at each period it has remained in the lead: it marks a perfection toward which other machines aspire.

LEWIS MUMFORD, *Technics and Civilization*

Use time, or time will use you.

OLD PROVERB

Introduction

"I KNOW WHAT TIME IS," said Augustine, "but if someone asks me, I cannot tell him." Things have not changed very much since then. The ordinary man (or woman) thinks he knows what time is but cannot say. The learned man, physicist or philosopher, is not so sure he knows but is ready to write volumes on the subject of his speculation and ignorance.

The ordinary man couldn't care less. What matters to him is that he can measure time. If, like the vast majority of the world's people, he lives in a rural society, his time is measured for him by natural events: daybreak, sunrise, high noon, sunset, darkness. He needs no more accurate division, for these are the events that demarcate his round of waking, working, eating, sleeping. The sequence of tasks fills the day, and when night falls and the animals are cooped or stabled, parents and children eat their evening meal and go tired to bed, to wake the next morning with the birds and beasts and start another day.

City dwellers measure time by the clock. Animals do not wake them; an alarm does. Their activities are punctuated by points on an abstract continuum, points designated as hours and minutes. If they have a job or class that starts, say, at nine o'clock, they try to get there *on time*. They have ap*point*ments, and these are fixed by points on the time scale.

Picture an immensely complicated and unevenly but often densely trafficked marshalling yard, with components shifting and shunting about in all directions; only instead of trains directed from without, we have people, sometimes directed but more often self-steering. That is the world of social and personal interaction, which works only because the member units have learned a common language of time measurement. Without this language and without general access to instruments accurate enough to provide uniform indications of location in time, urban life and civilization as we know it would be impossible. Just about everything we do depends in some way on going and coming, meeting and parting.

Indications, of course, are not enough. Knowledge of the time must be combined with obedience—what social scientists like to call time discipline. The indications are in effect commands, for responsiveness to these cues is imprinted on us and we ignore them at our peril. Punctuality (the quality of being *on the point*) is a virtue, lateness a sin, and repeated lateness may be cause for dismissal. The sense of punctuality is inculcated very early, indeed from infancy. Parents may feed their babies on demand, but their own schedules inevitably impinge on the consciousness of their children. As soon as children understand language, they pick up such notions as mealtime and bedtime. A child whose parents live and work by the clock soon learns that time is the most inexorable of disciplinarians. It passes slowly for children; but it passes, and waits for no one. It compels the laggard to hurry, for what one member of the family does with time affects the others. One of the most powerful notions to shape a child's consciousness is that of being late or of missing (the two notions are sometimes equated, which says something about the price of lateness)—missing a program, missing a plane, missing a meal, missing a religious service, missing a ball game, missing a party.

Most people operate within a margin of plus or minus several minutes. If they have a train to catch, they get to the station a few minutes early; likewise for appointments. For this range of tolerance, it is sufficient to check one's timepiece by radio and television announcements given to the latest minute. (It would be imprudent to rely on public tower clocks: they were more reliable when there was no such thing as radio and they really mattered.) Only the most precise persons will want to know the time to the latest ten-second interval as given by the telephone, or to the latest second, as given by certain radio stations that broadcast nothing but time signals; and few people will possess instruments, whether watches or clocks, that allow them to use this information in tracking the time.

For some purposes, however, hours and minutes are not enough. Astronomers were the first to want to measure time in seconds and fractions of a second, well before instruments existed that could do the job, even before such small divisions had been defined for general use. In the fifteenth century, observers began using time as a coordinate of location of celestial bodies, and where the indicators on clock dials could not furnish sufficiently accurate information, they counted the teeth of the turning wheels. The larger the wheels, the more the teeth and the finer the divisions.

From that point on, astronomy was always in the forefront of the demand for more precise and accurate timekeepers, and observatory "regulators" were typically the most discriminating available—the standard for all others. With the invention of the pendulum clock in the seventeenth century, it was possible for the first time to build timepieces accurate to less than a minute a day. A century later, observatory clocks could keep time within a fraction of a second a day, and by the end of the nineteenth century this variation had been reduced to a hundredth of a second or less. By this time, clocks were accurate enough to reveal irregularities in the motion of what had always been the timekeeper of final appeal, the earth itself. With the introduction in the twentieth century of quartz and then atomic regulators, scientists have abandoned all dependence on heavenly measures and established as the fundamental time unit the second, defined not as a fraction of a year but as a large number (over nine billion) of very rapid oscillations. The basis of time measurement has thus shifted from celestial to quantum mechanics. Astronomers meanwhile have continued to work with terrestrial days and years; but days and years are no longer uniform, and we now must intercalate leap seconds (as we continue to insert leap days) in order to make the calendar conform to the human experience of time and tide.[1]

Astronomers have been primarily concerned with tracking time over long intervals. By that I mean hours and minutes, days and years. They may want to measure such periods with great precision, even to small fractions of a second. Their work is such, however, that they need a timekeeper that can maintain accuracy more or less indefinitely, day in and day out. That is very different from time measurement over short intervals. The doctor who wants to count a pulse, the seaman who needs to time the log as it floats past from stem to stern, the horse breeder who wants to know how fast a mount has run a quarter-mile—all of these want a special kind of timekeeper, known as a *chronograph* (or sometimes *chronoscope*), whose primary function is to measure elapsed time. Such a timekeeper need not be accurate day in and day out, need not eliminate some of the sources of error that would be fatal to the performance of an astronomical regulator, but its user must be able to start and stop it at will; and since, for very small units of time, the slowness of human neural and muscular reactions produces more error than the timepiece itself, chronographs are often linked to electric or photoelectric triggers and synchronized with a camera that records both motion and duration, to a fraction of a second. (See Figure 1.)

The world of sports provides a good example of this kind of ever finer time chopping.[2] When Oxford and Cambridge met in March 1864 for the world's first dual track meet, the races were timed in quarter-seconds: the officials were presumably using watches beating 14,400 times an hour—not uncommon for English watches. In the decades that followed, stopwatches counting fifths became standard. This was the measure yielded by a so-called fast watch train—18,000 beats an hour. It was also long considered the smallest unit compatible with the responses of human timers.

Still, the desire to distinguish among close finishers provided a constant stimulus to finer timing. At Stockholm in 1912 the Olympic Games experimented with photographic-electric timers clocking tenths of seconds; then in 1924 in Paris they introduced instruments that could resolve hundredths. To what avail? There are few people outside the realm of organized religion who are as conservative as sports officials. For decades the International Amateur Athletic Federation refused to treat these results as valid. The human eye continued to pick the winner. Times were established by a comparison of handheld watches and adjusted in the direction of plausibility—that is, rounded up to the nearest fifth. If two of the watches happened to agree, that was treated as statistical proof of accuracy. The finer, electric readings were used only to rank the also-rans. Not until the 1932 Olympics were times given in tenths, and not until 1960 at Rome were handheld times abandoned and electric results accepted as official.[3] Even then, the winner was still chosen by the human eye, with predictable consequences. In the hundred-meter men's swim sprint, an American touched home first, if the electric clocks are to be believed; but the gold medal went to an Australian, who seems to have caught the judges' eyes. The Americans protested in vain: if one is to rely on human judges, it is very important that they never admit to error. But this was the last time that such a contretemps (if I may be permitted a pun) was allowed to happen. The Olympic Games have enough controversy in those many events whose scoring depends entirely on human judgment (diving, figure skating, gymnastics, boxing) without looking for trouble in areas where impersonal measurement can be used.

The introduction of automatic timers revealed the optimistic bias of human observation. Races as measured by the new technique invariably ran slightly slower than those timed by handheld chronographs. Clearly, human judges, fearful of underrating contestants and eager to share in the glow and glory of record-breaking performances, had a tendency to jump the gun, or rather the finish. The result has been two sets of record standards: one for handheld times, the other for automatic times. Eventually, as records are broken and times are brought further down, the older set will disappear.

The desire to avoid dead heats and make possible an ordering of even the closest finishes has led to the introduction of still finer units of measurement. In the 1970s the Olympic Games began using timers calibrated in hundredths of a second that were linked to automatic registers at start and finish. Then, in the winter Olympics at Lake Placid in 1980, the inevitable occurred: the victor in an hours-long cross-country ski race was separated from the second-place finisher by a hundredth of a second. The observer may reasonably ask: Isn't this really a dead heat? Is the margin of error for two pairs of readings taken hours apart significantly smaller than a hundredth of a second? Whatever the answer, the order of finish stood as timed, and one can confidently anticipate the introduction of timers calibrated in thousandths and ten thousandths with a view to avoiding this kind of chronometric and philosophical dilemma.[4]

The demands of sport, of course, are as nothing compared to those of some branches of science. The physicist who seeks to assign times to subatomic events is like the medieval astronomer whose subject matter was running ahead of his instruments. When one enters the world of subatomic particles, one leaves hundredths and thousandths of seconds far behind. This is the world of microseconds (10^{-6}), nanoseconds (10^{-9}), and picoseconds (10^{-12}). These are units invented for purposes of theoretical analysis. Episodes of short duration can in fact be observed and timed by recording and measuring their tracks, as in a cloud chamber, if velocity is known; or can be measured by converting the phenomenon being timed into vibrations (oscillations) and then counting them. Indeed, to the physicist, any stable oscillating phenomenon *is* a clock, and the best of them (the ones with the most stable and highest frequencies) are now the most accurate and precise measuring devices known to science. Today, when the finest measures are desired, such quantities as voltage, mass, and magnetic force are no longer measured in conventional ways but reduced to a frequency determination and then calibrated.[5]

ALL OF THESE extraordinary capabilities of tracking and measuring time in the long run and the short, which pervade and permit every aspect of our lives and work, go back to the invention of the mechanical clock in medieval Europe. This was one of the great inventions in the history of mankind—not in a class with fire and the wheel, but comparable to movable type in its revolutionary implications for cultural values, technological change, social and political organization, and personality.

Why so important? After all, man had long known and used other kinds of timekeepers—sundials, water clocks, fire clocks, sand clocks—some of which were at least as accurate as the early mechanical clocks. Wherein lay the novelty, and why was this device so much more influential than its predecessors?

The answer, briefly put, lay in its enormous technological potential. The mechanical clock was self-contained, and once horologists learned to drive it by means of a coiled spring rather than a falling weight, it could be miniaturized so as to be portable, whether in the household or on the person. It was this possibility of widespread private use that laid the basis for *time discipline,* as against *time obedience.* One can, as we shall see, use public clocks to summon people for one purpose or another; but that is not punctuality. Punctuality comes from within, not from without. It is the mechanical clock that made possible, for better or worse, a civilization attentive to the passage of time, hence to productivity and performance.

Much of what follows will be devoted to exploring these issues. First, however, I should like to clarify the technical character of the innovation. Without that, the rest is incomprehensible.

1. The mechanical clock was weight-driven. This made it (unlike the water clock) impervious to frost, which was no small matter in northern climes,[6] and kept it (unlike the sundial) working through the night and on cloudy days, an immense advantage in a region where it was not uncommon to go weeks without seeing the sun. To be sure, weights had been used before in horological devices, notably as a drive for secondary systems such as alarm mechanisms. But this was the first time they had been used to power the clock itself, where their constant force was a necessary (though not sufficient) condition of accurate timekeeping.

2. The mechanical clock transmitted the energy of the falling weight through a *gear train* (wheels and pinions)—what later generations were to call clockwork. Such trains were already known to the ancients and were occasionally employed in the Middle Ages as adjuncts to water clocks (more on this later); but now for the first time the timekeeping itself was effected by wheelwork. The gear train made possible a tight coupling of the power source with the rest, resulting in a more efficient transmission of energy and a more precise translation of time measures into time indications.

3. All of this was as nothing, of course, without the measuring mechanism itself. This was the Great Invention: the use of oscilla-

tory (continuous back-and-forth, to-and-fro) motion to track the flow of time. One would have expected something very different—that time, which is itself continuous, even, and unidirectional, would be best measured by some other continuous, even, and unidirectional phenomenon.[7]

This indeed is what the water clock does, or the sundial; and this is what contemporary Chinese horologists reasoned. Chinese water clocks were probably the most advanced timekeepers of that day, and to their builders their performance seemed logically to derive from their very nature: "There have been many systems and designs for astronomical instruments during past dynasties all differing from one another in minor respects. But the principle of the use of water power for the driving mechanism has always been the same. The heavens move without ceasing but so does water flow [and fall]. Thus if the water is made to pour with perfect evenness, then the comparison of the rotary movements [of the heavens and the machine] will show no discrepancy or contradiction; for the unresting follows the unceasing."[8]

Yet for reasons that are worth exploring, it has never been possible to keep anything moving at a continuous and even pace that even approaches the steadiness needed to track time. Instead, the secret of keeping an accurate rate has lain in generating some kind of repeating beat and counting the beats—that is, in summing series of equal, discrete parts (the digital principle). In general, the faster the beat, the smaller the parts and the more accurate the measure. The earliest clocks (c. 1300) had beats that lasted several seconds. Five hundred years later, the marine chronometer made do with two beats a second or five every two seconds. In our time, we have precision watches using quartz crystals that vibrate a hundred thousand times a second and more; and the most accurate timekeepers of all, the atomic clocks of the national observatories, use quartz crystal oscillators with frequencies of some 2.5 megacycles per second checked by cesium-beam resonators vibrating at 9,192,631,770 ± 20 cycles per second.[9]

All of this was prefigured in the first mechanical clock: all modern timekeepers, whatever their construction, are based on the same oscillating-digital principle. To be sure, the builder of that first mechanical clock had no idea what he had wrought. But that is often the way of technological advance: the plant cannot be envisioned from the seed.

In the mechanical clock as invented in the Middle Ages, the oscillatory timekeeper was combined with that part of the mechanism known as the escapement. Functionally these are two different things. The *oscillatory device* tracks the passing moments; it is what the Germans call a *Zeitnormal,* or time standard; we call it a controller or a regulator. It beats time. The *escapement* counts the beats (pulses) by blocking and releasing the wheel train at a rhythm dictated by the regulator. It does this in such a way that the wheel train moves equal distances (angular distances) in equal intervals and drives the hands or other indicators of the passing hours at the appropriate rate.

For today's physicist, the time standard is everything; the counting is a trivial problem to be left to the technicians. It would be anachronistic, however, to apply this invidious comparison to the process of horological invention in medieval Europe. Let me try to reconstruct hypothetically the range of possibilities open to the mechanicians of that era by way of illuminating the significance of their solution.

The task was to use the force of gravity to turn a wheel or wheels at a stable rate. This was a new kind of problem. So long as the timekeeper was a *clepsydra,* or water clock—that is, a vessel that takes in or gives out water in an even flow—one could avoid machinery altogether by reading the time directly from a scale marking the height of the water; or, if one wanted to drive wheels, one could use a float and toothed rod to communicate the thrust of rising water to a train. (See Figure 2.) This latter method went back to antiquity and was practiced in both Europe and Asia. Its principal limitation (aside from all the difficulties posed by low temperatures) was its physical weakness: for a variety of reasons that need not concern us here, it could not drive heavy mechanisms. One answer to that shortcoming was found by combining the force of gravity with the mechanical advantage of the wheel. Buckets were placed at the extremities of the spokes of a wheel and filled in turn by a stream from the clepsydra. When each bucket weighed enough, it tripped a release that allowed the wheel to rotate, thereby turning a drive shaft and bringing the next bucket into position. Depending on the size of the water buckets and of the wheel, such a system could drive very heavy trains indeed. (See Figure 3.) It was apparently invented in China, where it was used to power some extraordinarily complicated mechanisms, and it was probably known in Islam, though we have no explicit reference to its applications.[10] It does not seem to have been used in Europe.

By going over to weight drive, the Europeans transformed the terms of the problem within the context of their technology. The task now was not *to advance the reading or the mechanism regularly,* whether by continuous flow or thrust (float devices), but *to hold it back.* The reason is simple: an unbalanced force continuously applied will produce acceleration, and falling weights, if unopposed, will drive a wheel train ever faster until the weights reach the ground. The search was on, then, for some kind of controller that would curb the motion of a turning wheel so that it would rotate once in twenty-four hours. In the much-cited words of Robertus Anglicus, writing in 1271: "Clockmakers are trying to make a wheel that will move exactly as [*omnino . . . secundum*] the motion of the equinoctial circle."[11]

To this end, clockmakers presumably experimented with both braking and block-release devices. Among the former was the mercury-weighted wheel described in the *Libros del Saber de Astronomia* prepared for King Alfonso X of Castile in 1276. A drum-shaped wheel divided into segmental compartments and partially filled with mercury would be turned by a weight; the mercury, which would flow from one compartment to the next as the wheel turned, would act as an inertial brake on the speed of rotation. There is no evidence that such a clock was ever realized or that any other inertial or braking arrangement ever worked. Even with today's technology, it is very difficult to secure stable braking effects over time. Medieval craftsmen must have found it impossible.

It was the block-release, stop-go mechanisms that proved most effective, reliable, and accurate and hence swept the field. These mechanisms—we know of two varieties (see Figures 4, 5, and A.1)—are surely among the most ingenious inventions in history. There is no precedent for them, and their ability to combine several critical functions is testimony either to extraordinary conception or good fortune. First, by dividing time into discrete beats or pulses, they made it possible to count the passing moments and translate the count into units of time. Second, by blocking and releasing the train, they transmitted this count to the visual or auditory indicators (hands, dials, moving figures, bells). Third, also by blocking and releasing, they conserved and rationed the energy of the *prime mover* (weight or spring).

Multiplicity of function entailed variation in nomenclature. The English, French, Italians, and Spanish chose to call such a mechanism an "escapement" (French *échappement,* Italian *scappamento,* Spanish *escape*), thereby stressing the stop-go aspect. The Germans called it a *Hemmung,* a brake or impediment. This difference in sense and emphasis ordinarily does not pose any difficulty to students of the subject: a rose by any other name . . . so long as it is still a rose. But trouble arises when either term is applied inappropriately—when, for example, the mercury-flow inertial arrangement of the *Libros del Saber* is described as an escapement,[12] or when the same word is applied to very different kinds of mechanisms, such as the Chinese intermittent water-bucket escapement on the one hand and the European oscillating-beat escapement on the other. We will examine the consequences of this latter confusion in greater detail below.

One more aspect of nomenclature: the timekeeping unit of the early mechanical clock was both controller and escapement. The two were linked and influenced (controlled) each other, and the term "escapement" has understandably been applied to the whole device. Yet this terminology has contributed to a serious misunderstanding of the nature and significance of the new machine. Contrary to just about all the literature on the subject, the achievement of that unknown genius who built the first mechanical clock was not so much that he had used an escapement—momentous as this was—as that he had made use of oscillatory motion to divide time into countable beats. This was the great invention; this was the heart of the clock. By so doing he set time measurement on a new path and planted the seed of all subsequent improvements in chronometry. Nothing done elsewhere is comparable.

THE STORY that follows is divided into three parts. The first poses the question of how and why so seminal an invention occurred in Europe and remained a European monopoly for some five hundred years. This is not what an overview of the technological map would have led one to expect. Medieval Europe was anything but the scientific and industrial leader it would one day become; specifically in horology it stood far behind China and Islam. The invention of the mechanical clock was one of a number of major advances that turned Europe from a weak, peripheral, highly vulnerable outpost of Mediterranean civilization into a hegemonic aggressor. Time measurement was at once a sign of new-found creativity and an agent and catalyst in the use of knowledge for wealth and power. Part I, then, is primarily an exercise in cultural history, a study in comparative values and their implications for social action.

Part II is an essay in the history of science and technique. It tells the story of the improvement of the clock as instrument, from a big, crude, highly approximate, unreliable measuring device to a compact, portable, dependable, and long-lasting precision time-keeper. It took five hundred years to go from the turret clocks of the Middle Ages to marine chronometers and precision watches, from public time to private time, from dials with one hand to dials with three. The campaign to accomplish these gains enlisted the efforts of the finest craftsmen and greatest scientists in Europe. No other single project in applied science has ever drawn on so much intelligence and talent. And with good reason. Not only was much of commerce and navigation dependent on accurate timekeeping, but so, more and more, was scientific investigation. Every study of process, change, and movement needed a time measure in the denominator.

Part III—the longest—deals with the people who made clocks and watches and the way they made them. This is an exercise in economic history, the story of changing techniques of manufacture and modes of production. It is a fascinating story, the more interesting because we can follow it from beginning to end. The mechanical clock/watch, after some seven hundred years of domination, is now yielding to solid-state timekeepers using quartz-crystal controllers. Mechanical timepieces are still being made by the tens of millions every year, and they still have their uses and their devotees. But the quartz timekeeper is a superior instrument in terms of both precision and price, and it is bound to win out.

So we have a rare opportunity here to study the birth, maturity, and obsolescence of a major branch of manufacture. It is, surprisingly, a story that scholars have neglected: the books that have dealt with this subject as economic history can be counted on the fingers of both hands.[13] This is the more deplorable because this branch has not only been important in and for itself but for its influence on other branches of manufacture. One can think of few objects that have played so critical a role in shaping the character of life and work as clocks and watches. And no branch of production has done so much as this "nursery of the mechanical arts" to teach others the use of tools and machines and the advantages of division of labor.[14]

Furthermore, this is a branch of manufacture that differs from almost all others by its freedom from locational constraints. The reasons are obvious. First, the raw materials used in clock- and watchmaking represent a small fraction of total cost; hence there is no need to be near the source of supply. Second, the manufacturing process itself uses little fuel or energy; hence no need to be near coal or waterpower. And finally, the transport of the light final product is easy and relatively inexpensive; hence no need to be near the market. In short, clocks and watches can be made anywhere—anywhere, that is, where skilled hands can be guided and supervised by ingenious technicians and creative designers. The industry is therefore a marvelous laboratory for the study of the human (entrepreneurial and labor) factor in industrial achievement. This is an aspect of the larger story of economic development that many economists and economic historians have been reluctant to credit; all the more reason to give it careful consideration in those circumstances where it has been of special importance.

I   FINDING TIME

The question to ask is: Why clocks? Who needs them? After all, nature is the great time-giver (*Zeitgeber*), and all of us, without exception, live by nature's clock. Night follows day; day, night; and each year brings its succession of seasons. These cycles are imprinted on just about every living thing in what are called circadian ("about a day") and circannual biological rhythms. They are stamped in our flesh and blood; they persist even when we are cut off from time cues; they mark us as earthlings.[1]

These biological rhythms are matched by societal work patterns: day is for labor, night for repose, and the round of seasons is a sequence of warmth and cold, planting and harvest, life and death.

Into this natural cycle, which all peoples have experienced as a divine providence, the artificial clock enters as an intruder. For example, in ancient Rome:

> The gods confound the man who first found out
> How to distinguish hours. Confound him, too,
> Who in this place set up a sundial,
> To cut and hack my days so wretchedly

15

Into small pieces! When I was a boy,
My belly was my sundial—one surer,
Truer, and more exact than any of them.
This dial told me when 'twas proper time
To go to dinner, when I ought to eat;
But nowadays, why even when I have,
I can't fall to unless the sun gives leave.
The town's so full of these confounded dials . . .[2]

And yet the sundial is the most natural of clocks, for it simply registers the movement of nature's prime timepiece. In essence, it is a schematization of the tree that casts a shadow and thus tracks the passing day. Since our unhappy Roman thought sundials a plague, what would he have said about mechanical clocks, going night and day, sky cloudy or clear, keeping an equal beat and beating equal hours in all seasons? "By its essential nature," wrote Lewis Mumford, the clock "dissociated time from human events."[3] To which I would add: and human events from nature. The clock is a *machine,* a work of artifice, a man-made device with no model in nature—the kind of invention that needed planning, thinking, trying, and then more of each. No one could have stumbled on it or dreamed it up. But someone, or rather some people, wanted very much to track the time—not merely to know it, but to use it. Where and how did so strange, so *unnatural* a need develop?

1   A Magnificent Dead End

IN THE YEAR 1086 of our era, the boy emperor of China, or more probably his ministers, ordered an examination of the astronomical devices inherited from previous reigns and the construction of an astronomical clock that should surpass any that had been built before. He chose for this task a certain Su Sung, an experienced diplomat and administrator with scientific interests.[1] Su Sung gathered a team of officials, technicians, and craftsmen, naming as designer and superintendent of construction a certain Han Kung-lien, then "a minor official in the Ministry of Personnel." (In China there was no horological profession, and talent was to be found in unexpected places.) Two years later a working wooden model was finished; two years more, and the metal parts of the great clock were cast in bronze; four more years (1094), and the "explanatory monograph" was ready for presentation to the emperor, and presumably the clock with it.

Su Sung's clock, the third in a series of astronomical clocks beginning with Chang Ssu-hsun's in 979, was one of the marvels of the age.[2] (See Figure 6.) It was designed to reproduce the movements of the "three luminaries"—sun, moon, and (selected) stars—which were crucial to Chinese calendrical calculation and astrological divination.[3] It did this by means of an observational *armillary sphere*—that is, an assemblage of rings representing the paths of these bodies as they presented themselves to an observer on Earth—and a demonstrational celestial globe, each rotating on a polar axis and appropriately inclined to the horizon. These in turn were driven off a pair of vertical transmission shafts, one of which also bore a series of superimposed wheels six to eight feet in diameter. These wheels carried jacks, little manikins that revolved with wheel and shaft in measured pace and showed the hours, the "quarters" (*k'o,* each of which equaled fourteen minutes twenty-four seconds of our time), and the night watches by means of plac-ards. The whole mechanism, which must have weighed tons and occupied a tower about forty feet high, was powered by a water wheel designed to turn intermittently at a stable rate. Small buckets attached by pivots to the rim of the wheel were succes-sively filled from a clepsydra. When the weight of the water in the bucket was sufficient, it tipped a lever that released the wheel and allowed it to rotate, but only so far as to bring the next bucket into position. It was the clepsydra, of course, that kept time—that served, in other words, as the controller, while the fill-and-tip arrangement divided the flow of water into countable units and allowed relatively fine adjustment of the rate.[4]

Su Sung's astronomical clock apparently gave service for some few years. But in 1126 invaders from the north (the Chin Tartars) captured the Sung capital of K'aifeng and carried away what they could of Su Sung's clock. It is not clear from the account whether they were able to reconstitute it completely, water-wheel escapement and all; probably not. No matter; most of the mecha-nism "broke or wore out after some years." The Chin, for reasons unspecified, were then not able to replace the broken parts and re-store the clock. (Probably only Su Sung and Han Kung-lien could have done that.) All that remained was the heavy bronze armil-lary sphere, which must have been turned by hand, and this was badly damaged in a storm. Some years thereafter the Chin court had to flee before the invading Mongols. It was proposed that the armillary sphere be melted down "to make things, but the em-peror did not have the heart to destroy it. On the other hand its bulk was so large it would have been difficult to transport it on a cart, so in the end it was left behind."[5]

What happened to the marvelous sphere, we do not know. The

rings no longer turned smoothly, and this was enough to render it useless. Meanwhile, efforts by the fled Sung emperors to have the great clock duplicated failed: there were no astronomers to design it or mechanics to build it, perhaps because the Chin had taken off the best workmen to their capital in the north. The son of Su Sung could find no clue to the mechanism in the family books and papers. Soon nothing was left of Su Sung's clock but the latent memory.[6] The illustrated book he had prepared for the emperor was lost, then found in south China and reprinted in 1172; but apparently both original and reprint were then lost from sight, for when the Jesuits brought their clocks to China in the late sixteenth century, the technical achievements of Su Sung and his predecessors had been long forgotten. Then, in the seventeenth century, a scholar who happened to own a copy of the edition of 1172 brought it out again. This version, too, faded from public knowledge, and Su Sung's memoir had to wait until the beginning of the nineteenth century for its next appearance, which was followed by two or three further printings.

None of these, it is clear, achieved his resurrection as a horologist, for when scientist and historian of science Joseph Needham turned his attention to Chinese time measurement in the years after World War II, he brought to his study the traditional judgment of Chinese incompetence in this domain. This was the verdict of the Jesuits who first brought the mechanical clock to China in the sixteenth century. They said that all Chinese timekeepers were inaccurate, even the sundials, which were made as though all the world lived at thirty-six degrees of latitude, "neither more nor less";[7] and they have been echoed by any number of Western observers since. Thus Anton Lübke, author of "Altchinesische Uhren" and "Chinesische Zeitmesskunde," maintained that "the Chinese never made any discoveries comparable to those of Europeans in the technique of clockmaking."[8] And Mathieu Planchon, who wrote a standard history of the clock in the 1890s, stated: "The Chinese have thus not produced any mechanical clockwork properly so called—in this field they have only been bad imitators."[9]

Needham, author of an epic corpus of work on the history of Chinese science and technology, was the last person to let anyone, least of all a European, get away with that kind of scornful dismissal of Chinese talent and achievement. When historian of science Derek de Solla Price came to him in 1954 "with a bright idea about a Chinese mention of medieval mechanical water clocks," that was all the clue Needham needed. There followed what Price calls "the most hectic and intense scholarship" he had ever experienced:[10] within a year, Needham was able to lecture to audiences about the long-forgotten great tower clocks of emperors older than Kublai Khan. In Xanadu . . .

To appreciate the scope of this achievement, one must understand that Needham and his colleagues not only had to hunt up long-forgotten and widely scattered allusions to these astronomical water clocks, many of these references tantalizingly fragmentary, but to learn a new language—to tease out the meaning of superficially familiar ideographs used in unfamiliar senses. Of all these sources, the most valuable was Su Sung's book with its diagrams of the mechanism and detailed description of its operation. Nothing else like it was known. Withal there was a great deal of work to do before Needham and his associates could be reasonably sure they had worked out the meaning, and it was not until models of the timekeeping mechanism were actually built that some of the uncertainties could be resolved. These models were designed by John Combridge, a professional engineer in what was then the Engineering Department of Britain's General Post Office (now British Telecommunications), an amateur horologist so dedicated to his favorite subject that he learned Chinese in order to read the primary sources in the original.[11] There are still uncertainties: one can never be sure in reconstruction work how much is a faithful expression of the original design and how much a contemporary adaptation or emendation. Both Combridge and the Time Museum in Rockford, Illinois, have had to add to these models features that do not show in the record; the clocks will not work without them. Still, thanks to the devoted labor of these scholars and others, including a team in China, we have now filled in an important chapter in the history of technology, one that in Needham's words does much to qualify (undermine) "the opinion that a mechanical penchant was always characteristic of occidental, but not of oriental civilisation," and "that traditional Chinese culture was static or stagnant."[12]

Unfortunately, Needham, in his eagerness to restore the good name of Chinese horology, was not content with what must surely rank among the great detective coups of historical scholarship. He was convinced that there was more to these astronomical water-wheel clocks than the culmination of the hydraulic line of horological development. That would be a dead end, however magnificent. He preferred to see them, rather, as a beginning, as forerunners of the European mechanical clock.

Now this is optative history, the story of what should have been. There is, as Needham concedes, no evidence whatsoever linking the two horological traditions. There is not even a bond of logic or method between them. The Chinese measured time by the continuous flow of water; the Europeans, by the oscillating movement of a *verge-and-foliot*. Both techniques used escapements, but these have only the name in common. The Chinese one worked intermittently; the European, in discrete but continuous beats. Both systems used gravity as the prime mover, but the action was very different. In the mechanical clock, the falling weight exerted a continuous and even force on the train, which the escapement alternately held back and released at a rhythm constrained by the controller. Ingeniously, the very force that turned the scape wheel then slowed it and pushed it part of the way back—the effect that horologists call *recoil*: the scape wheel, in pushing the blocking pallet out of the way, by that very action turned the verge' and brought the other pallet in to slow it (the scape wheel), then push it back, thereby renewing the sequence. In other words, a unidirectional force produced a self-reversing action—about one step backward for three steps forward. In the Chinese timekeeper, however, the force exerted varied, the weight in each successive bucket building up until sufficient to tip the release and lift the stop that held the wheel in place. This allowed the wheel to turn some ten degrees and bring the next bucket under the stream of water while the stop fell back. (The closest analogy to this technique in mechanical horology is the *remontoir*, a device invented around 1600 for accumulating energy to a given trigger level and then releasing it to the train.) In the Chinese clock, then, unidirectional force produced unidirectional motion.

This comparison should not be seen as invidious, in either direction. Each technique had its advantages. The Chinese water-wheel clock, we shall see, was probably far more accurate than the early mechanical clocks. Its escapement mechanism was essentially independent of the train, hence unperturbed by its irregularities; and the use of the weight of the bucket as a trigger made possible a fine adjustment of the rate, for in that day—and indeed up to the very recent past—weighing was the most accurate and discriminating measure known. (Compare the experience of Galileo, who five hundred years later needed the finest possible time measurements in his experiments on acceleration; so he converted time into weight by running a clepsydra and weighing the outflow after each trial.)

On the other hand, the European clock had a much brighter future. Aside from its usability in all times and weather, it was susceptible of miniaturization, to the point of eventual portability. (There was no way a clepsydra would work if moved about.) The clock made possible, therefore, private as against public, general as against hieratic or royal time. And as we have seen, it held in germ performances far more precise than any water clock could provide. To be sure, some hundreds of years were to pass before this potential was realized. The point, however, is that this potential was embodied in the system—in the use of an oscillating controller—and that such a time standard was not available in a hydraulic timekeeper. The Chinese water clocks had come about as far as they could, whereas the mechanical clock marked the beginning of a new technology.

In any event, neither the systemic differences nor the absence of corroborating evidence discouraged Needham. Using a variant of *post hoc, ergo propter hoc* reasoning, he speculated that rumors of the Su Sung clock somehow filtered westward over a period of a hundred or more years and struck a spark. Needham calls this kind of free-floating inspiration a diffusion stimulus: "Perhaps we can imagine some early fourteenth-century scholar-craftsman puzzling over the tale that in the East they used a set of oscillating levers, tripping and holding back a wheel so as to regulate its turning."[13] The word "oscillating" is, of course, inappropriate here. It should be reserved for continuous to-and-fro motion, like that of a balance wheel or pendulum. It is not applicable to the kind of intermittent mechanism used in the Chinese water-wheel clocks, and its use by Needham in this sense may explain why he thought the rumor of the one could serve as stimulus and inspiration for the other.

To understand this figment, one has to place it in the larger context of Needham's philosophy of history. He makes a point of finding Chinese ideas and devices that can be presented as "anticipations or precedents or conjectured influences" in the one larger stream of what he calls ecumenical science. His concern has been not only to redress "the balance-sheet of indebtedness among the cultures of the Old World," but to search out those "nodal points of discovery and invention" that constitute permanent contributions to "the edifice of human knowledge." He is hostile to the idea of separate streams of culture. For Needham, "the whole march of humanity in the study of Nature [is] one single enterprise."[14]

This is a noble credo, but it should not be allowed to get in the way of historical accuracy. "Perhaps we can imagine . . ."? The historian may imagine what he will. Needham himself was persuaded. For him the only question was whether the Chinese escapement came to Europe "in person, or only as a rumour." Nor could there be doubt concerning the origins of the mechanical clock: it "owes its existence largely to the art of Chinese millwrights." This sublime conviction, this assertion of the hypothesis as fact, has inevitably had some influence on the subsequent writing of horological history. My own sense of the literature, though, is that most students of the subject have withheld assent, even comment. Joseph Needham is too great a scholar not to be taken seriously, but what does one do with his wishes become affirmation? Carlo Cipolla sees all this as fancy. I agree with him.[15]

Needham's preoccupation with this *fantaisie* is the more unfortunate because it has diverted him from the real issue: if the Chinese were so far ahead of the Europeans, if their water wheels embodied in germ the mechanical escapement, then why did the Chinese not invent it?[16] This is, in fact, Needham's kind of question—a variant of what Kenneth Boulding described as "the Needham problem": Why did science originate in Europe and not in China, which seemed so much readier for it?[17] The question is the more pertinent because horology is only one of several areas in which the technology of medieval China was ahead of that of Europe: it was China, after all, that gave us paper, gunpowder, movable type, porcelain, and other important and ingenious products.[18] Recall in this regard the wonderment of Marco Polo, who was no country bumpkin but a citizen of Europe's greatest trading city, Venice.

Clearly, the Chinese had every incentive to develop a truly mechanical clock, for in their climate they had the same problems with clepsydras that Europeans did, only more so. Temperatures often fell too low, and even when clocks did not freeze, they behaved very differently in cold weather and hot, in winter and summer. An early account tells us of an attempt to keep the water in a clepsydra warm by surrounding the instrument with heated water.[19] But this solution did not take hold, for reasons that do not appear in the record. (It is not hard to imagine difficulties posed by such an arrangement.) In later experiments the Chinese tried using sand or mercury instead of water, but sand does not flow smoothly enough or mercury cleanly enough to keep a good rate over time. Besides, mercury kills.[20]

Here, incidentally, one should distinguish both China and Europe from Islam. Muslims were concerned to know the time, if only for religious reasons, and they loved to build highly animated water clocks that told the hours by sound and spectacle— clocks far in advance of anything that medieval Europe could make. When Harun al-Rashid, the legendary caliph of Baghdad, wanted to impress Charlemagne, he sent him one of these wonder clocks. It was the admiration of the Frankish court, but that was all: no one had the courage or skill to copy it. In its way, then, it was just as representative of technological superiority as were the striking clocks the Jesuits used to win the favor of the Chinese emperor some eight hundred years later. But Muslim clockmakers had little trouble with cloudy skies or freezing temperatures; their sundials and clepsydras together gave them good service in all seasons. Hence, they had little incentive to invent a new technique of time measurement.[21]

Anyone looking, though, at the world's techniques of time measurement in, say, the eleventh century would have given odds that the Chinese would develop a mechanical clock well before the Europeans. Needham even permits himself a little rhetoric on this point: "The mechanised astronomical instrument was trembling on the verge [a pun?] of becoming a purely time-keeping machine. Inaudible echo must have answered 'A clock!' "[22] But the clock did not come. Chinese horological techniques stood still, then retrogressed. That the mechanical clock did appear in the West, and with it a civilization organized around the measurement and knowledge of time, is a critical factor in the differentiation of the West from the Rest and the rise of Europe to technological and economic hegemony.

NOW LET US SEE what light history, in particular Needham's own evidence, sheds on this paradox.

The first thing one has to remember is that it is not "natural" to want to know the precise time—that is, time as expressed in hours, minutes, and subminutes. We take such a need for granted, because hundreds of years of time discipline have inured us to it. But the historian would be guilty of gross anachronism if he assumed such an interest on the part of our distant (and not-so-distant) ancestors. To understand the invention of improved instruments of time measurement, one has first to ask: Who wanted them?

In T'ang and Sung China (eighth to twelfth centuries), almost no one wanted them—perhaps not even the astronomers. The vast majority of the population lived on the land and regulated life and work, as country dwellers everywhere are wont to do, by the diurnal round of natural events and chores. In places where the peasants farmed collectively, a drum might summon the people to labor in the fields, but this does not imply the existence of temporal obligations or points of reference. Such signals were presumably themselves irregular, dictated by nature, weather, and the varying requirements of agriculture, conforming not to schedule but to opportunity and circumstances. They were not so much a sign of punctuality as a substitute for it.

The pattern of work in the cities was little different. There, too, the craftsman awoke with the dawn and the animals and worked as long as natural light or oil lamps permitted. In the typical household workshop, one person, usually the newest apprentice, would "sleep on one ear," wake before the others, start the fire, get the water, then get the others up; and the same person would usually shut things down at night. Productivity, in the sense of output per unit of time, was unknown. The great virtue was busyness—unremitting diligence in one's tasks.[23]

This is not to say that Chinese society was indifferent to time or its measurement. Even villages, we are told, might have a clepsydra: any watertight vessel could presumably be adapted for this purpose and would keep approximate time so long as temperature permitted. Whether such devices were used to mark the actual time or simply the passage (duration) of time—as with sand clocks—we do not know. In the neighborhood of Buddhist monasteries, it was often the monks who provided an itinerant time service, usually combined with the collection of alms. And in larger agglomerations, there were the night watchmen, calling out as they made their way through the streets. We read that an apprentice charged with opening the shop should get up at the sound of the fifth watch, the last of the night; and that the K'ang Hsi emperor (1662–1722) prided himself on rising at that matutinal hour.[24]

It was in the largest places, as always, that time was most important. An Arab traveler of the mid-ninth century, Sulaiman al-Tajir, writes in an oft-translated passage: "Each city has four gates, at each of which there are five trumpets, which are sounded by the Chinese at certain hours of the day and night. There are also in each city ten drums, which they beat at the same time, the better to show publicly their loyalty to the emperor, as well as to give knowledge of the hours of the day and the night; and they are also furnished with sun-dials and clocks with weights."[25]

This was during the T'ang dynasty (618–906), when urban timekeeping seems to have been carefully regulated from above. In the capital at sunset, a drumroll of eight hundred beats signaled the closing of the gates to each of the walled neighborhoods that made up the city—no missing that kind of noise. And then, at the fifth and last night watch, drums beat again, first in the palace, then in all the streets, followed by the opening of the gates to the neighborhoods and markets. Such loud punctuation was designed to set strict bounds on urban movement and activity: the T'ang code provided for sixty-six blows to any subject who transgressed the curfew. The night was for officials and the military.[26]

That was one set of regulations, and it is hazardous to generalize from so little. China, after all, was not a static, unchanging society. In the long space of some thousands of years, it knew an alternation of tight centralization and political fragmentation, territorial expansion and invasion from without, native rulers and foreign conquerors, ecumenical peace and civil war. Besides, in no society are rules and practice congruent. Local records, for example, tell us of the development under the T'ang emperors of extramural quarters (*faubourgs*) that escaped some of these strict controls; and of the institution of a night market in the capital in the eighth century. Under the Sung (960–1279) and Mongol (1280–1368) dynasties, there flourished large, commercial cities, business rather than mere administrative centers, richer and more active than anything known in Europe at the time. These cities seem to have cracked the rigid bureaucratic mold, in part no doubt because the court found it necessary or desirable to grant foreign merchants and financiers exceptional opportunities as traders, purveyors, and fiscal agents.

These were the kind of people who must have been interested in time measurement. The sources, however, tell us little if anything on the subject, and that silence may in itself be evidence of indifference or satisfaction. Perhaps the public time signals were adequate. If so, there can have been little ordering of activity (appointments) around them, for their meaning was limited by the fact that in China the hours were not numbered serially. Instead they bore names, so that there was no unambiguous way to announce each hour by, say, the beats of a drum or clangs of a gong. On the other hand, the ordinary subject of the emperor had his own devices to know the hour, however roughly. Not the clepsydra: that might well serve for official time, but it did not lend itself to private use. It needed too much attention, too much space, or both. But there were fire clocks (graduated fuses) for intermittent timekeeping, and of course the sundial, with all its limitations.

In the long run, this kind of "rugged individualism" never took root. It is no accident that it was closely linked to alien operators and foreign trade. Chinese society, specifically the elite of mandarin literati, never trusted private enterprise and uncontrolled self-enrichment. Business was by definition "monkey business," and outsiders were potential subversives. Even in laxist times, the state never abandoned its right to tax, expropriate (usually via forced "loans"), or seize the fortunes of its subjects. The very success of these mercantile cities and their free-wheeling residents was reason to curb them. The richest trades were taken over by the state; the most active streams of commerce with the outside world were shut down. By the time Europeans came to China in some numbers in the sixteenth century, two hundred years of indigenous rule by the Ming had screwed the bureaucratic lid on tight.[27]

These European visitors, as we have seen, had only contempt for Chinese timekeeping, which seemed primitive by comparison with Western techniques. They reserved particular scorn for Chinese sundials, which they described as having a vertical *style* or *gnomon* (the shadow-casting indicator), hence as unadjustable for latitude; and as having an equinoctial scale (all hours equal) that was useless most of the year.[28] It is not clear why they found these dials so deficient. A vertical gnomon will make shadows whose angular movement across the dial will be more or less uniform from day to day; the sector covered will change as days lengthen or shorten, but the angles will map the solar hours. Unless the Chinese were marking time in temporal (unequal) hours, which does not seem to have been the case, these sundials should have given good service.[29] If they had not, it is hard to imagine that the Chinese could not have produced other, more satisfactory versions; they certainly had the knowledge to do so. It has been alleged that we have here another example of technological one-upmanship in the service of cultural and spiritual domination.

Be that as it may (we are perhaps too quick to project our own motives on others), it is obvious that even those Chinese who had no access to instruments could keep rough track of the time by keeping their eyes open and noting the changing light and shadows around them. For this purpose, timekeepers are where you find them. Thus, we are told that in modern Peking, houses of substance were built oriented to the cardinal points of the compass: on clear days their shadow on the paving stones would serve as a crude clock. Even unaligned houses could function this way for an experienced observer. One foreign resident writes that if he asked his servants the time (why didn't he consult his watch?), they would look about and say, "Perhaps near to eleven," or "Surely, now, at least half past four."[30] This, to be sure, was in the 1930s, when urban Chinese had long been sensitized to Western, numbered hours by contact with foreigners. Yet it may have been even easier to do this sort of thing a thousand years earlier, when timekeepers were rare and less accessible. One may assume, for example, that the periodic sounding of horns and drums came to be linked in the minds of listeners with the diurnal round of natural events—sun low, sun high; shadows short, shadows long; shadows west, shadows east—and must have imprinted upon them in the long run a sense of passing time independent of auditory signals.

Yet sensing or observing the passage of time is not the same thing as knowing and using it. The ordinary Chinese did not need to know the hour in order to do what had to be done. He knew what had to be done, and sometimes the hour impinged on his consciousness. In this context, such natural indicators as were available were good enough, for all their imprecision.

We shall see that in time measurement, good enough is the enemy of good.

The one group with a commitment to precision, at least in principle, consisted of the astronomers/astrologers of the imperial court, and their commitment was mitigated by their very function. The task of these seers was to study and predict the movements of the heavenly bodies and to compile *ephemerides* (tabular statements of these movements) that might serve as a guide to action. There was a time to sow and a time to reap, a time to fight and a time to stop fighting, a time to have sex with the empress and a time to couple with a concubine, and these times were written in the conjunctions of stars and planets and the motions of the sun and moon. In the words of Su Sung, "Those who make astronomical observations with instruments are not only organising a correct calendar so that good government can be carried on but also predicting good and bad fortune and studying the resulting gains and losses."[31]

The Chinese water wheel towers, then, were primarily intended not as clocks but as *astraria*—mechanisms for the study and display of the movements of heavenly bodies. The clockwork itself was simply a means to these ends, a way of facilitating observations and automating the display. To be sure, Su Sung's clock did show the time by means of those tablet-carrying jacks and the bells or gongs sounding at intervals. But these functions were accessory, and the manner of display was such that it would not have been possible for someone to know the time without approaching closely; even then, the reading would be subject to a margin of error of some minutes, because the jack in the window gave the time as of the last quarter.[32]

This says nothing, of course, about the accuracy of the mechanism itself or of the other water-driven clocks that preceded it. Here the Chinese "talked a good game," and one might be tempted to infer from their boasts a keen concern for accurate time measurement. For instance, of a very early bucket-wheel-drive armillary sphere of the fifth century, we are told that it showed transits at dawn and dusk that agreed with the heavens "like the two halves of a tally"; and of the preliminary model of the Su Sung clock, that it showed "perfect agreement" between the armillary sphere and the motions of the heavens.[33]

It would be a mistake to take such reports at face value. In the first place, there is an element of conventional hyperbole here: note the use of the tally simile, which went back to the Chou period (first millennium before our era) and had become a cliché. Second, the Chinese seem to have taken the best results as representative. This will come as no surprise: in matters of self-appraisal, it is human to err on the up side. Chinese instrument-makers and clockmakers were no different in this regard from swimmers or joggers or fishermen everywhere.

Such human failings aside, much of this optimism was implicit in the nature of these instruments. Since the Chinese water-wheel clocks were primarily astronomical devices, their accuracy was checked not by comparing the time with the heavens but a copy of the heavens with the heavens. The copy was the armillary sphere, with rings turning as the wheel—that is, at roughly half-minute intervals. Presumably two observers would work together on this: one would mark a celestial event—the rise, say, of a fixed star—and call it out to the other, who would immediately check the position of the appropriate ring to see whether it showed the event. Given the inherent error of observation, the character of exchange of information, and the intermittency of the mechanism itself, anything within a minute or two would have seemed to agree "like the two halves of a tally."[34] To be sure, this was no small achievement, and in conjunction with careful observation, it did permit consistent timekeeping. Yet this is not the same thing as *precision* timekeeping, which would have required the

comparison of time with sky (not "sky" with sky) and the maintenance of careful, subminute records of fast and slow. Such records
were not to appear until hundreds of years later, in Europe, with
the invention of the pendulum clock and the marine chronometer.

In any event, since contemporary testimony is often figurative
and rhetorical, we may want to conduct our own tests. After all,
we have a number of reconstructions of the timekeeping mechanism of the Su Sung clock, and we can rate these. This Combridge
did, with excellent results—a variance of only about a minute a
day. Needham used the results of an earlier, sand-driven minimo-
del to suggest that Su Sung's clock was not surpassed as a timekeeper until the invention of the pendulum clock by Christiaan
Huygens in the seventeenth century.[35] He may well be right. The
problem with such a leap of inference is that there are many discrepancies between even an accurate model and the Su Sung original. Aside from the difference in materials and the intrusion of
modern emendations, today's mode of verification and adjustment (checking by electric contacts against time signals) is far
more precise and effective than the ones available then.

On balance, however, there seems every reason to believe that
the Su Sung clock could perform well on a day-to-day basis—certainly far better than the early mechanical clocks that made their
appearance in Europe almost two hundred years later. How it did
over longer periods, we do not know. It is a horological commonplace that timekeepers do best under test conditions—that is,
when they have been freshly cleaned, lubricated, adjusted, and
rated. Afterward they invariably drift, and not always in the same
direction. In the context of Chinese technology, such drift would
come largely from dirt and corrosion (even bronze deteriorated)
and from daily and seasonal variations in temperature.[36] If we
were to apply here Needham's procedure of inference *a posteriori,*
we would be somewhat pessimistic. The model of the Su Sung
clock now standing in the Time Museum in Rockford has not
kept a stable rate, partly for mechanical reasons, partly perhaps
as a result of temperature problems.[37]

It is probably this difference between short run and long that
accounts for the contradictory reports that have come down to us.
The Chinese boasted of their peak performances, yet also com-

memorated their failures. The records seem to show that none of their great clocks kept good time over a long period; indeed, none of them lasted. The annalists report these disappointments in matter-of-fact declarative prose—reflecting perhaps the Chinese temperament and good annalistic manners, but also the functional reality. Given the calendrical-astrological objectives of these clockwork astraria, an accurate rate was desirable but not necessary. For horoscopes, the tolerable margin of error is relatively large. What does it matter if the timing of the winter solstice is off by an hour, several hours, or even a day? A great deal in principle; indeed, the very legitimacy of the emperor rested on the harmony of his decisions and actions with the patterns of the cosmos. In practice, though, there was room for error, so long as it was not patent. If the astronomers found an anomaly, the armillary sphere could be adjusted and the calendar corrected. The important thing was the appearance of knowledge, duly certified to the ruler by the court astronomers and proclaimed by him to the people. The criterion, in other words, was political rather than scientific. The astronomers understood that only too well: we are told that in the tenth century the representatives of the two imperial observatories, which were supposed to arrive at results independently and check them against each other, used to compare and reconcile their data in advance. They did not even bother to use their instruments to make observations, but rather copied out the positions from the tables of ephemerides. "Everyone knew about it, yet no one thought it strange." One hundred years later, this collusion was still taking place. When the new astronomer-imperial learned of it, he had some of the astronomers exposed and punished, but to no avail: "The deceptions continued as before."[38]

Su Sung himself ran into this problem in the course of his diplomatic career. He was charged with an embassy to the Liao kingdom in the north—in the words of the Chinese chronicler, to the Liao "barbarians." His task was to bring a gift to the Liao emperor, whose birthday chanced to fall on the winter solstice. But the Chinese calendar had the winter solstice a day early, so that when Su Sung brought his gift to the Liao court, the officials were at first unwilling to accept it. This is the story as the Chinese memorialist Yeh Meng-te (c. 1130) tells it:

As the barbarians had no restrictions on astronomical and calendrical study, their experts in these subjects were generally better [than those of the Sung], and in fact their calendar was correct ... Of course Su Sung was unable to accept it, but he calmly engaged in wide-ranging discussion on calendrical science, quoting many authorities, which puzzled the [Liao] barbarian [astronomers] who all listened with surprise and appreciation. Finally he said that after all, the discrepancy was a small matter, for a difference of only a quarter of an hour would make a difference of one day if the solstice occurred around midnight, and that is considered much only because of convention. The [Liao] barbarian [astronomers] had no answer to this, so he was allowed to carry out his mission [on the earlier of the two days]. But when he returned home he reported to the Emperor Shen Tsung, who was very pleased at his success and at once asked which of the two calendars was right. Su Sung told him the truth, with the result that the officials of the Astronomical Bureau were all punished and fined.[39]

Consider now the implications of the Yeh Meng-te account. "The barbarians," he writes, "had no restrictions on astronomical and calendrical study"; hence, "their experts in these subjects were generally better" than the Chinese—this, mind you, without the help of wonder clocks. In China the calendar was a perquisite of sovereignty, like the right to mint coins. Knowledge of the right time and season was power, for it was this knowledge that governed both the acts of everyday life and decisions of state. Each emperor inaugurated his reign with the promulgation of this calendar, often different from the one that had preceded it. His court astronomers were the only persons who were permitted in principle to use timekeeping and astronomical instruments or to engage in astronomical study.[40] His time was China's time.

In effect this was a reserved and secret domain. There was no marketplace of ideas, no diffusion or exchange of knowledge, no continuing and growing pool of skills or information—hence a very uneven transmission of knowledge from one generation to the next. Listen to the lament of an artisan of two thousand years ago who had spent his lifetime making astronomical instruments: "When I was young I was able to make such things, following the methods of division without really understanding their meaning. But afterwards I understood more and more. Now I understand it all, and yet soon I must die. I have a son also, who likes to learn how to make [these instruments]; he will repeat the years of my experience, and some day I suppose that he in his turn will understand, but by that time he too will be ready to die!"[41]

More than a thousand years later the same lament was heard. Tseng Min-chan was an astronomer-clockmaker of the early twelfth century. Only a generation had passed since the construction of Su Sung's wonder clock, but in southern China the secret of the bucket-wheel drive seems to have been lost, and Tseng was using floats in clepsydras to power a simple time display and strike the hours. A devoted stargazer, he used to lie on his bed and watch the heavens through a hole in his roof. But even in south China it gets cold, and one night he fell asleep and caught cold, dying soon after. "Alas!" wrote his grandson, "his knowledge was not handed down. Only the general constructional designs of his water-clocks and sundials were known to his son."[42]

Under the circumstances we can understand the repeated lapse of knowledge over long periods, so that each great clockmaker had to search in old records for the forgotten secrets of earlier reigns. By that time some of the meaning of the characters had been lost, and the texts were not always decodable. Trial and error led inevitably to false starts and dead ends, and changes came haltingly. When Chinese astronomer-clockmakers learned that iron and bronze were not suitable to some parts of water-wheel drives because of corrosion, it took them centuries to profit from their experience by trying wood. These were not stupid people—they just did not make astronomical clocks very often.[43]

In his memorial to the emperor, Su Sung complains of the loss of two important instruments that had enabled the ancients "to reach the greatest perfection in observing the heavens." With these gone, "I am afraid our results cannot be very correct." He then goes on to review the episodic character of the gains made by his predecessors:

> The system used in Chang Heng's instrument was not recorded in the official histories and [I-Hsings's] old instrument [the first fully evidenced example of the escapement clocks] of the K'ai Yuan reign-period [713–741] was already lost before the end of the T'ang dynasty [907]. At the beginning of the T'ai-P'ing Hsing-Kuo reign

period [976] a Szechuanese, Chang Ssu-Hsun [made] a plan and model [of an astronomical clock powered by a water-wheel and escapement] and presented it to the emperor T'ai Tsung. The emperor authorized its construction in the Palace and it was completed in the following year . . . After the death of Chang Ssu-Hsun the cords and mechanism went to rack and ruin and there was no longer anyone who could understand the system of the apparatus.[44]

Needham has not been unmindful of this pattern. He remarks that from Chang Heng (78–142), astronomer royal, mathematician, and engineer, the first in Chinese history to build a water-driven armillary sphere, to Matteo Ricci, the Jesuit missionary of the sixteenth century who first brought mechanical timepieces to China—that is, over a span of fifteen hundred years—only a half-dozen, perhaps only four, astronomer-clockmakers kept the great tradition alive in China or, more accurately, revived it at intervals. Needham presents this fact as something of a wondrous economy: "It is well worth noting how few men it took to span all the centuries of clockwork drive mechanisms."[45] He might have written that nothing better illustrates the constraints on experiment and the impediments to diffusion of knowledge in this domain than the paucity of successful practitioners over time.

One last comment about the annalist's account of Su Sung's mission: he writes that Su Sung, in an effort to persuade the Liao astronomers of the correctness of the Chinese calendar and his choice of date for presenting his gift, "calmly engaged in wide-ranging discussions on calendrical science, quoting many authorities, which puzzled the [Liao] barbarian [astronomers], who all listened with surprise and appreciation." Now it is hard to say, from this account, what exactly puzzled the so-called barbarians, but one thing that may have troubled them was the recourse to authority rather than to evidence. Reference to authority may have been decisive in the monopolistic astronomy of China; it was no doubt less effective in a competitive, emulatory context; and it must surely have inhibited curiosity, originality, and observation.

It also promoted the corruption of science by factional politics. A memorial by court official Wang Fu written only a generation after Su Sung, at a time when Su Sung's clock was presumably still working, treats the subject of astronomical water clocks without any mention of Su Sung. Needham suggests that partisanship is the explanation: "Su Sung's clockwork was associated with the Confucian Conservatives—Wang Fu was one of the Taoistic Reformers." When these came to power in 1094, a "committee of investigation" was established to scrutinize the court's astronomical instruments, and the very existence of Su Sung's clock was placed in jeopardy.[46]

Is that the way to run an observatory?

Is that the way to treat a clock?

2 Why Are the Memorials Late?

ALMOST FIVE HUNDRED YEARS were to pass before the Chinese would see clocks comparable in complexity and performance with Su Sung's masterpiece, by then completely forgotten.

In the fifteenth and sixteenth centuries the tiny Christian nation of Portugal, having cleared the land of the infidel Moor, pursued its crusade abroad and built an astonishing global empire on a web of sea routes. Portugal's population of about a million could scarcely furnish the cadres of communication and administration, but the Portuguese maximized their opportunities by occupying skillfully selected points of control whence they could draw in converts and commodities. Their enterprise was always a combination of propagation of the faith and the pursuit of trade, and it would be hard to say which of the two was more important. Together with an almost unslakable thirst for fame, these made up the three *G*'s that I, as a schoolboy, learned to identify as the prime movers of imperialism: God, Gold, and Glory.

One of these nodal points was the territory of Macao, a tiny tongue of land of about two square miles sticking out from a small island in the estuary of the Pearl River below Canton. The Portuguese who came to China in the mid-sixteenth century were allowed to build factories (that is, warehouses and sales offices) here in 1557. The aim of the Chinese authorities was to allow a modicum of carefully regulated trade while isolating these potentially dangerous "barbarians." There was to be no contamination by alien ideas or mores, and to emphasize this point the Chinese in 1573 built a wall that cut off the foreign compound from the rest of the island.

The Portuguese in turn wanted to use Macao as a bridgehead, not only to China but to Japan, and they moved energetically to evangelize in both areas. (Their impulse in this direction, which had earned them abiding enmity in India, was stronger than they.) In Japan their missionary activities enjoyed considerable success: the country had a tradition of borrowing and copying, and besides it was then in the throes of disorder and civil war, so that any organized opposition that might normally have impeded the importation of alien religious doctrines was momentarily disarmed. (Once the Japanese reestablished order and unity, they expelled the Portuguese and rooted out Christianity with unparalleled thoroughness.)

The Roman Catholic Church had no corresponding success in China. The Celestial Empire was intact, and the authorities had no intention of allowing any proselytizing by barbarians who, in contradiction of Asian values, practiced an exclusive religion and worshiped a god who would have no others before him. Saint Francis Xavier himself had not succeeded here, and his disappointment had been followed by failed missions of the Jesuit, Dominican, Augustinian, and Franciscan orders. The time-honored strategy of evangelization, going back to Roman times, was to win the favor of heathen rulers and, by their conversion, secure the Christianization of their people. In China this tactic was frustrated by the inability of the missionaries to penetrate the empire. How could they convert the emperor, or at least win his favor, if they could not get past the frontier?

The answer was to find something the emperor wanted and use that as bait.

In 1577 Matteo Ricci, a young Italian of noble family, then in his sixth year of service in the Society of Jesus, volunteered for missionary work in the East Indies. He was never to return home. After some years in Goa, headquarters of the Portuguese Indian empire, Ricci was dispatched to Macao, there to undertake the task of opening China to the true faith.

Ricci was no ordinary missionary. Handsome in appearance, gentle and suave of manner, he was Castiglione's ideal courtier, a type that may well have served as a model for his own education.[1] From the start he fit in with his hosts as though he were one of them, and so well did he study his Chinese language and letters that within less than a year he accomplished the extraordinary feat of conversing with the mandarins without interpreter, "which won their admiration and pleased them enormously."[2] This is the testimony of Father Alonso Sanchez, who knew Ricci at Macao from November 1582 to February 1583. The same Sanchez tells us that Ricci, besides being a good theologian, was also a great "astrologer" (he means astronomer). To be sure, Ricci brought with him to China an astronomical science that was Ptolemaic, earth-centered rather than sun-centered, and hence obsolete. But even in Europe the so-called Copernican Revolution was proceeding very slowly, and pioneers such as Galileo were still teaching their students Ptolemaic astronomy at the end of the sixteenth century.[3] European astronomy, moreover, for all the limitations of its theory and pretelescopic observations, was better than the late Ming variety. At least it was better at producing the kind of information the Chinese wanted. The Chinese emperor, Ricci wrote in 1605, employed more than two hundred persons "at much expense" to draw up the annual calendar; yet the calendar, which was symbol and expression of Chinese sovereignty and dominion, fairly swarmed with errors. Ricci revealed to the Chinese an astronomy that yielded more accurate predictions of the movements of celestial bodies, especially of eclipses. Nothing could have impressed them more, since, as we have seen, they regulated actions and decisions, big and small, by the conjunction of sun, moon, and stars.[4]

Ricci also brought with him globes and maps that were like magic casements opening on distant and unknown lands. Above all, he awed and delighted his hosts with his "self-ringing bells"—striking clocks, "such as never before had been seen in China." "These globes, clocks, spheres, astrolabes, and so forth, which I have made and the use of which I teach, have gained for me the reputation of being the greatest mathematician in the world. I do not have a single book on astrology, but with only the help of certain ephemerides and Portuguese almanacs, I sometimes predict eclipses more accurately than they do."[5]

By judicious gifts Ricci secured, first, permission to open a mission at Chao-ch'ing (just west of Canton) and thence to move step by step northward to the capital at Peking. His progress was slow and was repeatedly interrupted or reversed by jealous and suspicious officials, especially the much-detested eunuchs who protected the emperor from all contact with outsiders. But the fame of Ricci's wondrous mechanisms preceded him, preparing the way for each advance; so that in the end it was an impatient emperor who called for Ricci's presence (and presents): Where are the self-ringing bells?[6] And when the dowager empress showed an interest in her son's favorite clock, the emperor had the bell disconnected so that she should be disappointed. He could not have refused to give it to her, had she asked for it; but neither would he give it up, and so found this devious way to reconcile filial piety with personal gratification.[7]

Here, then, was China's second chance at a horological revolution. The Chinese, having failed within their own technology to advance beyond the clepsydra time standard to some kind of oscillating controller, now had the opportunity to copy and perhaps improve on European technique. The instruments that Ricci brought with him may or may not have been more accurate than those in use in China at the time; they were, as we shall see, very approximate timekeepers. But they were portable and held out the promise of the private knowledge of time—that is, time as an ever-present personal companion. They were also capable of performing an array of entertaining accessory functions: sounding bells, playing melodies, parading little automated figures. The Chinese, we know, "adored" them, and their possession quickly became a mark of status as well as a source of pleasure.

Why, then, did not the Chinese take up their manufacture? The market was there. The population of the kingdom was just about equal to that of all Europe, and income per head, we are told by bold comparativist econometricians, was almost as high as in Europe.[8] Moreover China never lacked for artful craftsmen capable of the most delicate and painstaking tasks. Nothing seems to have been wanting for successful imitation.

The answer in part is that the Chinese did try to copy the European machines. Not immediately. When Dutch ambassadors visited China in 1665, they were as unfavorably impressed by timekeeping practice as the Jesuits had been three quarters of a century earlier. "The Chinese," they wrote, "although ingenious and subtle, have almost no instruments for showing the hour, and those they have are so poor that you cannot trust them."[9] The emperor at that time was the great K'ang-hsi, just at the beginning of his long reign and still very busy consolidating his hold on the celestial throne. Once things quieted down, however, he turned to internal development and had a workshop for the making of clocks and watches installed in the palace. This was only one of a number of such studios devoted to luxury arts and crafts (compare the artists housed in the Louvre in the service of the contemporary French monarchy); but the clock-and-watch shop was surely one of K'ang-hsi's favorites. He alludes proudly to it in one of his poems (in China the art of poetry was cultivated by persons of education right on up to the emperor):

> The skill originated in the West,
> But, by learning, we can achieve the artifice:
> Wheels move and time turns round,
> Hands show the minutes as they change.[10]

K'ang-hsi apparently saw his clock manufactory as part of a larger effort to learn and emulate Western technology. "In the late Ming Dynasty [late sixteenth–early seventeenth century]," he wrote, "when the Westerners first brought the gnomon, the Chinese thought it a rare treasure until they understood its use."[11] So with clocks: the springs gave a great deal of trouble; but "since I have been reigning, having learned a great deal from the Europeans how to make these springs, I have had hundreds and thousands of clocks made, and all of them keep accurate time." "You are all still young," he wrote his children, "yet each of you has, by my generosity and favor, ten or twenty of these chiming clocks to play with [*come per gioco*]. Shouldn't you think yourselves very fortunate? Always remember with gratitude, then, the happiness abundantly accumulated and transmitted to you by your ancestors and your father."

Yet for all the emperor's solicitude and support, this palace workshop never became the nucleus of a home clock-and-watch industry. K'ang-hsi never intended it as such. It would not have occurred to him to bestow on the multitude the happiness "abundantly accumulated and transmitted" to princes and courtiers. Insofar as native Chinese did learn the art of clockmaking and set up on their own, to the point of constituting by the eighteenth century a separately identifiable trade, they were occupied essentially with repairing or imitating Western timepieces. Unlike the Japanese, the Chinese never developed a distinctive style of clock to their own tastes and needs. Their skills were mimetic, and while their cases eventually took on a distinctly Chinese appearance, the movements remained "at best poor copies of poor copies of European timepieces."[12] The difficult and supervisory work of the palace manufactory was always entrusted to European missionaries. When the British East India Company brought some clocks to China as gifts to the emperor and found that they had been damaged en route, it was to these horologists-in-the-service-of-God that it turned for assistance. The Catholic missions, of course, were quick to understand the opportunity and to assign to China clerics with training and experience as clockmakers and mechanics.[13]

Even so, the quality of the imperial workshop was not up to the best Western standards—in spite of K'ang-hsi's proud claims. When the Chinese court wanted accurate or complicated mechanisms, it turned to Europe, as did those subjects of the emperor who could afford to. During the quarter-millennium of Manchu rule, Peking purchased or accepted as gifts an extraordinary array of timekeepers. "The Imperial Palace," wrote Father Valentin Chalier in the 1730s, "is stuffed with clocks . . . watches, carillons, repeaters, organs, spheres, and astronomical clocks of all kinds and descriptions—there are more than four thousand pieces from the best masters of Paris and London, very many of which I have had through my hands for repairs or cleaning."[14]

The smaller and more portable of these treasures were among the targets of preference when French and British troops looted the great palace in Peking in 1860 in an orgy of rapine and destruction. The French arrived first, and the officer in charge claimed later that, after examining the imperial apartments, he had posted sentries to guard everything intact until the arrival of the British commander. A waste of time: his men had no intention of observing such fine courtesies. Spurred by the sight of coolies and peasants hastening through the gardens with ladders and ropes, they burst through and grabbed everything they could, dumping it in piles and tipping it into wagons for later disposal. With all their rush to grab, though, they could not resist the opportunity to play with the myriad automata, to bury their heads in the fragrant silks, to dress in the ornate mandarin costumes and the richly embroidered gowns of the court women.

"A large majority of them," wrote one witness, "were 'grown children' who were 'mainly tempted in the midst of all this unbelievable accumulation of wealth' by the extraordinary variety and number of mechanical toys and clocks, so that the whole area was 'one continuous symphony' with monkeys beating cymbals, rabbits rolling drums, birds singing, toy soldiers playing cornets and bagpipes, clocks chiming, and some four thousand musical boxes simultaneously tinkling their several tunes, and every now and then all this noise was 'drowned out by the easily amused soldiers roaring with laughter.' "

The British found most of the best things gone by the time they arrived, but made up for lateness with method and organization. They "moved in by squads, as if on fatigue duty, carrying bags, and were commanded by noncommissioned officers who, incredible as it sounds (yet strictly true), had brought jewellers' touchstones with them." One chaplain cut short his Sunday service to fill a mule cart with loot; the next Sunday he preached an "admirable sermon against covetousness." The jewels snatched then have long since disappeared into anonymous boxes and drawers. The fabrics, if they have survived, hang in strange closets and museum display cases. The clocks and watches turn up occasionally in European and American auctions, shamelessly proclaiming their violent provenance: a timepiece from the Summer Palace is *ipso facto* an object to be coveted and will invariably fetch a better price.[15]

Some forty years later, the suppression of the xenophobic Boxer movement gave an even larger alliance of Europeans and Japanese the opportunity to repeat the rapine of 1860. Again the palace was looted, chests and drawers were forced open, and all manner of treasures were plundered. Even so, when in the 1930s

Simon Harcourt-Smith did his catalogue of the timekeepers and other mechanisms of European manufacture remaining, he could still wonder at the now silent sounds and still movements of the hours—their passage "marked by a fluttering of enamelled wings, a gushing of glass fountains and a spinning of paste stars, while from a thousand concealed and whirring orchestras, the gavottes and minuets of London rose strangely in the Chinese air."[16]

S O WE ARE BACK TO SQUARE ONE: Why couldn't the Chinese, with all their interest and skill, do what K'ang-hsi thought they had done—that is, learn the Western artifices and even go beyond them?

The answer, I think, lies in the same attitude that had constrained horological development earlier: it was simply not important in China to know the time with any precision. Calendar dates mattered, but neither life nor work had ever been organized on the basis of hours and minutes. Carlo Cipolla puts it this way: "Foreign machinery could not be properly appreciated because it was not the expression of a Chinese response to the problems set by a Chinese environment." And he quotes a bibliographical catalogue of 1782, the *Ssu-K'u Ch'üan-Shu Ti-Yao* (The Index to the Grand Library), by way of conveying the Chinese sense of the significance of European horology: "In regard to the learning of the West, the art of surveying the land is most important, followed by the art of making strange machines. Among these strange machines, those pertaining to irrigation are most useful to the common people. All the other machines are simply intricate oddities, designed for the pleasure of the senses. They fulfill no basic needs."[17]

This is clearly a defensive statement. The dismissal of European devices as oddities suggests that we have here not only an expression of indifference but a deliberate rejection. This attitude was crucial to subsequent Chinese horological development. Everything was changed drastically by the simple fact that *once Ricci brought European clocks to China, progress lay not in the transformation of indigenous technology but in the adoption of an alien device.*

Let us go back to that fateful mission to get a sense of the terms of the challenge—for challenge it was. It will be obvious to the reader that Ricci's dazzling display of new knowledge and won-

drous objects was meant to do more than simply charm and se-
duce his hosts. Ricci was fishing not merely for compliments but
for souls. Implicit in this display of learning and technique was
the argument that a civilization that could produce a manifestly
superior science and technology must be superior in other respects, specifically in the spiritual realm. Clockwork was God's
work. (Conversely, God's work was clockwork.)

Today such an equation seems patently simplistic. We have,
after all, a long experience of technology in the service of violence
and oppression, as well as in that of improvement and progress.
We understand that techniques are two-edged instruments, available for both good and evil. It would be illogical, then, for a citizen of the twentieth century to infer spiritual qualities from material achievement. Yet to understand the Jesuit argument, we have
to take ourselves back to a more innocent age, caught up in the
intoxicating exhilaration of new knowledge and only too happy to
infer therefrom a more general virtue. Such sentiments were by
no means confined to Jesuits, who might be tempted to stretch
logic in the service of a higher cause. Take Leibniz, for example,
one of the great scientists and mathematicians of his age—though
to be sure a religious one—here congratulating the French government on its contributions to a science that would serve as spiritual inspiration to less enlightened peoples:

> The real discoveries that you have promoted are for all places and
> all time. A king of Persia will exclaim over the effect of the tele-
> scope, and a mandarin of China will be overcome with astonish-
> ment when he understands the infallibility of a geometrician mis-
> sionary. What will these peoples say, when they see this marvelous
> machine that you have had made, which represents the true state
> of the heavens at any given time? I believe that they will recognize
> that the mind of man has something of the divine, and that this di-
> vinity communicates itself especially to Christians. The secret of
> the heavens, the greatness of the earth, and time measurement are
> the sort of thing I mean.[18]

To be sure, Ricci himself was too smooth, too supple to con-
front the Chinese with a brutal alternative: Confucianism or
Christianity. He preferred to woo, to reconcile, to make conversion as comfortable as possible. To this end he composed in Chinese and published essays designed to assimilate the ethics of the

Chinese literati to Christian doctrine. This required him to treat as symbolism practices that others would have interpreted as idolatrous: the Chinese ideographs for ancestor worship, for example, became Ricci's word for the Mass. He also played down those aspects of Christian dogma that the Chinese might perceive as intrinsically mythical and hence incredible.

All of this was to bring down upon him the strictures of other churchmen, especially those of more rigid orders, and left Rome with an abiding anxiety concerning the reliability of the China mission: the Jesuits, it was widely alleged, had gone native. Yet that was just what made them more or less acceptable, sometimes even welcome, in a world of jealous courtiers and intriguing officials, a world that could not help but perceive them as competitors for influence, hence potentially subversive.

The abiding difficulty was that, try as Ricci and his successors might to soothe Chinese susceptibilities, they could not confine the challenge posed by Western science and technology to the theological sphere. China was the Middle Kingdom, the center of the world. The peoples round about had nothing to give to China but tribute. Ricci, for all his gentility and courtliness, came to China as a barbarian (by Chinese definition), so for the Chinese the very fact of Ricci's knowledge and wares turned the world upside down. The initial interest in Ricci's maps changed to annoyance and resentment when his hosts perceived how small a place China occupied, off toward the right-hand edge of the paper. Ricci immediately set about to draft a hemispheric projection that, while accurate, placed China in the center and exaggerated its area by comparison with all the lands around. This, we are told, obtained great favor and was widely circulated by the authorities, *ad maiorem gloriam . . .*[19]

There is a limit, though, to that kind of cleverness. The way in which the Chinese perceived the world and their place in it was premised on the integration of culture and civilization, and spiritual consistency compelled them to accept the painful logic of the Jesuit position. In the words of Mei Weng Tin (1632–1721):

> How can I detach my life from the Confucian wisdom
>   and devote my life to the Western theories?
> If I were to study astronomy without becoming a Catholic
>   our friendship would be insincere.
> How in this world can one go back on one's origins?[20]

Clocks, in other words, embodied far more than an argument for the validity of Christianity. They were an assault on China's self-esteem.

The classic response to this kind of challenge is to deny it or disparage it. The Chinese did both. Denial took the form of claiming priority for Chinese invention in this domain. Thus, Fang I-chih, an encyclopedist of the mid-seventeenth century, after reviewing some of the early history of Chinese water clocks and automata, comes to two contemporary artists, whom he credits with making watches only an inch across: "Their workmanship is splendid, but after all what is there really surprising in it?" Did not writers of the Sung Dynasty (tenth to thirteenth centuries) already speak of "revolving and snapping springs?" Were "the chiming watches of the red-haired people" any different? Western clocks, Chinese scholars argued, derived from the clepsydra—the Chinese clepsydra at that. In the words of a scholar-official of the late nineteenth century: "The automatically striking clock was invented by a [Chinese] monk, but the method was lost in China. Western people studied it and developed refined [time-keeping] machines." The same chauvinist went on to claim Chinese invention of the steam engine, too. As well be hanged for a sheep as a lamb.

Pre-shades of *Heavenly Clockwork!* Joseph Needham self-deprecatingly recalls that when his attention was first drawn to this claim by Wang Ling in 1946, neither of them was "disposed to take it seriously."[21] Second thoughts are not always best.

Other Chinese took note of the limitations of Western timepieces to depreciate their contribution. In the 1630s a censor by the name of Wei Chun wrote a pamphlet entitled "Ricci Has Invented Fables to Deceive Everyone," which argued among other things that as Western clocks were driven by springs whose force diminished as they unwound, they could not possibly keep good time.[22] It was necessary, he went on, to keep winding them and to correct them by checking their performance against a sundial; but if the sun was covered several days in a row, there was no way to correct them. True enough. A similar dismissal by Hsü Ta-shu compared European clocks unfavorably to Chinese clepsydras: "They spend in making them tens of ounces of silver. At least one can say of a water lift that it saves labor. But these clocks, which are so difficult to make and break down so easily, aren't they really nothing better than pure waste?"[23]

The most common response was to belittle the clock as a delightful but highly dispensable curiosity—that is, to treat it as a toy. Such a reaction was predetermined in part by the Chinese attitude to time and its measurement: if hours and minutes do not really matter, what purpose could these self-ringing bells serve, except to amuse? Yet there was more. As Carlo Cipolla shows, the clock was only one of a large array of European mechanisms and devices thus relegated to the status of oddities and trivia.

Some of this we should no doubt take at face value: the Chinese liked playthings. Admiral van Braam, leader of a Dutch embassy to Peking at the end of the eighteenth century, writes with some astonishment of a high-ranking mandarin who proudly showed him a little mill in a bottle. When the container was set upright, sand fell on the pallets of the wheel and turned it. The mandarin wanted to know whether van Braam had ever seen anything like it. Yes, he had, and much finer. Why, then, hadn't he brought any with him? "I observed in answer that as in our country they only serve for the amusement of children, we had not supposed that they would give the least pleasure or excite the smallest attention. He assured us of the contrary, and spoke in the language of a man who thinks himself the possessor of a wonder."[24] Any people that was prepared to pay up to five hundred ducats for kaleidoscopes—those cunning little viewers through which one can see an unending series of colorful shapes and patterns—obviously liked toys for their own sake.[25] It is only on this basis that we can understand the tens of clocks that K'ang-hsi gave each of his children, and the thousands—literally thousands—accumulated in the imperial palaces.

Yet this intense playfulness is not enough by itself to account for this trivialization of the clock. The Chinese may have sometimes been big children by European standards, but children learn very early to distinguish between toys and the real thing, between play objects and work objects. China's depreciation of the clock, among other Western mechanisms, was also a protective reaction, an effort to erase the discrepancy between self-perception and reality. Van Braam, I think, was right on the mark when he wrote that the Chinese "think that they hold the first rank among all created beings of this immense universe" and were not prepared to abandon so comforting a self-image. "It may be perhaps supposed that the sight of the masterpieces of art, which the Chinese receive annually from Europe, will open their eyes and convince them that industry is there carried farther than among themselves, and that our genius surpasses theirs; but their vanity finds a remedy for this. All these wonders are included in the class of superfluitics, and by placing them beneath their wants, they place them at the same time beneath their regard. If, for a moment, they fall into an involuntary fit of astonishment, they come out of it firmly resolved to do nothing to imitate that by which it was produced."[26]

So it was that when George Macartney went to China in 1793 bringing from the Court of St. James gifts (including some fine clocks and watches), greetings, and requests for representation and trade, the emperor could condescendingly turn him down. In an edict addressed to George III, the old Ch'ien-lung, then in the fifty-seventh year of his reign, set the record and the world right side up:

> The Celestial Court has pacified and possessed the territory within the four seas. Its sole aim is to do its utmost to achieve good government and to manage political affairs, attaching no value to strange jewels and precious objects. The various articles presented by you, O King, this time are accepted . . . in consideration of the offerings having come from a long distance with sincere good wishes. As a matter of fact, the virtue and prestige of the Celestial Dynasty having spread far and wide, the kings of the myriad nations come by land and sea with all sorts of precious things. Consequently there is nothing we lack . . . We have never set much store on strange or ingenious objects, nor do we need any more of your country's manufactures.[27]

China's assertive superiority and self-imposed autarky would have paradoxical consequences. On the one hand, they shut the door against the stranger; on the other, they gave him the desire and incentive to smash it in. Economic dependency in this supposedly imperialist encounter ran the wrong way. Europeans loved China's tea and porcelain, and were even ready to send furniture all the way from Europe to be lacquered and painted there. But since the Chinese had no interest in European fabrications, Western nations had no way to pay for these purchases except with specie. The British did their best to teach the Chinese to wear woolens, but climate and fashion were opposed: those Chinese who could afford costly fabrics preferred silk to scratchy broadcloth. Timepieces were among the few Western artifacts the Chinese were ready to pay for, and the profits of the "sing-song" trade—that is, the trade in chiming and automated mechanisms—were legendary. But there was a limit to how many of these handcrafted objects the Europeans could make and sell. They were commonly shipped in small assortments, immobilized a fair amount of cash, and few Chinese could afford them. Clocks and watches, then, could not in themselves come close to balancing the growing deficit on commodity account. The British had to make up the difference with silver (always painful) or, increasingly, with opium. The Chinese effort to block this import eventually brought the two nations to open conflict: the wretched Opium War of 1839–1842. This was concluded by the Treaty of Nanking (August 29, 1842), which compelled China to pay a large indemnity, to eliminate official middlemen (the Hong merchants), to open new treaty ports to Western traders and goods, to confirm the cession of Hong Kong to the British, and to accept the presence of a British representative rejected earlier by Ch'ien-lung. There was no mention of opium in the treaty, but the British did try to persuade the Chinese to legalize its sale on the ground of freedom of choice: if the Chinese were virtuous, they would not smoke it. And if officials were virtuous, they would not let it in. None of that was very logical or consistent, but vice and greed take their pretexts where they can. In the meantime, some of England's most respectable merchant houses (Barings, Jardine-Matheson) owe part of their fortune to this infamous trade, just as some of the best families in New England owe theirs to slaving. Time has a way of laundering these gains.[28]

IT IS HAZARDOUS to generalize about social values and attitudes, since there are always exceptions. Here, too. A clerk of the Grand Council of the Ch'ien-lung emperor, Chao I by name, writing in the mid-eighteenth century, tells of one such: the minister Fu-heng. Chao I begins by noting the usefulness of Western clocks and science: "Self-striking clocks and watches [that hang from one's neck] and tell the time all come from the West. A clock can strike according to the hour. The watch has a needle which moves as time passes and can express twelve hours.[29] They are both extremely ingenious. At the present time the imperial astronomers, when they watch the stars and make calendars, all employ Westerners, and from this we may deduce that the Westerners' methods may be said to be finer than the old methods used in China."

But then Chao I turns to the limitations of such mechanisms: "Clocks and watches often have to be repaired. Otherwise the gold thread inside [the mainspring? the chain drive?] will break or they go too fast or too slow, and then you cannot get the correct time. Therefore among the court officials there are some who possess these things, but they still forget meetings—or, [to put it another way], those in the court who never miss meetings are the ones who do not own clocks."

Fu-heng, though, was an aficionado of timepieces: "Fu-heng's house is one which has clocks and watches, so much so that among his servants is not even one who does not hang one on his body. They can mutually check the time against each other and never be off the mark. One day when there was a formal imperial audience, [Fu-heng's] watch had still not noted that the time for the audience had arrived, but the imperial retinue had already gone in, and the Emperor had already taken his seat. Suddenly [Fu-heng] appeared in a great state of agitation and kowtowed at the foot of the throne. For the rest of the day he went around in a most upset, untranquil frame of mind."[30]

Here, then, was an object lesson—affectionately (derisively?) told, but a lesson nonetheless. Clocks and watches had their weaknesses, could let the user down; so when all was said and done, the people who never came late were the ones who did not trust them. How did this clockless majority make sure they came to audiences on time? In the same way that Russian peasants of the late nineteenth century learned to "catch" a train: the courtiers got to the palace hours early (at midnight for a predawn ceremony) and waited for the drums to beat, the bells to ring, and the gates to open. Nothing is more symptomatic of a clockless, unpunctual society into which time intrudes as an unwelcome stranger. The historian Ray Huang, writing of court etiquette under the Ming Dynasty in the fifteenth and sixteenth centuries, tells us that the morning audiences especially were vexatious, trying the patience of all the officials, who had to stand about in the open in bad weather as in good, but also of the emperor himself. Some remedy was found by reducing the frequency of these ceremonies; by permitting more appropriate dress on inclement days; and by allowing officials to be accompanied by umbrella bearers.[31] But all these were at best palliatives, and the whole system rested on the assumption that an official's time—all his time—belonged to the emperor, who could do with it (and waste it) as he pleased. The only way to correct this waste—which the Chinese could not even perceive as such—would have been to recognize time as private property. This was not a simple matter in a system where material possessions were also held on loan, as it were, from the emperor. The position of the mandarin was analogous to that of the apprentice: both were servants and their time was their master's. Under the circumstances, it was not easy to inculcate a sense of time as something to be tracked, measured, saved.

The K'ang-hsi emperor, who loved his Western clocks and sought to naturalize their manufacture in his palace, put the alternative very well; but then his time was his own:

> Red-capped watchmen, there's no need to announce the dawn's coming.
> My golden clock has warned me of the time.
>
> By first light I am hard at work,
> And keep on asking, "Why are the memorials late?"[32]

3   Are You Sleeping, Brother John?

IT IS ONE OF THE MISFORTUNES of scholarship that there was only one word for clock in the western Europe of the Middle Ages: (*h*)*orologium*. This generic term referred to every kind of timekeeper, from sundial to clepsydra to fire clock to mechanical clock. So when, in the late thirteenth century, we get an unprecedented spate of references to clocks, we cannot be sure *prima facie* what kind of device our sources are talking about. Not until the fourteenth century do we get our first unmistakable reports of mechanical clocks—namely, the tower clock with astronomical dial built by Roger Stoke for Norwich Cathedral (1321–1325); the highly complicated astronomical mechanism that Richard of Wallingford initiated at St. Albans around 1330 and that took thirty years to build; and then, completed in 1364, Giovanni de' Dondi's astronomical clock, the marvel of its time. The latter two were described by their authors in such detail that we have been able to make working copies in our own day.[1] (See Figure 7.)

The obscurity of what Needham calls "one of the most important turning-points in the history of science and technology" has been a fruitful source of legend and speculation. For some centuries it was common to attribute the clock to the canon Gerbert, who later became Pope Sylvester II (999–1003), the Pope of the

Millennium. Gerbert was indeed a savant in his generation. He had learned mathematics and astronomy in Spain, perhaps at the feet of Jewish and Muslim scientists there, and had taken away with him a fund of knowledge and technique that reemerged in the fourteenth century with the work of Richard of Wallingford and Giovanni de' Dondi. The historian and monk Richer (tenth century), who was one of Gerbert's students, tells us that his master built a globe, also an armillary sphere for the planets and another to show the motions of the stars.[2]

Gerbert, then, presumably had the knowledge and skill to build a mechanical clock. But *could* is not *did*. There is no contemporary proof of Gerbert's inventing such a device and some reason to think otherwise: if the oscillating controller and mechanical escapement were known as early as the year 1000, why do we have to wait another three hundred years to see clocks appear in the belfries and towers of city halls and churches? Surely, moreover, if so remarkable an invention had been coming slowly into use, it would have left some literary or pictorial trace, if not some physical remains. Some have suggested that if Gerbert did indeed build a mechanical clock, it must have been suppressed by the church, which might have seen it as the illicit fruit of intercourse with infidels or as the cunning product of some dark, Faustian compact. Gerbert did eventually acquire the reputation of a sorcerer and heretic—to the point where, in the sixteenth century, militant Protestants exploited his "infamy" to calumniate the papacy.[3] But this amounts to heaping speculation on speculation. Would the church have wanted to suppress a device so useful in its own management of time? Could it have? I am not a medievalist and am in no position to answer such questions. But the matter is worth investigating.

In the meantime, we are left with over two hundred fifty years of near silence, followed by a rush of ambiguous noise. The resulting uncertainty has given negative encouragement and support to a most unexpected interpretation of the invention and development of the mechanical clock. This is the thesis of Derek de Solla Price, coauthor with Joseph Needham and Wang Ling of *Heavenly Clockwork* and specialist in the history of ancient and medieval scientific instruments. Price argues that the machines of Wallingford and Dondi were the first mechanical clocks, that the timekeeping components served simply as drives for astronomical devices of a type going back to antiquity, and that timekeeping for its own sake was an unanticipated by-product of this experiment in automation. "The escapement, which originally gave perfection to the astronomical machine, was also found useful for telling time, and as social development led to an increased social awareness and importance of time reckoning, simplified versions of this part of the astronomical device were made and became widely used as mere time-tellers."[4] Price is hard on "simplified versions" and "mere time-tellers": on another occasion he refers to "degeneration in complexity" and describes the later fourteenth century as a time when "tradition of escapement clocks continues and degenerates into simple time-keepers." To cite his by now well-known metaphor: "The mechanical clock is nought but a fallen angel from the world of astronomy!"[5]

Now, it is no doubt true that terrestrial timekeeping is less elevated than heavenly clockwork, but "degeneration" does seem a mite strong. So do Price's strictures against the other wisdom on this subject, which he finds "unsatisfying, misleading, and often false." Earlier students of horology, he warns us, were on the wrong track: "On no account must we take the easy way out which abandons the history of the clock and talks instead about the history of time-measurement. It is most unfortunate that such a term was ever coined."[6] "What did not happen was that man wanted to measure time and so devised new ways of doing it. What did happen is that in the course of following an old trend, not quite yet extinct, he developed quite sophisticated techniques, important for their technological brilliance, that gave him for the first time the possibility of doing something he had not wanted before it was readily available. This product, timekeeping, caught on, and it is due to this ancient fashion that time became a matter of the deep philosophical and scientific importance it has today."[7]

Price's thesis, if true, would imply a most intriguing paradox. Here you have two societies, Europe and China, thousands of miles apart, both of them building extraordinary machines to imitate the movements of heavenly bodies, both of them automating these planetaria/astraria by means of clock drives. In both cases, the clock is an accessory, and neither society cares much about time measurement for its own sake. Yet one society, the European, abstracts the time function from its device and starts building a civilization based on pure (simple) timekeepers, whereas the other, the Chinese, does not.

Unfortunately, in both logic and evidence this is an unconvincing, indeed a most surprising thesis. (The exclamation point in Price's allusion to fallen angels is well justified.) In logic: the normal sequence of technological development runs from simple and rudimentary to complex and refined. To quote Price himself, "Historically speaking we expect that the further back we delve, the more primitive and simple the technology becomes." Any alleged deviation from this rule should put the historian on his guard.[8] And in evidence: we are as sure as we can be, short of possessing the remains of a very early machine, that there were mechanical clocks before Wallingford and Dondi, hence that simple timekeepers preceded complex, clock-driven planetaria and astraria.[9]

It is a fact that no early escapement clock seems to have survived, but then, neither has any medieval water clock. If anything, we should be less surprised by the disappearance of the former than of the latter. The first mechanical clocks were crudely fashioned and liable to break down at any time. They needed continual care, frequent overhauls, and substantial replacement every ten or twenty years. They were made of brass or iron, valuable metals at the time, and we may be sure that clockmakers who repaired them were not inclined to treat discarded parts or machines as junk to be abandoned. Rather, just as roofers today routinely keep and recycle the copper gutters and sheets they replace, so medieval clockmakers must have treated used brass and iron as valuable "perks" of the trade.

We are thus thrown back on literary evidence, ambiguous at times because of the generic character of the term *horologium*, none of it definitive in itself but collectively decisive. For one thing, there is the abrupt increase in the frequency of references to clocks toward the end of the thirteenth century. Suddenly clocks are news because clocks cost money. Cathedral chapters mention them in their accounts; itemize their repairs; pay people to watch them and keep them going; hire all manner of specialists to replace wheels, paint dials, carve wooden figures. A new profession makes its appearance, that of the clockmaker or *horologeur*.[10] There is simply nothing like this for the earlier period. The late British scholar C. F. C. Beeson argued, I think correctly, that this in itself was indicative of a new device.[11]

For another, the new clocks and their associated bells were often sited in high places—the better to hear them. But towers are no places for a water clock: no one hauls water any higher than he has to, and lofty exposures make it very difficult to keep water from cooling and freezing. Do not imagine, either, that medieval clockmakers were placing clepsydras on ground level and using them to drive or trigger mechanisms thirty or more feet above. Such an arrangement is not inconceivable, but would have been extravagantly costly in the context of medieval metallurgy. We would have heard about it, if only about its breakdowns. The fact is that contemporary accounts make no mention of water in connection with this new generation of timepieces—no concern for freezing, no reference to leaks or evaporation, no hint of rust or corrosion—nothing. Under the circumstances, all efforts to salvage the clepsydra connection, however ingenious, must be rejected as highly improbable. Indeed, on the principle of Occam's Razor, their very ingenuity makes them suspect.

What we do have in the contemporary sources is a clear sense of excitement and pride. These great clocks were, like computers today, the technological sensation of their time. When a poet like Dante looks to the clock and its wheel train for vivid similes, you know that he is speaking to established and conspicuous sensibilities. Thus in canto 24 of the *Paradiso* (written between 1316 and 1321):

> And like the wheels in clock works, which
>   Turn, so that the first to the beholder
>   Seems still, and the last, to fly.
>
> E come cerchi in tempra d'oriuoli
>   si giran si che il primo, a chi pon mente,
>   quieto pare, e l'ultimo, che voli.

Again, it takes excessive ingenuity to see here anything but the wheel train with reduction gearing characteristic of the mechanical escapement clock.

By the time we get to Wallingford and Dondi, then, the mechanical clock was in its third or fourth generation—at least.

Dondi himself must have learned much of what he knew about these devices from his own father, an astronomer and clockmaker. Both Wallingford and he provide us with careful descriptions of their wheelwork and gear ratios, but neither finds it necessary to say anything about the character and construction of the controller-escapement—that is, the timekeeping heart of his machine. Indeed Dondi explicitly dismisses the mechanical clock as a commonplace, the making of which "will not be discussed in such detail as the rest, because its construction is well known, and there are many varieties of them and, however it is made, the diversity of methods does not come within the scope of this work." His own clock, he says, beats at the usual two-second rate; and anyone who is not capable of making a "common clock"—"by himself and without written instructions"—should not attempt the rest.[12]

To sum up: the Wallingford and Dondi masterpieces, far from being the first mechanical clocks, made use of an already established technique. What is more, the use of falling weights as power source made it possible to impart steady drive to more complex mechanisms than could be worked by a clepsydra. It was the clock, in other words, that facilitated and thereby fostered the automated planetarium or astrarium, not the reverse. Indeed, the simple mechanical clock, by opening for the first time serious possibilities of precision timekeeping, eventually laid the basis for modern astronomical science. So much for "fallen angels"!

The clock did not create an interest in time measurement; the interest in time measurement led to the invention of the clock.

Where did this demand come from? Not from the mass of the population. Nine out of ten Europeans lived on the land. "Labor time," to quote the medievalist Jacques Le Goff, "was still the time of an economy dominated by agrarian rhythms, free of haste, careless of exactitude, unconcerned by productivity—and of a society created in the image of that economy, *sober and modest,* without enormous appetites, undemanding, and incapable of quantitative efforts."[13] Town and city life, to be sure, was different. The city dweller has no natural sequence of tasks to rhythm his day. The very uniformity of his occupation makes him time-conscious; or if he is moving about, the irregular pattern of his contacts imparts a sense of haste and waste. But urban centers developed late in the Middle Ages, from about the eleventh century on, and already before that there was an important timekeeping constituency. That was the Christian church, in particular the Roman branch.

It is worth pausing a moment to consider this temporal discipline of Christianity, especially of Western Christianity, which distinguishes it sharply from the other monotheistic religions and has not been adequately examined in the literature on time measurement. In Judaism the worshiper is obliged to pray three times a day, but at no set times: in the morning (after daybreak), afternoon (before sunset), and evening (after dark). A pious Jew will recite his prayers as soon as possible after the permissible time; but if circumstances require, he has substantial leeway in which to perform his obligation. Today some of the starting times of worship are given on calendars to the minute, thanks to astronomical calculations. In ancient and medieval times, however, nature gave the signals. The animals woke the Jew to prayer, and the first of the morning blessings thanks God for giving the rooster the wit to distinguish between day and night.[14] The evening prayer could be recited as soon as three stars were visible; if the sky was cloudy, one waited until one could no longer distinguish between blue and black. No timepiece or alarm was needed.

Islam calls for five daily prayers: at dawn or just before sunrise, just after noon, before sunset, just after sunset, and after dark. Again, none of these requires a timepiece, with the possible exception of the noon prayer. I say "possible," because high noon is easily established in sunny climes by visual means. Besides, insofar as the local religious authorities wanted to set times for prayer and used clocks for the purpose, they could easily make do with the sundials and water clocks of the ancients. In most Islamic countries, the sun usually shines and water rarely freezes. Moreover, in Islam as in Judaism the times of prayer are bands rather than points, and local tradition determines how much the prayers may be delayed without impairment. In both religions prayer is a personal act, without clerical or congregational mediation, and worship, with some exceptions, need not be collective and simultaneous.

Christianity, especially monastic Christianity, differs from both. The early Christians had no standard liturgy; the new faith was not yet a church. Usage varied from place to place, and prayer was as much a function of opportunity as of obligation. Insofar as the Nazarenes were still Jews, they built on the practices of the older faith, with its morning and night recitations (Deut. 6:7, "when thou liest down and when thou risest up") or its triple office (Dan. 6:11, "he kneeled upon his knees three times a day"). But then they added their own devotions, in part to give expression to those praises and supplications that had no place in the Jewish service, in part to distinguish themselves from the "obdurate" Hebrews. By the early third century, Tertullian, acknowledging the impracticality of the Pauline ideal of ceaseless prayer (1 Thess. 5:17), recommended daily prayers at set times: in addition to the morning and evening prayers prescribed by the Law, there would be devotions at the third, sixth, and ninth hours. These were the points that divided the daytime into quarters,[15] and Tertullian asserts that they were recognized as temporal punctuation marks by all nations: "they serve to fix the times of business and they are announced publicly."[16] Very convenient: that way there was no problem of knowing when to pray, since civil time signals would serve to summon the faithful.

The setting of prayer times by the clock was no small matter. It represented a first step toward a liturgy independent of the natural cycle. This tendency was much reinforced by the introduction of a night service, which apparently went back to the earliest days of Christianity, when the Jewish followers of Jesus, having celebrated the Sabbath, met again on Sunday for nocturnal devotions. The choice of hour had some precedent in scripture:

> I have remembered Thy name, O Lord, in the night . . .
>
> (Psalms 119:52)

> At midnight I will rise to give thanks unto Thee . . .
>
> (Psalms 119:62)

> I rose early at dawn and cried;
>   I hoped in Thy word.
> Mine eyes forestalled the night-watches,
>   That I might meditate in Thy word.
>
> (Psalms 119:147–148)

Scriptural precedent, though, is more often sanction than cause. The early Christians had good prudential reasons for coming together in the night while Caesar slept; also a most potent spiritual motive, namely the hope of salvation. The Gospel speaks of the Bridegroom's coming at midnight (Matt. 25:6), which led the church in Constantinople to institute a midnight office. Yet such precision was the exception, indeed was deliberately avoided. Uncertainty was preferable, because more compelling. The Lord will come, it is written, "at an hour you do not expect" (Matt. 24:42–44). "If he comes in the second watch, or if in the third," blessed are the servants who are watching and waiting (Luke 12:37–38). Nocturnal devotions, then, appropriately called vigils, were a spiritual watch for the second coming (the *parousia*) of the Lord.[17] Pliny the Younger wrote of this practice to the emperor Trajan at the beginning of the second century: "They are wont to come together before the light."[18]

For hundreds of years there were no rules, only practices. Rules came with monasticism—with the formation of a regular clergy (that is, a clergy subject to a *regula,* or rule) whose vocation it was to pray and pray often, and in so doing to save that multitude of the faithful whose worldly duties or inconstancy prevented them from devoting themselves entirely to the service of God. The innovator here was Pachomius in Upper Egypt in the early fourth century: against the prevailing eremitic individualism, his new order instituted a minute regulation of the collective praying, working, eating, and sleeping day. "It was there that for the first time we see realized the practice of an office in the strict sense, recited every day in the name of the church, *publicum officium,* at set hours."[19] Among the services: vigils, the *officium nocturnum* that was later merged with and called matins. From Egypt the practice spread to Palestine, Syria, Mesopotamia, and Europe.

Still, rules varied—"they were still feeling their way."[20] Temporal prescriptions, for example, may have been looser in the Eastern churches, where the natural diurnal cues continued to play an important role.[21] It was in the West, in the Rule of Saint Benedict, that the new order of the offices found its first complete and detailed realization: six (later seven) daytime services (lauds, prime, tierce, sext, none, vespers, and compline) and one at night (vigils, later matins). As the very names indicate, most of these were designated and set in terms of clock hours. Hence the very term "canonical hour," which eventually became synonymous with the office itself: one "recited the hours."[22]

This was around 530. In the centuries that followed, the Benedictine rule was adopted by other orders, including the great houses grouped around the Vatican and Lateran basilicas, thereby ensuring the eventual normalization of the canonical hours throughout Western Christendom. Progress in this direction was uneven owing to the physical insecurity of a violent age; in many parts of Europe, monastic life was disrupted for long periods by recurrent invasions and internecine strife. Besides, each house had its own interpretation of the Rule: we are talking here about customs (*consuetudines*), and there is nothing so idiosyncratic as custom.

Beginning in the tenth century external pressures eased, and the foundation of the Cluniac order (910), with its almost exclusive devotion to prayer, was the first sign of a general monastic revival. Cluny was followed by others, in particular the Cistercians (beginning of the twelfth century), under whom work regained the place it had held alongside prayer in the original Benedictine discipline. The very nature of these foundations, as expressed by the idea of an order, pressed them toward uniformity of practice and observance, and their reformism found expression not in the latitudinarianism often associated with the idea of reform today, but in the restoration of discipline. Discipline in turn had at its center a temporal definition and ordering of the spiritual life: *omnia horis competentibus compleantur*—all things should be taken care of at the proper time.

To be sure, one should not interpret the new discipline to mean an absolute uniformity of practice. We are still dealing here with uses and customs, and some of the confusion and contradiction among accounts and analyses of the monastic *horarium* is no doubt due to these conventual, regional, and national variations. Within each house, though, time discipline was taken seriously, and the abbot himself or his representative was personally responsible for its accuracy and enforcement. "Nothing, therefore, shall be put before the Divine Office," says the Rule.[23] Nothing was so important as the round of punctual, collective prayer.

Why was punctuality so important? One reason was that lateness—"God forbid!"—might make it necessary to abridge an office, in particular matins: "Let great care be taken that this shall not happen."[24] Another, I think, was that simultaneity was thought to enhance the potency of prayer. That would also explain the requirement that devotions be chanted aloud: to sing along is to sing together. That indeed was the point of community: the whole was greater than the sum of the parts.

Multiplication of simultaneous devotions—this was the way of salvation for all. Indeed, there were those who would have revived the Pauline ideal of continuous prayer (in relays presumably): thus Benedict of Aniane in the early ninth century and, even more, the monastery at Cluny in the tenth. (The latent purpose—or, if you will, the objective consequence—was, in conjunction with ascetic diet, to promote a state of light-headedness conducive to enthusiasm and hallucinations, or, euphemistically, to illumination and visions.)

The performance of such a demanding sequence, in particular the recitation of the nocturnal office after a period of sleep, imposed a new and special kind of temporal servitude. Unless some member of the congregation were ready to stay awake through the night and watch the clock—a precarious resort, as anyone who has stood sentry duty knows—it was only too easy to oversleep. In Roman times, some sympathetic or coreligionist member of the night watch may have served as waker; but with the fall of the empire, urban services broke down and watches became only a memory. To replace them, the medieval church would learn to make alarm mechanisms. Otherwise no one would ever have gotten any sleep, for fear of failing in his duty and jeopardizing not only his own salvation but that of others. Hence the instructions of one of the Villers Abbey fragments (1267–1268): "You must do the same when you set [the clock] after compline, so that you may sleep soundly."[25]

This religious concern for punctuality may seem foolish to rationalists of the twentieth century, but it was no small matter to a monk of the Middle Ages. We know, for one thing, that time and the calendar were just about the only aspect of medieval science that moved ahead in this period. In every other domain, these centuries saw a drastic regression from the knowledge of the ancients, much of it lost, the rest preserved in manuscripts that no one consulted. Much of this knowledge was not recovered until reimported hundreds of years later via the Arabs and the Jews in Spain or, still later, from Byzantium. But time measurement was a subject of active inquiry even in the darkest of the so-called dark ages. One has only to compare Isidore of Seville's rudimentary notions of time in his *De Temporibus* (615) with Bede's enormously popular textbook, the *De Temporum Ratione* (725)—written in the peripheral, tribal battleground that was Anglo-Saxon England—to realize the progress made in this field.

In large part this progress reflects the church's continuing concern to solve and systematize the dating of Easter and the other so-called movable feasts. These dates were established in accordance with the lunar as well as the solar calendar—like the Jewish calendar, but different. The principles of calculation, the science known as the *computus,* were sufficiently complex to give rise to multiple solutions, which came eventually to divide different Christian rites from one another. The task of extrapolating these dates into the future was particularly difficult, so much so that a thousand years later even so brilliant a mathematician as Carl Friedrich Gauss was not able to reduce the calculation to a comprehensive algorithm.[26]

It was in this area that Bede made his greatest contribution, and the rapid diffusion of his work on the continent testifies to its superiority and interest. Certain monasteries became centers of training and calculation (thus Sankt Gallen and Auxerre) and produced a substantial literature on the subject that was avidly copied elsewhere. The great volume of tables, charts, discussion, and diagrams that can be found today in any major manuscript collection testifies to the vigor and creativity of this effort.[27]

Most of this literature deals with dating, but calendrical concerns invariably spilled over into the area of time measurement, and vice versa. Indeed, I would argue that it was precisely this that made European astronomy and the computus so different: the practitioners were interested not only in the moon and the seasons, but in the day and its divisions. In particular, these same monks wanted to know the division of the day into light and darkness, the better to set the hours of the liturgy. The best of them, Gerbert for example, were quite aware that day (and night) did not grow and diminish at an equal rate from week to week, and they worked out the schedule of changing proportions (what they called a *horologium*) by measuring day and night at the solstices and adjusting from there. Gerbert offers one correspondent advice on how to take this measure: use a clepsydra, he says, and collect the water separately for night and day; then pour them together, and if the sum makes twenty-four (equinoctial) hours, you know you have it right.[28]

This combination of measure and calculation made possible the construction of *horologia* giving night and day for every day in the year. We have one tenth-century table, for example, which gives the division not only by hours (*horae*), but by points (*puncta*, five to the hour) and *ostenta* (twelve to the *punct*). That made each *ost* equal to one of our minutes, and the clocks of the day could not measure that accurately; so the figures in the table were given to the nearest third of a *punct*, that is, four *osts*.[29]

Time mattered to such experts as Gerbert, but it also mattered to the ordinary monk, for whom getting up in the dark of the night was perhaps the hardest aspect of monastic discipline. Indeed, the practical meaning of "reforming" a house meant first and foremost the imposition (reimposition) of this duty. The sleepyheads were prodded out of bed and urged to the office; also prodded during service lest they fail in their obligations. Where the flesh was weak, temptation lurked. Raoul Glaber (early eleventh century) tells the tale of a demon who successfully seduced a monk by holding out the lure of sweet sleep: "As for you, I wonder why you so scrupulously jump out of bed as soon as you hear the bell, when you could stay resting even unto the third bell . . . but know that every year Christ empties hell of sinners and brings them to heaven, so without worry you can give yourself to all the voluptuousness of the flesh."[30]

The same Glaber confesses to two occasions when he himself woke late and saw a demon, "come to do business with the laggards."[31] And Peter the Venerable, Abbot of Cluny in the twelfth century, tells the story of Brother Alger, who woke thinking he had heard the bell ring for nocturns. Looking around, he thought he saw the other beds empty, so he drew on his sandals, threw on his cloak, and hastened to the chapel. There he was puzzled not to hear the sound of voices lifted in prayer. He hurried back to the dormitory. There he found all the other monks fast asleep. And then he understood: this was all a temptation of the devil, who had awakened him at the wrong time, so that when the bell for nocturns really rang, he would sleep through it.[32]

These, I suggest, are what we now know as anxiety dreams. They clearly reflect the degree to which time-consciousness and discipline had become internalized. Missing matins was a serious matter, so serious that it has been immortalized for us by perhaps the best known of children's songs:

> Frère Jacques, Frère Jacques,
> Dormez-vous? dormez-vous?
> Sonnez les matines, sonnez les matines,
> Ding, ding, dong; ding, ding, dong.[33]

4 The Greatest Necessity for Every Rank of Men

"I DLENESS," WROTE BENEDICT, "is an enemy of the soul."[1] The fixing of a daily schedule of prayer was only part of a larger ordering of all monachal activity, worldly as well as religious. Indeed, for monks there was no distinction between worldly and religious: *laborare est orare*—to work *was* to pray. Hence, there were rules setting aside times for work, study, eating, and sleeping; rules prescribing penalties and penance for latecomers; rules providing explicitly for the maintenance of the clock and its nightly adjustment, so that it would wake the sacristan at the proper time. Note that at this stage, it was not the clock that worked the big bells. As "Frère Jacques" tells us, the clock merely rang loudly enough to get the bell ringer out of bed.

These were not necessarily clocks in our sense of the term. Many of them did not indicate the time or run continuously. Rather, they were what we now know as timers and associate with three-minute eggs or film developing. But these were timers that ran for hours. They were set to run during the night and served only to trip the alarm; to use the medieval terminology, they were *horologia nocturna* or *horologia excitatoria*. It is now generally agreed, moreover, that some of them made use of an escapement-type mechanism to produce a to-and-fro motion of the hammer(s) beating on the bell, and that this mechanism was often weight-driven.

It is this mechanism, probably, that was the forerunner of the clock escapement.[2] We have already met the device: the wheel train ends in a *scape wheel* (also called the *crown wheel,* after the shape of the teeth), whose rotation is alternately blocked and released by pallets on a staff (the *verge*) pivoting to and fro.[3] In the alarm version, one puts a knocker on the end of the verge, and this oscillating knob strikes a bell. In the timekeeping version, the verge is T-crossed by a rod that swings back and forth with it (called the *foliot,* perhaps because of its "mad" motion). This foliot is an inertial controller: by moving weights along it, one can change its moment and thus the beat of the mechanism. (See Figures 5 and A.1.) The action, in other words, is the same in both versions, and it would not have taken much to go from one to the other.

Etymology makes clear what was happening. Before the invention of the weight-driven mechanical clock, remember, the clepsydra and sundial were both known as *horologia,* and this generic term was subsequently applied in the vernacular to the new machine as well. Thus we get French *horloge,* Italian *orologio,* Spanish *reloj.* But new things often call for new names: the English called the new device a clock; the Dutch and Flemings, a *klokke.* And what is a clock, but a bell? (Compare medieval Dutch *clokke,* German *Glocke.*) Even the French, who stayed with the old name, changed their word for bell at about this time, from *sein* or *sain* (from the Latin *signum*) to *cloche.*[4] Something new had come on the scene. Seen ontologically and functionally, these timekeeping machines began as automated bells.

Bells, bells, bells. Big bells and small. Monasteries were beehives of varied activity, the largest productive enterprises of medieval Europe. Brothers, lay brothers, and servants were busy everywhere—in the chapel, the library, the writing room (scriptorium), in the fields, the mill, the mines, the workshops, the laundry, the kitchen. They lived and worked to bells. The big bells tolled the canonical hours and the major changes, and their peal carried far and wide, not only within the convent domain but as far as the wind could take it. And the little bells tinkled insistently throughout the offices and meals, calling the participants to attention and signaling the start of a new prayer, ceremony, or activity. All of this was part of a larger process of depersonalization, deindividuation. Monastic space was closed space—areas and corridors of collective occupancy and movement—so arranged that everyone could be seen at all times. So with time: there was "only one time, that of the group, that of the community. Time of rest, of prayer, of work, of meditation, of reading: signaled by the bell, measured and kept by the sacristan, excluding individual and autonomous time."[5] Time, in other words, was of the essence because it belonged to the community and to God; and the bells saw to it that this precious, inextensible resource was not wasted.

The bells, in short, were drivers—goads to effective, productive labor. It is this larger role, going far beyond reveille, that may account for the higher standard of punctuality enforced by the new monastic orders of the eleventh and twelfth centuries. The Cistercians in particular were as much an economic as a spiritual enterprise (they would not have recognized a difference). Their agriculture was the most advanced in Europe; their factories and mines, the most efficient. They made extensive use of hired labor, and their concern for costs made them turn wherever possible to labor-saving devices. Their Rule enjoined them, for example, to build near rivers, so as to have access to water power; and they learned to use this in multifunctional, staged installations designed to exploit power capacity to the maximum. For such an undertaking, timekeeper and bells were an indispensable instrument of organization and control; and it may be that it was the proliferation of this order throughout Europe and the expansion of its productive activities that stimulated the interest in finding a superior timekeeper and precipitated the invention of the mechanical clock. The Cistercian abbeys of central Europe must have had their hands full getting satisfactory performance from clepsydras.[6]

Whatever the inspiration, it seems clear that in the century or two preceding the appearance of the mechanical clock, there was a substantial advance in the technique of hydraulic timekeeping and a concomitant diffusion of the new methods and devices. For the first time we see the temporal discipline of the cloister explicitly linked to the *horologium:* thus the instructions of William, abbot of Hirsau in the eleventh century, on the duty of setting the clock each night (to take account of the unequal temporal hours);

and the several provisions of the Cistercian Rule (early twelfth century) on the care of clocks and bells. From these and similar references, still occasional but too frequent to be dismissed as exceptional, we may infer that the bell-ringing clepsydra became in this period a feature of the "well-tempered" monastery. The strongest corroboration, in my opinion, is the language of Robertus Anglicus in the passage cited earlier (p. 10) on the search for a mechanical clock. "Clockmakers," he tells us—*artifices horologiorum*—are trying to make a wheel that will make one turn in a day. Who were these *artifices* if not technicians (mechanics) who had made a specialty of the clepsydra, in particular the bell-ringing clepsydra, and were led by their experience of wheelwork to experiment with new kinds of timekeepers? I am not one of those who give credence to the existence of a guild and street of clockmakers (presumably water-clock makers) in Cologne in the late twelfth and early thirteenth centuries—after all, why Cologne? But I do think that Robert's designation of horological specialists, *artifices horologiorum,* is unconscious testimony to the presence, if not of a trade, at least of an established group of producers and, by implication, of a corresponding market.[7]

The monastic clergy may have provided the primary market for timekeepers and the principal stimulus to technical advances in this domain, but the church alone cannot account for the popularity and development of the new device, which for all its limitations rapidly drove the clepsydra from the scene. For one thing, clerical demand by itself was probably insufficient to sustain what rapidly became a major craft. For another, the nature of time measurement as practiced by the church was incompatible with the technological possibilities and characteristics of the new instrument.

Consider the new sources of demand. These consisted of, first, the numerous courts—royal, princely, ducal, and episcopal; and second, the rapidly growing urban centers with their active, ambitious bourgeois patriciates. At the very beginning, in the thirteenth and early fourteenth centuries, princes and courtiers may well have accounted for the greater part of secular demand for timekeepers. Typically they were the wealthiest members of society, the more given to luxury expenditure because they did not earn their income. (It is always easier to spend other people's money.) The preceding centuries, moreover, had been an era of sustained increase in wealth and power: population was growing, and with it the area under cultivation; trade also, and with it the yield of duties and taxes. These new resources nourished central authority and enabled it to enforce that condition of order that is in itself the best encouragement to productive activity. (One should not exaggerate, of course. Europe was still a perilous place by today's standards, but security had immensely improved over what it had been when Northmen, Magyars, and Saracens were raiding everywhere with impunity and the law of the strongest and most violent prevailed.) One can well understand, then, how after centuries of reconstruction and growth the rulers of Europe seized upon and delighted in the new bell-ringing clocks, these wondrously ingenious instruments, costly to build and maintain, but well worth it for their plangent ubiquity—the ideal, quotidian reminder of and symbol of high authority.

In the long run, though, the future of the infant clock industry lay with the bourgeoisie—originally and literally the residents of the *bourgs* (in colloquial American English, the *burgs*). Along with the crown, indeed in alliance with it, the town was the great beneficiary of the agricultural and commercial expansion of the high Middle Ages (eleventh to fourteenth centuries). Sleepy villages were becoming busy marketplaces; administrative centers and points of transshipment and exchange were growing into nodes of wholesale and retail trade and craft industry. The more successful residents of these new cities quickly came to constitute a new elite, an urban patriciate possessed of great wealth and a sense of power and self-esteem that rivaled that of the older landed elite. They were able, further, by shrewd cooperation with the crown and the construction of an urban military base, to win substantial autonomy for their municipalities, which were organized by collective agreement among the residents and by contractual arrangement with or concession from higher authority into self-administering communes. These had their own fiscal resources, so that when mechanical clocks appeared on the scene, the cities of western and Mediterranean Europe could afford to build them as complements to or successors to the cathedrals—a symbol of a new secular dignity and power and a contribution to the general welfare.

Why the general welfare? Because, just like the monastery, the city needed to know the time even before the mechanical clock became available. Here, too, necessity was the mother of invention.

We have already noted the contrast between the "natural" day of the peasant, marked and punctuated by the given sequence of agricultural tasks, and the man-made day of the townsman. The former is defined by the sun. The latter is bounded by artificial time signals and the technology of illumination and is devoted to the same task or to an array of tasks in no given sequence. The spatial compactness of the city, moreover, is an invitation to serial engagements: with careful planning (that is, timing), one can multiply oneself. To be sure, the medieval town long remained half-rural. Everybody who could, kept a *basse-cour* of chickens, roosters, rabbits, and other useful livestock; so that some of the natural time signals heard in the countryside were heard in the city as well. Still, it is one thing to receive or perceive the time; another thing to track and use it. The two environments differed radically in their temporal consciousness.

This difference was growing. (It was not to contract until the nineteenth century, with the coming of the railroad and the penetration of the country by the rhythms and servitudes of the city.) As commerce developed and industry expanded, the complexity of life and work required an ever larger array of time signals. These were given, as in the monasteries, by bells: the urban commune in this sense was the heir and imitator of the religious community. Bells sounded for start of work, meal breaks, end of work, closing of gates, start of market, close of market, assembly, emergencies, council meetings, end of drink service, time for street cleaning, curfew, and so on through an extraordinary variety of special peals in individual towns and cities.

The pressure for time signals was especially strong in those cities that were engaged in textile manufacture, the first and greatest of medieval industries. There the definition of working time was crucial to the profitability of enterprise and the prosperity of the commune. The textile industry was the first to engage in large-scale production for export, and hence the first to overflow the traditional workshop and engage a dispersed work force. Some of these workers were true proletarians, owning none of the instruments of production, selling only their labor power. They streamed early every morning into the dye shops and fulling mills, where the high consumption of energy for heating the vats and driving the hammers encouraged concentration in large units. These workers—called *ciompi* in Florence, "blue nails" (stained by dye) in Flanders—were poorly paid, overworked, potentially troublesome and mobilizable. Other branches of the manufacture could be conducted in the rooms and cottages of the workers. Employers liked this so-called putting out because it shifted much of the burden of overhead costs to the employee, who was paid by the piece rather than by time; and the workers preferred it to the time discipline and supervision of the large shops. They could in principle start and stop work at will, for who was to tell them what to do in their own home?

The bells would tell them. Where there was textile manufacture, there were work bells. Artisans in other places might work the traditional day from sunup to sundown, but Brussels had its *joufvrouwenclocke* at dawn, another work clock (called the *werckclocke*) a little later, a *drabclocke* in the evening for weavers and twisters, among others, and a *lesteclocke* for tapestry workers, cobblers, and whitesmiths. Sometimes these bells were public, installed by the municipal authorities in a church tower, perhaps rented, or in a belfry erected for the purpose. This was the case in Amiens in 1335, where the king granted the request of mayor and aldermen "that they might be permitted to issue an ordinance concerning the time when the workers of the said city and its suburbs should go each morning to work, when they should eat, and when return to work after eating; and also, in the evening, when they should quit work for the day, and that by the issue of said ordinance, they might ring a bell which has been installed in the Belfry of the said city, which differs from the other bells." Sometimes the bells were private, the property of the employer. In Ghent in 1324 the abbot of St.-Pierre authorized the fullers "to install a bell in the workhouse newly founded by them near the Hoipoorte."[8]

These work bells inevitably gave rise to conflict. Part of the problem, no doubt, was implicit in the effort to impose time discipline on home workers. In principle, payment by the piece should have taken care of the matter, with workers responding to wage incentives. In fact, the home workers were content to earn what they felt they needed, and in time of keen demand, employers found it impossible to get them to do more, for higher pay only reduced the amount of work required to satisfy these needs. The effort to bring the constraints of the manufactory into the rooms and cottages of spinners and weavers made the very use of bells a focus of resentment.

Meanwhile in the fulling mills and dye shops the bells posed a different kind of problem, especially when they were controlled by the employer. Consider the nature of the wage contract: the worker was paid by the day, and the day was bounded by these time signals. The employer had an interest in getting a full day's work for the wages he paid, and the worker in giving no more time than he was paid for. The question inevitably arose: How did, indeed how could, the worker know whether bell time was honest time? How could he trust even the municipal bells when the town council was dominated by representatives of the employers?

Under the circumstances, workers in some places sought to silence the *werckclocke*. At Thérouanne in 1367 the dean and chapter promised "workers, fullers, and other mechanics" to silence "forever the workers' bell in order that no scandal or conflict be born in city and church as a result of the ringing of a bell of this type."[9] But few places gave in so completely, and the years after the Black Death of 1347–1350 saw repeated trouble on this score. The plague had sharply reduced the population of the cities and towns, some of them by well over half, and a skeleton labor force was exploiting its enhanced bargaining power to demand concessions from employers and authorities. Among other things, they turned the very bells that bound them into tocsins of revolt. The decrees of these years make it clear what was at stake: the heaviest penalties were reserved for such *lèse-majesté*. At Commines the fine was sixty pounds (an enormous sum) for anyone ringing the bell as a call to assembly; and for sounding a call to revolt, the punishment was death.

Such efforts to eliminate the work bells never achieved success: as soon suppress the system of wage labor. Besides, once the workday was defined in temporal rather than natural terms, workers as well as employers had an interest in defining and somehow signaling the boundaries. Time measurement here was a two-edged sword: it gave the employer bounds to fill and the worker bounds to work. The alternative was the open-ended working day:

We'll always be weaving cloth of silk,
And shan't be better dressed for it.
We'll always be poor and bare
Always hungry and thirsty.
We'll never be able to earn enough
To eat better.
Bread, we have to share,
A little in the morning and in the evening less.
And we are in great misery,
But the man we work for
Gets rich on our wages.
We're up a good part of the night
And work all day to make our way . . .[10]

It was not the work bells as such, then, that were resented and mistrusted, but the people who controlled them; and it is here that the chiming tower clock made its greatest contribution. It provided regular signals—at first on the hour, later on at the halves or quarters—which necessarily limited the opportunities for abuse. Of course, with the appearance of the dial (from the word for day), it was possible for all interested parties to verify the time on a continuous basis.

This was not the end of the matter. As new clocks were built, discrepant time signals gave rise to new issues of discord: Why are we obliged to start work earlier than they? Or to stay later? Perhaps it was with this kind of conflict in mind that Charles V of France decreed in 1370 that all clocks in the city should be regulated on the one he was installing in his palace on the Ile de la Cité. He thereby affirmed the primacy of royal power, but such decrees could not solve the problem. The early tower clocks were far too crude and inaccurate to synchronize, even approximately: "C'est l'horloge du palais; / Elle va comme il lui plaît." Almost two hundred years later, another Charles V, Holy Roman Emperor, was to spend the last years of his life trying among other things to make his clocks sound together. He was still trying when he died.

This change in the technique of time measurement and signaling was associated with an equally drastic change in the units of measurement. The church, remember, kept temporal hours that changed with the season, and as the church kept time, so did the rest of society. The punctuation marks were, in addition to the natural diurnal events (sunrise, sunset), the liturgical offices. Thus, the Parlement of Paris met at the hour of the first mass in Sainte-Chapelle and remained until the bell for none. In Bruges court cases ran until noon, and appeals until vespers.[11] In Liège a citizen condemned for a debt had to pay or give security by sunset. The millers of Paris ceased work on Sunday from the announcement of the holy water in the chapel of St-Leufroy to the ringing of vespers. Spinsters of silk quit work in the summer when the bell of Ste-Marie-des-Champs called to alms, and carpenters stopped on Saturday when the big bell of Notre Dame sounded none. In summer (defined as from Easter to Saint Rémy's day) the tanners of Paris worked from sunrise to sunset. But what about wintertime, when the sky was often cloudy and the sun obscured? They worked as long as it was light enough to tell two similar coins apart, the *livre tournois* and the *livre parisis*. Another regulation provided that work would begin as soon as it was light enough to recognize someone in the street.[12]

These are obviously nonclock standards. Whether, as some historians have believed, they are also preclock—that is, whether they are evidence that as yet no clockwork turned in Paris or clock bells chimed—is another story. Nothing was so conservative in the Middle Ages as hours of business and terms of labor. Any change in these was sure to injure some vested interest, while the hazards, costs, and limitations of artificial illumination made employers and workers alike reluctant to work days of fixed length. The fact that earlier usages persisted well into a new timekeeping era is testimony to the difficulty of changing so fundamental a way of thinking about and ordering life and work.

In the long run, however, change was implicit in the new mode of measurement. Whereas variable (temporal) hours were easily measured by the clepsydra, they were incompatible with mechanical turret clocks. Not that the flow of a water clock can easily be made to vary with the season; but the reading is easily adjusted to the calendar. All that is required is a series of scales, either marked on the clepsydra itself or on measuring rods, for the different times of year. (The ancient Egyptians learned to use a similar system with their shadow sticks and vertical gnomons.)

The mechanical clock is ill-adapted to such use. Its beat is regular and it marks its beat. To be sure, one can allow for variable

Plate I. Table clock by Steffen Brenner, Copenhagen, 1558. Height 21 cm. The movement is made of iron; the case of bronze, gilt brass, and silver. *Sonnerie* hours and quarters, plus alarm. Five dials. On the four sides: (1) astrolabe dial, sun hand, dragon hand with rotating moon (showing phases) at tip; number rings for temporal (unequal) hours, 2 × 12, and the astrological houses, 1–12; (2) age of moon, moon phases, duration of moonlight in hours and minutes; position of sun in zodiac; (3) length of day and night; correction dial in center; (4) days of the week, with associated planets; alarm dial in center. On top: Calendar and holy days; year-to-year dominical letters, movable feasts, golden number; minute hand.

A word about the very rare dragon hand: its function is to show the position of the moon's nodes (the points on the celestial sphere, 180 degrees apart, where the moon's path intersects the sun's). Used in conjunction with the sun hand, which tracks through the year along the ecliptic, such a dragon hand makes it possible to predict eclipses. Since the line of nodes retrogrades relative to the stars, the dragon hand turns counterclockwise, sweeping the calendar and zodiac scales and completing its revolution in some 18.6 years.

Brenner was court clockmaker to Frederick II of Denmark. This table clock is surely one of the most beautiful and mechanically ingenious astronomical clocks ever made—in its architecture, far ahead of its time.

Plate II. Table "tabernacle" clock, Augsburg, 1600, no signature. Height 52 cm. Hour and quarter striking; also alarm. Movement is of brass, except for epicyclic gearing (iron). Case is of gilt bronze, brass, silver. Dials are of silver, partially enameled. Astrolabe dial has dragon, sun, and moon hands, age and phases of moon; dial below gives days of week and associated planets. Opposite side: hours and minutes, sectors for length of day and night; below, dial with hand for adjusting length of day and night. On sides: two calendar dials, each covering half a year, giving saints' days, dominical letters, the golden number, the epact (age of the moon on January 1, used in fixing the date of Easter), and the date of Easter for the years 1600–1687. Below, two small dials for regulating the striking and going rates and setting the alarm. A typical, high-quality astronomical clock, multiplying functions and information at the price of reliability and accuracy.

Plate III. Clock in the form of a Turkish vessel. Augsburg, c. 1585. Height 45 cm. The movement is made of iron; the case is of gilt bronze and copper. *Sonnerie* hours and quarters. Clock in center. The eyes of the Turk on top are linked to the going train and move from side to side. On the hours, he lifts his arm, which probably held a sword originally; at the same time, the two oarsmen make rowing movements. On the quarters, the monkey on the prow moves and the front oarsman turns his head back and forth. This is a clockwork toy, perhaps made for the Turkish tribute.

Plate IV. Table automaton clock, by Nikolaus Schmidt, Augsburg, c. 1580. Height 43 cm. The movement is made of iron; the case is of gilt bronze and copper. The eyes of the elephant move with the clock balance; the legs are soldered in position. (Did the raised leg move at one time?)

Animal clocks were very popular at the time, especially those showing exotic beasts (lions, camels, and mythical birds, as well as elephants). But this one resembles nothing so much as a rook (castle) in an Indian chess set.

Plate V. The Tower of Babel: a "Blois enamel" watch. Movement by Matthis Wentzel, Strasbourg, 1636; case attributed to Jean Toutin, inventor of the technique of painting on enamel. Diameter 47 mm. A superb example of a vitreous polychrome miniature. The case is painted back and side, inside and out. The dial is also painted. Pre–balance spring; one hand.

Plate VI. A miniature copy of the Velasquez portrait of Philip
IV of Spain (1621–1665): a "Blois enamel" watch with movement by Edme Burnot, Brussels, c. 1650; case unsigned. Diameter 62 mm. A very rare combination of polychrome painting
and floral enamel in relief. On the reverse a comparable copy
of the Velasquez portrait of Maria Anna of Austria, Philip's
wife. Philip spent much of his reign and the resources of his
kingdom trying to assert Spanish dominion over the Low
Countries—in particular, over Protestant Holland; hence the
link to a watch made in Brussels.

Plate VII. The Holy Family: a "Blois enamel" watch by Salomon Pairas (Payras) of Blois, c. 1650. Painting unsigned. Characteristic *bassine* case, painted all sides, in and out. The choice of a religious subject was common and signaled, I think, to others the loyalties and style of the wearer. Another favorite theme was Greek mythology, which conveyed a different message.

Plate VIII. The high-fashion clock of the mid-eighteenth century: all packaging and presentation. The case here is Meissen porcelain; the clock itself, by the distinguished Ferdinand Berthoud, is incidental. The secret of making hard-paste porcelain, reserved to China for a thousand years, had been discovered at Meissen in the early years of the century, and now polychrome porcelain figures were all the rage. It goes without saying that a clock like this was almost untouchable. Even the winding called for the utmost care, and dusting or moving was just about out of the question. The wonder is that a fair number of these have survived, though almost never without some damage, however slight.

hours by either changing the rate or the reading. The clock will beat faster or slower as the weights on the foliot are closer or farther from the center of oscillation. In principle, then, a mechanical clock could be adjusted as necessary to take account of the changing and different lengths of day and night hours. Such adjustments, though, are at best approximate even with small, easily accessible chamber clocks. They would have been far more difficult with the large tower clocks that marked the passage of time for the general public.

Changing to a variable reading was even harder. It can be done. The Japanese, who retained their variable hours (in every way comparable to the temporal hours of ancient and medieval Europe) even after learning of and adopting the mechanical clock in the sixteenth century, resorted eventually to the ingenious device of movable numerals on the dial of their timekeepers. But they were adjusting these numerals (moving them closer together or farther apart) on chamber clocks; and even so, the results cannot have been very accurate, if only because the changes were made only about once a fortnight.[13] (See Figure 8.)

Medieval Europe did not have this recourse. Remember that many of these early clocks were public, deliberately sited in towers and spires, tens of feet above ground. They usually had no dial, hence no numbers to move about—hardly an option in any case. Instead, as we have seen, these were automated bells. Since there was no practicable way to render the *detent* (release) for the bell train independent of the regular beat of the time train, these clocks perforce marked equal hours; and as they marked, so increasingly did the urban society that depended on them. Perhaps the earliest recorded example of the new, secular time standard comes to us from Sarum, England, where a regulation of 1306 stated that "before the clock of the Cathedral had struck one no person was to purchase or cause to be purchased flesh, fish or other victuals."[14]

The introduction of equal hours and the habituation of urban populations to public time announcements had profound consequences for the European mentality. Medieval man, it has been observed, was innumerate as well as illiterate. How much reckoning could he do in a world that knew no uniformity of measurement? Units of distance were linked to physical characteristics that varied as people do (the English *foot,* for instance, and the

French inch, called a *pouce,* which means thumb); while weights typically were converted to volume standards (a *bushel* of grain) that inevitably varied from place to place and mill to mill. Even the learned were not accustomed to using numbers. The calculation of the calendar, for example—a crucial aspect of liturgical discipline—was confined to specialist computists. The schools offered little if any training in arithmetic, and the very persistence of roman numerals was both symptom and cause of calculational paralysis.

All of this began to change in the twelfth and thirteenth centuries—just as one would expect. This was a period of growing trade, and he who trades must reckon. So must clerks and functionaries who count taxes and expenditures, and these were years of rapid development of royal power and government apparatus. It was no accident that arabic numerals came in at this time, or that books that had once resorted to metaphor now gave numbers, however erroneous, of armies, treasures, buildings, and the like.[15]

It was the urban, commercial population that seems to have been quickest to learn the new language and techniques. Arithmetic was the province above all of the unlettered speakers of the vernacular (as opposed to Latin). Many of these learned arithmetic in the shop or on the road, but even before they entered trade, they learned to count by the bells of the clock. Not by the old church bells ringing the canonical hours; these did not mark equal units and hence did not lend themselves to addition and subtraction. But the new bells and the calculations they made possible (how long until? how long since?) were a school for all who listened and began to organize their lives around them.[16] Meanwhile the church clung to old ways and, so doing, yielded the rhythm of life and work to the lay authorities and the bourgeoisie. Equal hours announced the victory of a new cultural and economic order. Here indeed was an unintended consequence: the monks had wrought too well.

THE EARLY TURRET CLOCKS were very expensive, even when simple. Wrought iron and brass needed repeated hammering, hence much labor and much fuel. The casting of the bells was a precarious operation. The placement of the mechanism usually entailed major structural alterations. We shall see later that the construction and installation of a tower clock might take months, if not years; that teams of craftsmen and laborers had to be assembled on the site and there lodged and boarded; and that the task of subsequent maintenance required the permanent attendance of a resident technician, repeated visits by specialized artists, and an endless flow of replacement parts. Constructing a clock was not the same as building a cathedral, a project so costly that it engaged the surplus resources of an entire community, to the point of extenuation.[17] But it did entail a substantial and continuing commitment, usually by the public authority, in the name of the common weal.

These costs increased substantially as soon as one went beyond simple timekeepers to astronomical clocks and automata. The medieval accounts show this process clearly: the sums paid to painters and woodcarvers bear witness to the growing importance of the clock as spectacle as well as time signal. The hourly parade of saints and patriarchs; the ponderous strokes of the hammer-wielding jacks, the angel turning with the sun, the rooster crowing at sunrise; the lunar disk waxing and waning with the moon—all these movements and sounds offered lessons in theology and astronomy to the up-gazing multitude that gathered to watch and wonder. The clock as pageant was an imitation of divine creation, a miniaturization of heaven and earth. As such it was a source of immense pride to the kings and communes that built it, a challenge to all kingdoms and cities around. The show clock was to the new secular, urbanizing world of the later Middle Ages what the cathedrals had been to the still worshipful world of the high Middle Ages: a combination of sacrifice and affirmation, the embodiment of the highest skills and artistry, a symbol of prowess and source of pride. When Philip the Bold of Burgundy defeated the Flemish burghers at Rosebecke in 1382 and wanted to punish those proud and troublesome clothiers, he could do no worse (or better) than seize the belfry clock at Courtrai and take it off to his capital at Dijon.

This symbolic (totemic) role of the clock goes far to account for the rapid diffusion of these instruments in western and central Europe. A show clock was a matter of prestige, an edifying spectacle for residents and visitors alike. In this sense, clocks were the secular analogue to the religious relics that had long been the

most potent attraction to pilgrims and travelers. Cost was a secondary consideration, and the multiplicity of autonomous chartered communes with substantial tax revenues of their own
provided a ready demand. (Such communes, be it noted, were a
uniquely European phenomenon, not to be found in Islam or
East Asia.)

These public clocks, moreover, were only the top of the market.
They are the ones that history knows best, but we know only a
fraction of what was made. In this regard, the records are misleading: they have preserved the memory of a spotty, biased selection and largely omitted the smaller domestic clocks made to
private order. As a result, it was long thought that the first mechanical clocks were turret clocks and that the smaller domestic
models were the much later product of advances in miniaturization. Yet there was no technical impediment to making chamber
clocks once the verge escapement had been invented. Indeed, Antonio Simoni has persuasively argued that since the mechanical
clock was a development of the timer alarm, itself made to chamber size, small clocks must have preceded the big turret machines.[18]

Whichever came first, the one logically implied the other, so
that we may fairly assume that both types of clock were known
and made from the start. It so happens that the first literary allusion we have to a mechanical clock refers to domestic timepieces.
This goes back to the late thirteenth century, in Jean de Meung's
additional verses to *Le roman de la rose.* Jean, a romantic poet of
curiously worldly interest, attributes to his Pygmalion a fair array
of chamber clocks:

> Et puis faire sonner ses orloges
> Par ses salles et par ses loges
> A roes trop subtillement
> De pardurable mouvement.

> And then through halls and chambers,
> Made his clocks chime
> By wheels of such cunning
> Ever turning through time.[19]

From the middle of the fourteenth century, chamber clocks
show up in inventories and accounts. Chancellery records of Aragon, recording clock purchases by the royal family, mention more

than half a dozen master clockmakers, producing simple and complicated timepieces, transportable (from the middle of the century) as well as fixed.[20] And the inventory of the personal possessions of Charles V after his death in 1380 mentions among other things a clock all in silver "without iron," with two silver weights filled with lead, apparently made toward the beginning of the century for his ancestor Philip the Fair (died 1314).[21]

These were, of course, the furnishings of kings, which have a way of being remembered. There were many more clocks, surely, that were made, went out of order, were cast aside, and disappeared unnoticed. We may reasonably infer this from the numerous references to clockmakers. The clock lists show pieces in some very small and unexpected places, no clocks in some important cities, and only a handful of clocks in the greatest centers. A handful of clocks in London and Paris? There must have been ten times that many and more by 1400. If a clockmaker could make one, he could make many, and there were probably dozens of clockmakers active in Europe by the end of the fourteenth century.

This rapid emergence of a new profession was at once a strong force for improvement in quality and reduction in costs, hence a stimulus to demand, and the best sign of the popularity of the new device. Few inventions in history have ever made their way with such ease. Everyone seems to have welcomed the clock, even those workers who toiled to its rules, for they much preferred it to arbitrary bells. Indeed, one of the themes of contemporary observers was the usefulness of the clock to people of all walks of life: *summe necessarium pro omni statu hominum* was the way Galvano Fiamma, chronicler of Milan, put it, when he proudly marked the restoration in 1333 (?) of a clock that not only struck the hours but signaled each one by the number of peals.[22] And this in turn recalls an earlier inscription on a clock installed in 1314 on the bridge at Caen: "Je ferai les heures ouir / Pour le commun peuple rejouir" "I shall give the hours voice / To make the common folk rejoice").[23]

Even the poets liked the new clocks. That is the most astonishing aspect of these early years of mechanical horology, for no group is by instinct and sensibility so suspicious of technical innovation. Here, moreover, was an invention that carried with it the seeds of control, order, self-restraint—all virtues (or vices) inimical to the free, spontaneous imagination and contemplation so prized by creative artists. Yet it would be anachronistic to impute these ideals to the thirteenth and fourteenth centuries; they came much later. The medieval ideal was one of sobriety and control, along with due respect for worthy models. Besides, it was surely too soon to understand the potential of the new device for forming the persona as well as for dictating the terms of life and work. Instead, the availability of this new knowledge gave all a sense of power, of enhanced efficiency and potential, of ownership of a new and valuable asset; whereas we, living by the clock, see ignorance of or indifference to time as a release from constraint and a gain in freedom. When we go on vacation, we want to be able to put our watch in the drawer and not look at it until we return to the "real" world: that is the essence of what the American armed forces call R & R (rest and recreation). Everything depends, I suppose, on where one is coming from.

In any event, the early celebrators of the clock were no mere poetasters. The great Dante Alighieri praised the "glorious wheel" moving and returning "voice to voice in timbre and sweetness"—*tin tin sonando con si dolce nota* (almost surely a reference to a chamber clock, unless Dante had a tin ear)—therein echoing the pleasure that Jean de Meung's Pygmalion took in his chiming clocks a generation earlier.[24] And a half-century later Jean Froissart, poet but more famous as historian, composer of "love ditties," sang in his *L'horloge amoureuse* (1369) a love song to the new machine:

> The clock is, when you think about it,
> A very beautiful and remarkable instrument,
> And it's also pleasant and useful,
> Because night and day it tells us the hours
> By the subtlety of its mechanism
> Even when there is no sun.
> Hence all the more reason to prize one's machine,
> Because other instruments can't do this
> However artfully and precisely they may be made.
> Hence do we hold him for valiant and wise
> Who first invented this device
> And with his knowledge undertook and made
> A thing so noble and of such great price.[25]

11　KEEPING TIME

The mechanical clock began as a crude, imprecise, unreliable instrument. It took four hundred years to turn it into an accurate timepiece.

It proved much easier to do what seems at first thought a harder task—that is, adapt the principles of clockwork to complex mechanisms for tracking the heavenly bodies or driving ingenious automata. Within less than a century after the invention of the mechanical clock, we have the masterpieces of Richard of Wallingford and Giovanni de' Dondi; more important, we have a rapidly growing array of show clocks—the legendary Strasbourg cathedral tower clock is a spectacular example—combining fanciful astronomical and temporal indications with a pageant of moving figures, historical, mythical, and symbolic. These mechanisms called for esoteric mathematical calculations, clever and sometimes original mechanical arrangements, and extraordinary craft skills. Even so, the essential problem, that of linking different rates of circular (hence angular) motion by appropriate systems of gearing, was an old one and had been solved as far back as ancient Greece.[1] By using the right ratios of wheel teeth and pinion leaves, along with a mix of round and not-so-round wheels, one could track the sun through the zodiac (easy), reproduce the nineteen-year cycle of the epact (harder) and the movable feasts of the church calendar (harder still), trace the epicyclical orbits of the planets (very hard), and so on, with sometimes amazing theoretical accuracies.

Consider, for example, the moon drive of Richard of Wallingford's machine. This was theoretically capable of producing a lunation of 29 days, 12 hours, 44 minutes, 7.64 seconds—only 4.6 seconds off what was then thought to be correct and 4.74 seconds off the modern value.[2] That represents an error of only 1.8 parts in one million. But the lunar train could have done that only if the clock drive had been perfect. It was anything but, not only because of the limitations of the verge escapement but because of the very complexity of the mechanism. All of these additional functions called for extra wheels and pinions, special gearing, epicyclical trains; and each articulation was another locus of friction and imprecision. (It is a general rule of machinery that precision and efficiency vary inversely with the number of mobiles.) Not that this inaccuracy bothered Richard of Wallingford. He even made his clockwork disconnectable, so that the machine could be run backward and forward at will. That is what planetaria are for.

By comparison, to keep time within half a second a day—long the standard for a marine chronometer—one had to hold error within a limit of one part in 172,800 (less than 10^{-5}). To meet the eighteenth-century standard for astronomical regulators, one had to hold variance to one-tenth of a second per day—that is, one part in 860,400 (almost 10^{-6}). More to the point, these limits were not hypothetical, not the potential performance of ingeniously calculated gear ratios, but operative constraints. Either the timekeeper kept to these bounds, or the sun and stars signaled its failure. Its merit lay in its performance, not in its cunning. There is no harder test.

Most antiquarian horologists and many historians of science have been so impressed by the ingenuity of complex astronomical machines that they have undervalued or ignored the achievement of precision timekeeping. They have missed the real story.

5   My Time Is My Time

F ROM THE BEGINNING there was a demand for smaller clocks that could be installed indoors. These, too, were necessarily weight-driven and had to be hung fairly high if they were to run for a reasonably long period; they were suited only to spacious residences. Technology, then, as well as cost, limited demand to princes, courtiers, and the richest of bourgeois. Not until clocks could be made cheaper and more compact could they become a characteristic furnishing of a well-to-do home.

Miniaturization was limited by tools and materials. Turret clocks were made of iron, and the iron of that day, smelted in a bloomery furnace, was of uneven composition and hardness, hence poor matter for fine work. Their wheels were cut so roughly that only heavy weights—up to a thousand pounds—could drive the mechanism: not the kind of thing one could put in a house. Chamber clocks called for better wheels, preferably made of brass, which was far more homogeneous and workable than iron and more accurate in small scale. (Steel, it should be remembered, was a by-product of ironmaking, very costly, also uneven in character, and in its harder states almost unworkable in the fine. In those days, steel was largely reserved for grindable cutting tools:

swords, knives, razors, surgical instruments.) Good, close cutting was more important in these small clocks than in large ones, because irregularities in small wheels translate into bigger errors on the dial.

Miniaturization opened the way to portability—not in the fourteenth-century sense of a mechanism that could be taken down, dismantled, transported, reassembled, and mounted, but in the sense of an instrument that could be moved about even while it continued running. This was the kind of clock that could be taken from room to room or that, when smaller still, could be worn on the person. Such a device required an internal prime mover in the place of trailing weights. And whereas weights could be made as big as the mechanism required, an incorporated prime mover was necessarily limited in size and strength. The device adopted, the coiled spring, could never have worked had not the clock train already been so reduced in weight and improved in articulation that it required relatively little force to drive it.

Spring-driven clocks appeared on the scene toward the beginning of the fifteenth century—about one hundred fifty years after the invention of the mechanical clock. The principle was an old one, but well-behaved springs were hard to make and even then proved frustratingly short-lived. Their fabrication quickly became a highly specialized art that clock- and watchmakers were only too happy to leave to experts, as we shall see when we talk about the clock- and watchmaking industries. Spring drive posed a problem that had not been encountered before. Weights imparted equal force to clock trains throughout the course of their fall: each drop was too short to allow more than the tiniest acceleration, and each drop was the same as the one before. The force of a spring, however, diminishes as it unwinds, so that some way had to be found to equalize the force over time. The answer was the *fusee wheel,* an intermediary wheel between the mainspring and the wheel train, conical in shape (something like a top), which the mainspring pulled round and round by means of a cord (later a chain).[1] (See Figure 9.) At the start, when the mainspring was fully wound and at its strongest, it pulled on the narrow part of the fusee, where it had the least mechanical advantage; and as it wound down, so did the cord on the fusee wheel, so that the increasing mechanical advantage compensated for the diminishing force of the spring. The fusee, like the spring itself, was a known device. When it was adapted to clockmaking, we do not know—presumably very soon after the introduction of the spring. The earliest recorded clock with fusee—the so-called Burgundy clock of Philip the Good—dates to 1430.[2]

The introduction of mainspring *cum* fusee made possible the truly movable domestic clock and eventually, with further miniaturization, the portable timepiece that we know as a watch. Myth has it that the watch was invented around the beginning of the sixteenth century by a certain Peter Henlein (alias Hele) in Nuremberg, an old center of metalwork and instrument- and clockmaking, as well as of trade and finance. It was a city that had the workers, the craft tradition, the international contacts, and the market, so the myth has some basis in fact.[3] But there is no hard evidence in its favor, in spite of the repeated appearance in the antiques marketplace of watches signed diversely Henlein and Hele, the better to gull those amateurs who want to buy the first watch ever made.[4]

Italian historians have disputed the German claim to priority. They point out, persuasively in my opinion, that the invention of the watch was implicit in the small table clock. As soon as the clock was small enough to be carried on the person, someone was bound to do so; and as soon as someone did, the very usefulness of the object would have called forth copies and improvements.[5] (See Figure 10.) In this sense, there was probably no invention, just a silent transition occurring in several centers more or less contemporaneously, probably in the last quarter of the fifteenth century.[6]

Once the fashion of wearing watches took hold, makers vied for smallness. Very soon watches were so tiny as to be placed in the hilt of a dagger (Francis I of France paid a small fortune for two of these in 1518) or in a finger ring (Elizabeth of England wore one that not only told the time but served as an alarm: a small prong came out and gently scratched her finger). Some of these early miniatures were about as small as anything made since.[7] Needless to say, these *tours de force* were meant primarily to impress and impose, not to tell the time.

The small timekeeper (portable clock or watch) proved to be a revolutionary instrument. By its very nature it stimulated horological technique, for miniaturization is a school for skill. The tinier the mechanism, the greater the penalty for irregularity or inaccuracy of shape or cut. Only the finest artists could work in the small, and in the long run they tended to separate themselves from the smiths and cannon founders who made the big tower clocks. (A well-trained watchmaker could and did make both clocks and watches; this was not true of most clockmakers.) The separation grew with the growth of demand and a concomitant tendency toward the division of labor.

This technical stimulus was reinforced by the proliferation of timepieces and diffusion of ownership. Where once clocks had been the conspicuous consumption and privilege of an exalted few, two centuries of technical advance and production experience had now made them available to a widening circle of bourgeois. As we know, none of these domestic and personal timekeepers were very accurate, so that their very multiplication was an invitation to comparison. Whose watch or clock was right—or rather, closest to right? Comparison in turn entailed emulation. The way to minimize these errors was to make better timepieces. Here the market played its role: some clocks and watches ran truer than others, and users soon learned which makers did the best work. Consumer preference did the rest. The finest artists could not work fast enough to fill their orders. The "botchers" spent their days turning repair jobs into annuities.

In the prevailing state of the art, meanwhile, the most effective measure was to adjust one's clocks and watches as often as possible to some more reliable timekeeper. The most accurate was the sun, and many a watch was made with a miniature sundial inside the cover, along with a compass so that the gnomon could be aligned north-south. In Europe, of course, especially north of the Alps, the sun does not always shine, and often the best one could do was to set by some turret clock with a reputation for accuracy—although these too were proliferating and creating their own babel. The search for a reliable referent gave poets the opportunity for a new figure of speech; thus John Suckling in the epilogue to *Aglaura* (1637):

> But as when an authentic watch is shown,
> Each man winds up and rectifies his own,
> So in our very judgments.

Even more profound were the consequences of miniaturization for society and culture. Where people had once depended on the cry of the night watch, the bell of the church, or the turret clock in the town square, now they had the time at home or on their person and could order their life and work in a manner once reserved to regulated communities. In this way, privatization (personalization) of time was a major stimulus to the individualism that was an ever more salient aspect of Western civilization. The public clock could be used to open markets and close them, to signal the start of work and its end, to move people around, but it was a limited guide to self-imposed programs. Its dial was not always in view; its bells not always within hearing. Even when heard, hourly bells are at best intermittent reminders. They signal moments. A chamber clock or watch is something very different: an ever visible, ever audible companion and monitor. A turning hand, specifically a minute hand (the hour hand turns so slowly as to seem still), is a measure of time used, time spent, time wasted, time lost. As such it was prod and key to personal achievement and productivity.

It was also the best defense against the heightened disquietude of an age of transition. To be sure, all times are times of transition, but the pace of change varies and with it the uncertainty that spoils our ease and eats at our sense of peace and well-being. Historians have argued that the period from the fourteenth to the sixteenth century was an age of anxiety, when old boundaries seemed to have dissolved and old moorings to have broken. In a perceptive article entitled "Anxiety and the Formation of Early Modern Culture," William Bouwsma links this unease to the tribulations and perils of urban life.[8] In contrast to what were remembered as the sweetness and freshness of the countryside, its orderly round of natural tasks and the rightness of human and social relations, the city seemed a pit of disorder, dirt, and confusion, even of intrigue and danger. The source of all this evil, of course, was money: "Where money is at stake or some personal interest, one finds no relative or friend who prefers you to himself and has not forgotten his conscience."[9]

Much of this perception was false: the country was never so sweet and pure as nostalgia had it, and the city never so bad that it did not continue to draw those hungry for employment, wealth, power, stimulation, or sensation. Indeed, it was this flow of immigrants that alone enabled cities to subsist; for in that day, deaths in urban areas so far exceeded births that cities would have shrunk to nothing without continual reinforcement.

Still, in matters of the psyche perceptions are decisive. The repugnance for city life in an age of urbanization reflected an underlying tension between *was* and *ought,* between the actual and the normative. This tension was in turn reinforced by the intense awareness of the brevity of life and the imminence of death. This was a society that had experienced and could not forget the great pandemic known as the Black Death (1347–1350), which wiped out between a quarter and a third of the population of Europe, and an even larger proportion in crowded urban areas. The Europeans of these centuries saw death as standing close by, ever ready to take them—who knew when? And indeed the Black Death was followed at irregular intervals by further epidemics that we have forgotten only because the first of the series was so devastating in its effects.

The imminence of death made the use of time a crucial concern. As Francesco Datini, merchant of Prato, grew old, his wife, who knew better than anyone the old man's love of possessions, began to fear for his salvation: "In view of all you have to do, when you waste an hour, it seems to me a thousand . . . For I deem naught so precious to you, both for body and soul, as time, and methinks you value it too little."[10] And Francesco in turn gave similar advice to one of his partners, single-mindedly chasing riches in Spain: "You take no account of time and remember not that you must die . . . You count on your fingers and say : 'In so much time I shall have made so much, and shall have so much time left, and when I am rich indeed, I shall go back to Florence and take a wife.' "[11]

Sacred time! Time to save, in both senses of the word "save," for time saved (from pursuit of material things, from frivolity, from the sins of the flesh) and well spent was the coin of salvation. The clergy—those old keepers of the clocks, ringers of bells, masters of the calculus of purgatory and redemption—waxed eloquent on this point, since it was precisely here that their own regimen intersected with, translated into, a spiritual discipline for the layman. Thus Thomas a Kempis in the *Imitation of Christ* (1417–1421): "It is sad that you do not employ your time better, when you may win eternal life hereafter. The time will come when you will long for one day or one hour in which to amend; and who knows whether it will be granted?"[12] And a century later Teresa of Avila (1515–1582), visionary, ascetic, but also founder of convents and monasteries, organizer, and strict disciplinarian: "If we find ourselves unable to get profit out of a single hour, we are impeded from doing so for four."[13]

This emphasis on time thrift, on diligence in prayer and virtue, was a favorite theme of sermonizers because it was a potent one. It is hard for the skeptics and doubters of our secular and secularist age to appreciate the dread that then gripped small and great people alike, but we must believe them when they talk to us. Petrarch, poet and humanist, abhorred sleep because it reminded him of death, and shunned his bed because it hinted at the grave. Describing his thoughts of hell, he wrote: "Terror grips my heart / Seeing the others I tremble for myself / Others urge me on, / My last hour may be now."[14] And how could one appease this gnawing care, this horror before the unknown? How else but by the careful use of time? Time, Petrarch wrote the emperor Charles IV, whose reign began with the Black Death, is "so precious, nay so inestimable a possession, that it is the one thing that the learned agree can justify avarice."[15]

In an age of clocks, time thrift means clock watching. The timepieces of the fifteenth and sixteenth centuries were often *memento mori,* of skull form or engraved with motto reminding the user that every moment was precious, for every tick brought him closer to his final reckoning. (See Figure 11.) Yet what's sauce for the goose is sauce for the gander, and time saved for the sacred is just as easily saved for the profane. Most timepieces dispensed with reminders of death; they were instruments for living and doing more. Merchants already understood that time was money. While Francesco Datini was old and wealthy enough to worry about his end, his young partner in Spain was counting the days that were making him rich. "In the morning when I get up," wrote Leon Battista Alberti in 1433, "the first thing I do is think as though to myself: what am I going to do today? So many things: I count them, think about them, and to each I assign its time." He went on: "I'd rather lose sleep than time, in the sense of the proper time for doing what has to be done [*la stagione delle fac-*

cende]." After all, one can catch up with sleeping, eating, and the like on the morrow, but not business. The important thing was "to watch the time, and assign things by time, to devote oneself to business and never lose an hour of time." And returning to the subject: "He who knows how not to waste time can do just about anything; and he who knows how to make use of time, he will be lord of whatever he wants."[16]

For better or worse, this was a new kind of man, one who became more and more common with the growth of business and the development of a characteristically urban style of life. For reasons that would take too long to examine here, the new species became particularly prominent north of the Alps, where the clock and watch industry concentrated. Italy, country of Dante and Dondi del'Orologio, declined over the next two hundred years to a position of chronometric laxity and horological dependence. Alberti was Italian and Catholic, but by the late sixteenth century, the typical watch wearer was North European and Protestant.[17]

This shifting balance of production and use of timekeepers was clearly both symptom of and factor in the larger displacement of the center of commercial and industrial activity from southern to northern Europe over the course of the sixteenth and seventeenth centuries. This loss of centrality by the once dominant Mediterranean is, as all historians know, one of the classic themes and problems of modern European historiography. It has drawn the attention and exercised the analytical powers of some of the greatest scholars, among them the man who was probably the premier social scientist of the twentieth century, the German Max Weber. It is ironic that Weber, who spent a lifetime seeking out and studying those characteristics that set Europe and especially Calvinist Europe apart and created capitalist man, never hit upon this chronometric aspect. He would surely have seized upon it with delight, for—to paraphrase his own formula—what the clock was to the cloistered ascetics of the Middle Ages, the watch was to the in-the-world ascetics of post-Reformation Europe.

The latter, in turn, usually bought their timepieces from Protestant makers. Calvin himself, so impatient of ornament and distraction, accepted the watch as a useful instrument and thereby enabled the jewelry trade of Geneva to save itself by reconversion. Augsburg, the leading German center of clock and watch manufacture, was a city divided between Catholics and Protestants in a Catholic countryside: in the period 1500–1700, of 189 master clockmakers whose religious affiliation is known (out of 284), 165, or 87.3 per cent, were Protestant.[18] Meanwhile in France, overwhelmingly Catholic but with a small, active community of *réformés,* a disproportionate share of the leading watchmakers of the sixteenth and seventeenth centuries were Protestants.[19] No one to my knowledge has done a quantitative sample, but a quick look at the standard biographical dictionary of French clock- and watchmakers by Tardy reveals the high frequency of makers with Old Testament names: the Davids and Daniels and Isaacs and Samuels that were then characteristic signs of Protestant faith.[20] So when Louis XIV reversed a near-century of tolerance and revoked the Edict of Nantes in 1685, he drove two hundred thousand *réformés* from the country and devastated the French watch industry. Some of the best of these refugees went to England, where the trade needed little help; but others went to Switzerland, where they did much to establish the mountain manufacture that would one day dominate the world.[21]

The multiplication of clocks and watches also made possible a new organization of all those activities that depended on meeting and parting, on coming and doing together. Two areas where time measurement had great potential were transportation/communication and war. The limits of a technology based on horse traction, combined with the frequent stops imposed by frontiers and tolls, meant that movement of persons and goods over long distances was a stuttering succession of stages, often accompanied by opening or transshipment of cargo. Even so, the volume of traffic grew rapidly from the sixteenth century on, the more so as advances in the techniques of coach construction enhanced the demand for passenger transportation. Not that these coaches were comfortable, certainly not by modern standards; but a closed vehicle, for all its jarring and jouncing over miserable roads, was better suited than horse-and-saddle to the needs and possibilities (talents) of most travelers.

The effort to speed the movement of passengers, information, and goods entailed the establishment of scheduled departures, designed to allow as closely as possible for the arrival of feeder carriers and the completion of customs and similar formalities.

Thus, courier regulations of the sixteenth century in Augsburg called for mail to Venice to be in by eight hours on Saturday and for return mail from Venice to be in by twelve hours. Other routes had their own timetables.[22] Ideally these links implied the existence of a commonly accepted time standard that was known to and could be tracked by all interested parties. In fact, such a standard did not exist in the sixteenth and seventeenth centuries. Different countries and places began their day at different hours—some at sunrise, some at sunset, some at noon, some at midnight. Basel began at noon, but designated that as one o'clock, so that its clocks always ran an hour ahead of those of its neighbors. Some places divided the day into twenty-four consecutive hours (so-called Italian time); this was very hard on bell-ringing clocks. Others marked it off into two rounds of twelve hours each ("German hours" in Bohemia; "French hours" in Italy).[23] Even a good clock or watch was not enough to tell the time, and the experienced traveler never went abroad without his conversion tables. In the long run, of course, the needs of commerce and communication were such as to encourage standardization, and usage shifted gradually toward the diurnal pattern we know today: midnight to midnight, twelve hours A.M., twelve hours P.M. Meanwhile, every locality continued to have its own "true time" as marked by the sun. It was not until the coming of the railway in the nineteenth century that a faster, denser traffic compelled the establishment of regional and national time zones; and not until the end of the nineteenth and start of the twentieth that international agreements reduced these to a global system.

This raises the question of how transportation could work to time in the earlier period. The answer lies largely in the very slowness of vehicles: when it took a day to cover thirty miles, local horary discrepancies did not make that much difference. Today we have to make similar adjustments as we go from one time zone to another. To be sure, we go a thousand miles before we have to reset our watches; but at the speed we travel, we have to do it about as often as our ancestors of four hundred years ago reset theirs. That timekeepers were useful to the traveler of those centuries is clear from the appearance of a new kind of portable timekeeper—the so-called *coach watch,* a large (generally 90–140 millimeters in diameter), readable device, often with a swivel ring for easy hanging, almost invariably fitted with an alarm to prevent oversleeping.[24]

The clock/watch changed military tactics more slowly, slower indeed than one might have expected. In principle, no activity was so in need of timing as moving troops, for there was a limit to the number of men who could be controlled by direct visual or auditory commands. In the sixteenth and seventeenth centuries five to six hundred was the upper bound, and it was this that dictated the size of the battalion, the primary unit of maneuver.[25] But armies were composed of many battalions, and masses of men, dispersed over the landscape, would have been far more effective if their moves could have been synchronized.

The early timepieces, however, were far too approximate and discordant to admit of such coordination. "Give me a watch [that is, a clock with dial]," says Shakespeare's Richard III as he withdraws to his tent on the eve of Bosworth. And to his aide: "Ratcliff, about the mid of night come to my tent, and help to arm me." Ratcliff comes late to wake him: "The early village-cock Hath twice done salutation to the morn; Your friends are up, and buckle on their armour." So much for Richard's watch, which would not have behaved much better in Shakespeare's time. Meanwhile, in the other camp, Richmond is already up: "How far into the morning is it, lords?" "Upon the stroke of four." (Their clock is working.) "Why, then 'tis time to arm and give direction."[26]

That was the way of war in those days: armies waited out the night (one had to see to fight) and rose at dawn for battle. But commanders and troops alike preferred to get an early start. In 1479 the marchese of Mantua, then on campaign in Tuscany, complained in a letter to the cardinal of Mantua that his clock (or possibly watch) had broken down and that he could not get it fixed: would the cardinal send him one that "would show the hours even if it didn't ring."[27] Ordinary soldiers could not afford such an instrument nor did they have cardinals to write to (they couldn't even write), but some took their own roosters with them to wake them in good time.[28] The actual fighting was often preceded by parleys and boasts and salutes: *"Messieurs les Anglais, tirez les premiers!"* That was chivalry, the kind that survived in the Wild West, at least the Wild West storybooks: "Draw!") In such a world, nature was the clock—not a dictator of timing but a cue. Combat began and ended without regard to hours and minutes. Attacks were improvised, sallies the work of a moment. Armies fought as long as they could, which meant until one side won or night fell. Commanders relied on visual and auditory signals (banners, horns, shouts) to direct their troops. The field of battle was a welter of confusion and truncated perspectives.

Over time, the increasing wealth of the state and the introduction of new armament—field artillery, in particular—tended to increase the size of armies, the range of firepower, and the tactical complexity of combat. These trends, moreover, were reinforced by coalition politics: it was always harder to command or coordinate the separate armies of independent princes, and more than one battle was lost because someone arrived too late, not always inadvertently. Here the watch had an obvious contribution to make, and it was no personal whim that made Prince Eugene of Savoy and the duke of Aremberg engage the great English watchmaker Henry Sully to accompany them on the Flanders campaign of 1708. As well or better take a watchmaker than a personal surgeon; at least watches could be made to run again.[29]

By the end of the eighteenth century, a good watch was standard equipment for any army or navy officer, as some of the best timepieces that have come down to us testify. But common, too, were ordinary "onions": in the first decade of the nineteenth century, Swiss mountain watchmakers followed the French imperial armies all over Europe selling their wares, down to the cheapest *tocante* that ran only long enough for the seller to move on.[30] This diffusion of ownership no doubt made it possible to coordinate actions on the staff and corps, perhaps even regimental, levels. The matter still remains to be investigated. My guess is that it was not until reasonably accurate and dependable watches could be mass produced—that is, not until the American Civil War—that knowledge of time was sufficiently diffused to permit the synchronization of movement on the company level. It was Waltham's sales of its lowest-priced model to military personnel during the Civil War years that turned red ink into black, rescued the company from repeated reorganization, and laid the basis of its prosperity.

These were typically sales to private individuals; institutional arrangements lagged. When in the late 1860s the Swiss G. F. Roskopf invented the first reliable cheap watch (it sold for twenty francs, which equals about seventy-five of our dollars or what was then about four days' pay of a skilled worker), some European command officers wrote directly to buy them in quantity for their subordinates: thus this Colonel Tyvolowitch, who ordered 240 of them for the Russian Twenty-third Infantry in Helsingfors. Bureaucracies tend to be conservative in these matters, especially where money is concerned. Besides, the tradition that officers (hence gentlemen) should pay for their own equipment went back to the Middle Ages and died hard.[31]

As late as World War II the American government—by far the most generous in matters of equipment—issued timepieces only to those command and operations personnel who could not do without them; pilots, navigators, and selected combat officers. The assumption was that the wise officer or soldier would buy his own, which even the lowliest buck private could do for less than a month's pay. Of course a GI watch was not the same as a civilian timepiece: the former usually had a *hacking device*—a means of stopping the mechanism completely and starting it again at the press of a button—the better to synchronize within a second or two; the latter was just a timepiece. Indeed its most valuable use may well have been at night, for sleeping rather than for working. It often had radium-coated hands that glowed in the dark. How else would a soldier know he had another hour or two before reveille? One is reminded of the liberating effect of the clock on the medieval worker: it set off the time that was his from the time that was his employer's.

6 Of Toys and Ornaments and Serious Things

THE INTRODUCTION of spring drive and fusee opened an era in the history of time measurement. A new industry was born: the manufacture of what the Swiss and French call small-size timekeepers (*l'horlogerie en petit volume*).

All kinds of possibilities now lay open. For the first time the watch and clock trade ramified, specializing by country and moving in directions dictated by differences in needs, in taste, and in the organization of production. Of these lines of development, the most important in the long run was the pursuit of precision—the effort to build ever more accurate and reliable instruments. This will be our main concern in the pages that follow, but before taking it up, we should look a moment at alternative ideals. These absorbed the energies and imagination of some of the finest artists (art in the old sense of "arts and crafts") and inspired the production of some of the most highly prized masterpieces in the history of clock- and watchmaking.

Two such ideals deserve special notice: first, the pursuit of further miniaturization, aiming to reproduce in the small the highly complex mechanisms of the great public show clocks, with their astronomical indications, automated figures, and carillons; and, second, the development of the timepiece, especially the watch, as ornament or jewel, with primacy given to the container rather than the contents.

Both tendencies have persisted to our own time. Swiss makers continue to turn out a very few highly complicated mechanical watches a year, some of them selling for prices in six figures, obviously for a very limited clientele of rich aficionados—for the kind of person "who has everything." And the watch (and occasionally clock) as jewelry continues to adorn the glossy Christmas advertisements of high-fashion magazines. Some of the most beautiful objects made by Fabergé at the beginning of this century or by Cartier in the interwar years were clocks for milady's dressing table or bedroom mantel. I would not for a moment depreciate the technical ingenuity and esthetic value of these *objets d'art* (in both senses of the word "art"). Even less would I want to underestimate their profitability, and hence their significance for the growth and prosperity of the clock and watch industries. Still, from the horological perspective they are a diversion and, in some manifestations, another magnificent dead end.

It was the Germans, with their multiplicity of courts and princelings, who took the first and, on balance, more expensive path. Nowhere were rulers so eager to compete in patronizing arts and crafts; nowhere was consumption so conspicuous; nowhere were guilds stronger and more uncompromising in their standards. Another consideration was the obligation beginning in the mid-sixteenth century of the Holy Roman Emperor and some of the other German princes to pay tribute to the Ottoman Turks by way of dissuading them from further aggression. The Germans called this the *Türkenverehrung,* the Turkish honorarium, but a tribute by any other name remains a humiliation. In any event, it generated a continuing demand for important and elaborate timepieces, because the sultan, his vizier, and such provincial officials as the pasha at Ofen (alias Buda) simply delighted in these toys, which they were in no position to maintain and hence had to replace at short intervals.[1] Add to all this a German fascination with science, and one can understand the extraordinary multifunctional chamber and table clocks that were produced in Augsburg, Nuremberg, Ulm, and similar centers from the sixteenth to the eighteenth century. These, often built in the form of mimetic,

miniature towers (called *tabernacle clocks*), afford all the usual temporal indications, plus information on the movements of stars and
planets (often presented in the traditional astrolabe format), with
clock striking and sometimes automata thrown in. The ornamen-
tation is typically luxuriant, the chasing and engraving superb,
the detailing impeccable. From the modern point of view, to be
sure, the exuberance of these pieces is not always pleasing. Even
so, they are much admired and sought after by today's collectors,
who see in them an extravagant embodiment of long-lost skills.
Very few collectors can afford them.[2] (See Plates I–IV.)

In sacrificing time measurement to complexity and versatility,
such clocks went against functional logic. They cost a fortune,
kept a mediocre rate, and were difficult if not impossible to maintain. The purchase price, in effect, was only the beginning: the
buyer (or recipient), if he meant to have his clock go, committed
himself to pay an annuity to its maker or some other craftsman
for upkeep and repair. Where skilled attention was not easily
available, the clocks found their way into storerooms and trea-
suries, such as the Topkapi Serail in Istanbul. If, then, from one
point of view these marvelous machines were the acme of the
clockmaker's art, from another they were a technical sidetrack
and a sumptuary quagmire. They kept alive and nourished, as
did the great show clocks of an earlier era, the principle of automation, which would find important and widening application in
a later age; and for that alone we should be grateful to those who
made them. But in the shorter run they spelled the death of the
traditional German clock industry.

With time the demand for these extravagances, always sensitive
to political circumstances, diminished. As Germany grew stronger
and Turkey weaker, the tribute came to an end (peace of Zsitva-
torok, 1606); and then that long civil war that we know as the
Thirty Years' War (1618–1648) devastated large areas of central
Europe and impoverished much of the rest. Princes were more interested for the moment in guns than in clocks. Instead of adjusting to such changes, guild masters continued to prefer complex
chefs-d'œuvre to more commonplace productions; opposed innovations in the technique and organization of manufacture; then
complained of and did their best to suppress the competition of
nonguild workers and country clockmakers (*Landmeister*), whom

they denounced as *Pfuscher*—botchers and bunglers. By the mid-eighteenth century the German clock and watch industry had shrunk to a shadow of its former prosperity. Recruitment of new talent dwindled as skills gravitated to more expansive trades, and this tendency was accentuated by the effort of the guilds to shelter what was left of the market by restricting access to newcomers: the best way to become a clock- or watchmaker was to be the son of one or marry his widow.[3]

This was not the end, however, of the production of supercomplex, superornamented clocks. There were still mandarins in China, maharajahs in India, pashas in the Ottoman empire—most of these richer than ever. But guild labor (*das alte Handwerk*) could no longer compete with the industrial mode, and the craftsmen of Augsburg and Nuremberg were replaced in the eighteenth century by such assembler-makers and merchant entrepreneurs as Christopher Pinchbeck, James Cox, Henri-Louis Jaquet-Droz, William Ilbury, and William Anthony. All of these worked out of London, which was the world's greatest commercial and shipping center, but they bought their pieces as well as made them, which meant that more and more of these came from Geneva and Neuchâtel.[4] In Europe, meanwhile, the market for these *tours de force* had largely (though never entirely) disappeared. The taste for baroque sumptuousness and rococo swirls yielded to neoclassical sobriety. Warm gave way to cool; toys, to instruments. Besides, the astronomical-astrological indications were no longer useful.

The second path—the timepiece as decoration or jewel—found its highest expression in French watches of the late sixteenth and seventeenth centuries. (See Plates V–VII.) Under the wings of perhaps the richest court in Europe, the watchmakers of Blois and Paris went from delightfully engraved cases and dials, to champlevé and high-relief enamels, to polychrome paintings on enamel that were masterpieces of the miniaturist's art.[5] This last technique, invented by an artist named Jean Toutin in the 1630s, produced exquisite cases, painted overall, in and out, that have never been matched. Their high cost reflected not only the hours of painstaking work but the high risk of failure: each case had to be fired again and again as new colors were added, in furnaces whose temperature had to be carefully controlled so that earlier work would not melt. The loss to spoilage and breakage must have been maddening.[6]

By comparison with these French watches, the work of other European centers of this period is cleanly simple at its best, and crude (naive?) at its worst. The Germans, British, and Dutch preferred to decorate the metal itself, using *taille-douce* engraving or high relief (cast from masters and then chased). But we shall see that already in the first half of the seventeenth century, makers in these countries were importing cases from France, by way of enhancing the attractiveness and value of their movements. (The *movement* is the working part of the timepiece. All the rest is packaging and presentation.) How important this international division of labor was, it is impossible to say. It was clearly facilitated by the tendency in the second half of the seventeenth century to make rough movements to standard sizes, and later by the appearance of paint shops in Geneva turning out little enamel disks and plaques by the hundreds, which then could be inserted in gold, silver, or brass case frames of suitable size. This obviously simplified the work enormously: no longer was it necessary, as when enameling *en plein,* to apply the colors to the inside, outside, and sides of the case at the same time; and this simplification eventually (late eighteenth century) gave rise to specialization in the painting of the decorative scenes, with one artist doing nothing but faces, another clothes and draperies, another background.[7] It also reduced the risk, for the little disks were far easier to handle in the furnace and damage to a painting would not entail the loss of the whole case.

But such process innovations came in later, and in the seventeenth century the Blois enamels, like the German *Prunkuhren*, cost a fortune and had a tiny market. Even at cheaper levels, moreover, the French makers were more concerned with appearance than substance, with the case rather than the movement. (This attitude persisted; see Plate VIII.) I do not mean to imply by this that French makers were not also making simple watches of simple appearance. They were, just as the Germans were producing modest clocks alongside their wonder pieces. My point is just that their peak effort lay in the direction of ornament, because this is what their richest customers valued and wanted. In a mode of hand technique and limited output, highly skilled craftsmen who thought of themselves as artists saw high-cost, luxury work as the most dignified, honorific, and profitable. Again, the effect was to confine demand to a fortunate few, who then set the style for others, and to diminish awareness of the watch as a tool, an instrument of planning and self-discipline.

The point, which is one of emphasis (everything is relative), is reinforced by a comparison with British styling. British makers produced some elegant watches and, as noted above, imported continental cases for those customers who would and could have them. But the characteristic London watch of the seventeenth century is of modest appearance, sometimes adorned with simple floral or geometric motifs, sometimes devoid of ornament. These latter are known as puritan watches, as they were presumably made for that pious fraction of the middle class that in the best Calvinist tradition looked askance at decoration and accepted the watch only as a useful device.[8] (See Figure 12.) No group anywhere was quicker to adopt the watch as monitor, and hence to focus on precision, simplicity, and reliability as against versatility, ingenuity, or appearance. The makers catering to this market, themselves often Puritans, were in turn impelled to stress the same virtues. How else could they differentiate their unadorned product from that of their competitors?

Which brings us to our main line of horological development.

THIS MAIN LINE was the perfection of clocks and watches as instruments of measurement. Users in such eminently mercantile and bourgeois societies as Holland and England pressed for greater accuracy. The more numerous the users, the greater their dependence on time measurement and the less the tolerance for error or disagreement. Remember those verses of John Suckling: bring in an "authentic" (that is, reliable) timekeeper, and everyone present adjusts his own. If Suckling spoke for his countrymen, we may fairly infer that Englishmen wanted to know the right time.

To this generalized demand should be added that of two special interests.

The first was that of the astronomers. They were, as noted earlier, less interested at first in the clock as timekeeper than as an accessory, a means of automating their planetaria and astraria. Now they came to realize that the clock was an invaluable, even indispensable tool of observation. The pioneer here, according to the great Danish astronomer Tycho Brahe, was Bernhard Walther of Nuremberg (c. 1430–1504), who hung the drive weight on his clock at the moment of Mercury's rising, then reckoned his minutes by counting the number of teeth of the hour wheel that passed until the rising of the sun. Walther was using, in other words, a clock without a minute hand because it was not accurate enough to warrant showing minutes. But it could give a reasonably good reading over short intervals, and by counting the teeth, Walther was getting fractional information far more precise than the dial could show.[9]

It was a beginning. A century later, better clocks made possible more accurate and systematic work. The key innovation was the technique of star mapping developed at Cassel by William IV of Hesse and his astronomer Christoph Rothmann in the late sixteenth century.[10] This called for location by two spherical coordinates: *declination* (elevation above the celestial equator) as determined by any of several sighting instruments, and *right ascension* (the angle measured by the time elapsed between the star's passage of the meridian and that of the sun or other reference star). Note that William and Rothmann, like their contemporary Tycho Brahe, were working with the naked eye; the telescope would not be invented until after the turn of the century, and Galileo would seize upon it as the open sesame to the heavens only in 1609.

In the meantime, it is astonishing what could be seen and done without it—provided one had a precise instrument of time measurement. One can follow Brahe's efforts to improve performance in this regard. He did better and better, thanks in large part to the use of a clock with an enormous wheel over a meter in diameter with over a thousand teeth; but he never did well enough to be satisfied. From 1563 to 1570 he mentions minutes in his journals only twice; in 1577 he speaks of a new, more accurate clock showing minutes; in 1581 he speaks of seconds. But in 1587 he complains about the discord among his four clocks (there is nothing harder than making clocks agree) and says he cannot get closer than plus or minus four seconds; and he uses some version of the clepsydra, with mercury or other fluid, for his finest measures of time elapsed between two star transits.[11] The mechanical clock, in other words, still left much to be desired.

Yet there were clocks and clocks. It is no accident that William of Hesse engaged in 1579 as his court clockmaker the finest horologist of the day, the Swiss Jost Bürgi, a man whose conceptions and realizations were generations in advance of those of his contemporaries. Bürgi gave him what he needed, a clock that would mark seconds as well as minutes, that often varied (if we are to believe William) no more than a minute a day.[12]

The second major source of pressure for accurate time measurement—very much analogous to the needs of astronomers—was the search for a way to calculate longitude at sea. Here, as in star mapping, the problem was to find the second of two spherical coordinates. The first, latitude, was easily established in northern waters, and with reasonable accuracy, by measuring the altitude of the North Star above the horizon. The farther north the observer was, the higher the polestar; the farther south, the lower. The earliest oceanic navigators made no effort to translate this into latitude, but they knew from experience that one degree of altitude was equal to about fifty miles of distance north-south, and that was all they needed to know to reckon their position on the vertical scale.[13]

Once ships sailed below the equator, star sights would not help. The polestar was no longer visible, and there was no southern substitute. But one could use the sun, whose altitude also varies with latitude. There was a complication, though: because the apparent annual path of the sun around the earth (the *ecliptic*) is not parallel to the plane of the equator, the sun moves with the seasons higher and lower in the sky. This angular deviation is known as the sun's *declination,* and it has to be taken into account when reckoning latitude from the sun's noontime altitude. Tables of declination had been prepared from the thirteenth century, and the invention of printing (the Gutenberg Bible dates from 1455) came at the right time to make them widely available to navigators, who were just then beginning to sail the southern seas. These tables, plus some kind of sighting instrument (quadrant, mariner's astrolabe, cross staff) and the ability to do a little arithmetic, were all a good pilot needed.[14]

Admittedly, good pilots in this sense were not to be found in the early days of oceanic navigation. "Celestial navigation," writes Samuel Eliot Morison, "formed no part of the professional pilot's or master's training in Columbus' day, or for long after his death." And he cites Pero Nunes, Portuguese-Jewish mathematician, some forty-five years after Columbus' discovery: "Why do we put up with these pilots, with their bad language and barbarous manners; they know neither sun, moon nor stars, nor their courses, movements or declinations; or how they rise, how they set and to what part of the horizon they are inclined; neither latitude nor longitude of the places on the globe, nor astrolabes, quadrants, cross staffs or watches, nor years common or bissextile, equinoxes or solstices?"[15] By the end of the sixteenth century, however, these skills were standard features of navigational training, and the English, who had once learned from the Spanish and Portuguese, were now setting an example to the rest. Their greatest contribution was the backstaff, invented at the start of the seventeenth century by Captain John Davis of Sandridge. This was a cross staff so modified as to enable one to measure the sun's altitude while facing away from it—an inestimable boon in low latitudes. It was not to be surpassed until John Hadley's invention of the quadrant some 140 years later.[16]

Longitude was another matter. Since the earth turns continually on its axis, there is nothing visible from one longitude that is not visible in the course of the day from every other: along a given parallel one sees the same sun, the same moon, the same stars. The only difference is that one sees them at different times: the sun rises, for example, some three hours earlier in New York than in San Francisco. But this means that the earth is a clock (indeed, the original clock) and that longitude can be inferred from time differences. All that is needed is (1) to compare the time of observation of a given celestial event (for example, an eclipse) at a place of known longitude with the observed time at site; or (2) to keep constant track of the time at a place of known longitude and compare that with local time. We shall see that solutions of both types were devised and employed, but that in the long run the use of an accurate chronometer to keep reference time (that is, the second solution) proved more convenient and prevailed.

The problem became a matter of moment in the fifteenth century, when Europeans first entrusted themselves to the ocean sea; and it assumed the highest political and commercial urgency in the sixteenth, when costly sailings to the New World and the Indies became routine and paid for themselves many times over in bullion, spices, and other precious cargo. These voyages took months and exposed both passengers and crew to rigors (inadequate diets, foul water, filth, vermin) that astronauts of the twentieth century, with their carefully calculated and controlled environments and ingeniously packaged and balanced food supply, would find unbearable. One survivor (the word is justified) of the nine- to ten-thousand-mile run from the Philippines to Mexico, the line of the so-called Manila galleons, described it as "the longest, and most dreadful of any in the World . . . as well because of the vast Ocean to be crossed, being almost the one half of the *Terraqueous* Globe, with the wind always a-head; as for the terrible Tempests that happen there, one upon the back of another, and for the desperate Diseases that seize People, in a voyage of 7 or 8 Months, in a variety of latitudes, sometimes near the Line, sometimes cold, sometimes temperate, and sometimes hot, which is enough to Destroy a Man of Steel; how much more, then, an ordinary Man of Flesh and Blood, who at Sea had but indifferent Food."[17] More than thirty galleons were lost in the two hundred fifty years of the official Manila crossing; as many more gave up the trip and limped back to port with heavy loss; and no ship made harbor without a huge quota of deaths from scurvy, for "scurvy never fails on the way from the Philippines."[18] The historian of the Manila-Acapulco crossing, William Lytle Schurz, estimated property losses over these years at some sixty million pesos, a huge sum that translates into several billions of today's dollars.[19]

One may well ask why anyone should voluntarily put up with such risk and torment. The answer, we know, was money: from captain on down, the wages were fabulous: so great was the gain on oriental goods that could be delivered intact in Mexico, that traders smuggled them aboard in water jars if necessary. What was a little thirst compared to pieces of eight? Besides, with all those storms one could count on God's rain for refills. During the course of the voyage, we are told, the sailors would swear, in between prayers and novenas, never to do it again; but as soon as they landed in Acapulco and saw the bonus that came with the return trip, they were back on board.

The Pacific crossing was the worst; but the long voyages around the capes Horn and Good Hope were not much better; and the wealth (and hence the power) of all maritime nations came to rest in some degree on the ability to make these deadly runs. Yet without an accurate means of measuring longitude, there was no sure way of knowing where one was and how far and long one still had to go. An error of longitude could mean a missed landfall; days, even weeks, of wandering in emptiness; crews lost to disease and thirst. Once the able-bodied complement fell below the minimum, the survivors could no longer even bring the ship about and were at the mercy of waves and sky. Vessels driving rudderless before the wind and bearing cargoes of treasure and corpses to some unknown shore were the stuff of legend—tales of such ghost ships as the *Flying Dutchman,* doomed never to rest, never to reach harbor. The risk of disaster, as we have seen, was greatest on the high seas, where the distances seemed endless and the target was often a small dot of land. But every body of water had its perils, especially at night: the invisible shoal or rock, the unexpected landfall. The Spanish, for example, lost a large proportion of their galleons in Philippine waters; they never made it to the Pacific. English navigators were wont to accuse the Spanish crews of excessive timidity, of never letting the vessel run at night, even before a favorable wind. They should have been more willing to take chances in order to shorten the crossing. Perhaps—but their fright is understandable.[20]

Before the invention of more precise astronomical and chronometric methods, navigators had no better way of measuring longitude than by so-called dead reckoning (French *estime*)—a more or less informed guess of the ship's location based on running observations of its speed and course. When one realizes that in the beginning the best that navigators could do to measure speed was to throw an object overboard from the bow and see how long it took to float past the stern; that for want of suitable timers, seamen had to rely on some subjective verbal standard—the time, say, of an Ave Maria (the hazards of the sea made sailors the most religious of men); that even after the introduction in the seventeenth century of the minute glass, the log (French *loch*), and the knotted cord (which gave us nautical speed in knots), speed readings could never be frequent or timely enough to take account of the changes in wind and current; that measures of speed and course varied, not always randomly, from one crewman to another; and that each day's reckoning built on the cumulative error of previous estimates—then one can well understand that such calculations were often wildly inaccurate.[21] The introduction of improved methods of reckoning, moreover, did not ensure their adoption; it is always a mistake to infer general practice from best practice. French seamen—at least some of them—were still not using the log-and-line to estimate speed as late as the latter part of the eighteenth century, almost two hundred years after its introduction. A book of 1766, the *Abrégé de pilotage* of Coubard and Lemonnier, notes that ordinarily one made one's reckoning "by watching the water pass along the side of the ship, by taking account of the qualities of the ship, the force of the wind, the way in which it fills the sails; whether one is close-hauled or not, running or tacking, which helps one judge whether the water seems to be flowing faster than it really is because of tide or currents; if the vessel is in ballast, newly greased, etc. . . ."[22] In these circumstances it was better to sail alone than in a fleet, where every pilot had his own estimate.

To be sure, sailors were nothing if not resourceful. They almost always overestimated their speed: it was more prudent to think land near than to think it far. Since they could measure latitude with some accuracy, they headed straight to the known latitude of their destination, then ran the parallel east or west. The voyage might take longer that way, but better safe than sorry. And always they sounded the depths, for nothing was so reassuring as the leadsman's cry, "No bottom, Sir!" Meanwhile, they watched carefully for natural clues to location: the changing color and sound of the water, the presence of birds and shore-hugging fish, the reflection of green vegetation on the white clouds overhead. The crews of the Manila galleons went wild on sighting the *señas*—the first signs of the approaching North American continent: fungous *aquas malas* hundreds of miles out; then frolicsome seals cavorting in the water; then, close to shore, rafts of matted grass. The first seaman to see a sign of land received a money prize, and the day was given over to a party that included a trial of passengers by crew. No one escaped without a fine: someone had to pay for those months of close confinement and peril, with their inevitable harvest of quarrels and resentments.

It was almost uncanny how far and straight *le sens marin*—what Samuel Morison called "that intangible and unteachable God-given gift"[23]— could take a captain and his ship, especially along a parallel. Andrés Bernáldez, a man of the church and host to Columbus, wrote that "no one considers himself a good pilot and master, who, although he has to pass from one land to another very distant without sighting any other land, makes an error of 10 leagues, even in a crossing of 1,000 leagues, unless the force of the tempest forces him and deprives him of the use of his skill."[24] That would be an error of one percent—potentially fatal if the target was an isolated point, otherwise superb. A generalization based on a sample of one? Columbus was admittedly extraordinary, and he had a fairly easy run from east to west with good following winds. He was also very lucky, and Fortuna was the sailor's best friend. She did not smile on everyone. Some navigators made horrendous mistakes and lived to tell the tale, while others paid for small errors with their lives and those of their companions. "No kinde of men of any profession in the common wealth," wrote geographer and publicist Richard Hakluyt, "passe their yeres in so great and continuall hazard of life; . . . of so many, so few grow to gray heires."[25]

The pernicious effects of ignorance of longitude were multiplied by the consequences for cartography. The map, after all, was the primary medium for the transmission of information and experience in matters of navigation—just as the book was in other areas of knowledge. In the international contest for access to the riches of the Indies, maps were money, and secret agents of aspiring powers paid gold for copies of the carefully guarded Portuguese *padrons*. Bad measurements made bad maps, though, and many a ship spent precious days searching for land that showed only on paper. Cartographers had a dearth of accurate information and a plethora of guesses to go by, so that even contemporaneous maps differed in detail. In the white, empty spaces of the great oceans, islands appeared and disappeared as sailors reported their "findings" and revised those of their predecessors. Or they stayed on, in spite of decades of vain efforts to find them —legendary lodes to lure the greedy, or false harbors for ships in distress.

(Sometimes these unsuspected errors did no harm; with a little bit of luck, they might even help. Thus, Columbus was encouraged in his hope of quickly reaching the Indies by a monumental underestimate of the size of the world. Conversely, the kings of France long enjoyed the thought that their kingdom was larger than it actually was. When French astronomers recalculated the longitudes correctly in the seventeenth century and drew a new map, Louis XIV chided them for robbing him of one sixth of his domain. Presumably he was smiling when he said that; the sources do not say. In another country, the king might well have decided to shorten by a sixth these learned bearers of unwelcome tidings.)

Cartographic inaccuracies, with all the dangers they entailed, persisted into the nineteenth century, largely because astronomical methods of ascertaining longitude, the only ones then available, were so unreliable. Not in principle, but in practice. The techniques of observation were rudimentary, and the complicated calculations gave every occasion for error, even to those who understood the method. Samuel Morison, biographer of Columbus, cites Jean-Baptiste Labat, missionary to the Antilles in the early eighteenth century and "the earliest writer . . . who gives the position of Hispaniola correctly": "I only report the longitude to warn the reader that nothing is more uncertain, and that no method used up to the present to find longitude has produced anything fixed and certain."[26] In the last analysis, moreover, only navigators could correct this plague of errors; for competent astronomers were few, and fewer still were those among them prepared to sacrifice their comfort and risk their lives on long voyages to distant and unknown places.[27]

L ONGITUDE, THEN, was the great mystery of the age, a riddle to seamen, a challenge to scientists, a stumbling block to kings and statesmen. Only such will-o'-the-wisps as the fountain of youth and the philosophers' stone could match its aura of tantalizing promise—and longitude was real. To be sure, so closely guarded a secret of nature inevitably acquired magical overtones. Sebastian Cabot, an Italian navigator in the service of the English crown, told Richard Eden on his deathbed "that he had the knowledge thereof by divine revelation, yet so, that he myght not teach any man." That was a safe brag, but Eden would have no part of it: "I thinke that the good olde man, in that extreme age, somewhat doted, and had not yet even in the article of death, utterly shaken off all worldly vayne glorie."[28]

As is usual in such circumstances, the demand for a solution tended to peak in response to disaster, regardless of cause and effect. It is not at all certain, for example, that a good instrument for measuring longitude would have saved the Spanish Armada from the high winds and seas that drove its fleeing remnants onto the rocks of Ireland and the Orkneys. Philip III, though, may have thought so, for soon after ascending the throne in 1598, he offered a princely reward to the "discoverer of the longitude": a perpetual pension of six thousand ducats, a life annuity of two thousand ducats, and a cash prize of one thousand more. The money attracted a fair number of schemers and dreamers, and the Spanish crown paid out large sums, partly out of hope, partly no doubt *pour encourager les autres.* To no avail: Spain simply did not have the pool of knowledge and talent to solve the problem.

Spain was not alone. In the century that followed, the Dutch, French, and British announced in turn their own rewards. The British offer, by act of Parliament in 1713–14, was the most important, both for the size of the prize and its notoriety: sums of up to £20,000 (equals over five million of our dollars) would go to whoever, of whatever nationality, discovered a way of finding the longitude within one degree.

Attracted by the challenge, the money, the prospect of fame, the most learned men and skillful technicians in seventeenth-century Europe turned head and hands to the task. No project had ever mobilized so much talent; the list of names reads like the cast of a Hollywood spectacular on the history of science: Galileo, Pascal, Hooke, Huygens, Leibniz, Newton. And this was only the first team. Alongside them worked an array of patient astronomers, whose observations provided the data base for systematic analysis and prediction of the movements of heavenly bodies; and some of the most creative mathematicians of the day (the brothers Bernoulli, La Hire, later on Leonhard Euler), who turned to astronomical, mechanical, and horological problems as tests of the power of newly devised techniques. Finally there were the master clock- and watchmakers who, whether working on their own or in collaboration with these savants, deserve to share the credit with them. The most important were Salomon Coster in The Hague, Isaac Thuret in Paris, the Fromanteels and Thomas Tompion in London, in the seventeenth century; and George Graham, John

Harrison, John Arnold, and Thomas Earnshaw in London, Henry Sully, Pierre Le Roy and Ferdinand Berthoud in Paris, in the eighteenth. Do not think, moreover, that such teamwork was the product of a simple division of labor between the scientist's theoretical imagination and the craftsman's skilled hand. Theirs was a less specialized age than ours, and the unity of science and craft had not yet been ruptured. The greatest scientists did much of their own manual work, grinding lenses and devising and making instruments; while the best of the craftsmen possessed surprising theoretical knowledge and conceptual power. It was this partnership that finally gave us precise timekeepers to the measure of the stars and the sea; and toward the end, when the scientists thought they had done as much as they could, it was the craftsmen who persisted and completed the task.

7   My Time Is Your Time

THERE IS A LAW of error that may be stated as follows: *small errors do not matter until larger errors are removed.* So with the history of time measurement: each improvement in the performance of clocks and watches posed a new challenge by bringing to the fore problems that had previously been relatively small enough to be neglected.

In the first two hundred fifty years of clockmaking, the verge escapement reigned supreme. The available alternatives (for example, the "strob" escapement used by Richard of Wallingford) were abandoned; at least they have left no trace in the written or archeological record. Such gains in accuracy as were achieved were derived from improvements in the materials employed (better steels, in particular) and from more regular cutting and fitting of moving parts. These gains were substantial enough to warrant the introduction in the fifteenth century of a minute hand in both tower and chamber clocks.[1] Watches lagged behind: because smaller, they were less accurate. But they, too, were made to run better and truer, and in the seventeenth century (one hundred years of progress), we begin to see half-hour and then quarter-hour divisions on watch dials. These markings reflected not only the improvement of the instrument but the tighter requirements of social time.

Now it was the turn of the escapement to draw the attention of horologists. The verge had to be a source of error because the oscillator, whether foliot or balance wheel, had no inherent frequency. Linked as the oscillator was to the verge itself and through the verge to the movement, its beat necessarily varied with the force transmitted by the wheel train. Even an unpracticed ear can hear the irregularity of the beat of an early-seventeenth-century spring-driven timepiece. A cardiac specialist would recommend immediate surgery.

Fortunately, a clock or watch is not a human heart. An uneven beat was normal for a verge escapement, not a sign of impending collapse. Changes in rhythm, moreover, partially canceled out, and what trend remained could be reduced by altering the tension of the mainspring. Over a twenty-four-hour period, then, these timepieces might give a reasonably good account of themselves, with variances as small as a quarter-hour for clocks, a half-hour for watches.

Such performances, though, were deceptive. It is in the nature of an average—such as time over a twenty-four-hour period—to conceal the range of deviation of its components. These early verge timekeepers looked better from day to day than they were from hour to hour. Even at their best, moreover, they were not accurate enough to meet the needs of astronomers or navigators. The next task, then, was to find or invent an oscillator with its own inherent, stable frequency and link it in such a way to the escapement mechanism that it would not be disturbed by the remaining irregularities of the wheel train. The answer was found in two marvelous devices: the *pendulum* for standing clocks and the *balance spring* (French *spiral réglant*) for movable or portable timepieces.

The first major advance in this direction was Jost Bürgi's invention of the cross-beat escapement (1584). This used two linked, flexible swinging arms as regulator, alternately blocking and releasing the teeth of a single escape wheel. Bürgi combined this with another device of his invention, a remontoir, which replaced the fusee as a way of equalizing the force of mainspring on movement.[2] Instead of allowing the mainspring to work through the movement on the escapement, Bürgi used its power to wind up a smaller spring at short intervals; this smaller spring then gave near-identical impulses to the swinging arms of the cross-beat.

(See Figure 13.) The result was a significant gain in precision—
between one and two orders of magnitude. This and other inventions earned Bürgi the plaudits of contemporaries: he's practically
another Archimedes, boasted his patron, William of Hesse.[3]

Yet Bürgi is little known today, essentially because his creations, however ingenious, are not those that later generations built
on; they were anticipations more than innovations. His escapement *cum* remontoir was copied by few other makers and never
found general acceptance. It was hard to make and even harder to
adjust; presumably its advantages did not outweigh the additional cost, at least not for ordinary domestic use. In any event it
was made obsolete, indeed consigned to oblivion, by the pendulum clock, conceived by Galileo in 1637 or earlier and successfully
realized by Christiaan Huygens by late 1656.[4] Working hand in
hand with clockmaker Salomon Coster in Amsterdam, Huygens
simply eliminated the balance-wheel regulator and substituted a
pendulum hanging freely from a cord or wire, hence largely isolated from disturbance by the train. (See Figure A.1.) The improvement in performance was spectacular, down from at least
fifteen minutes' variation for a verge with balance to ten or fifteen seconds a day; and the device was so easy of application that
not only were upright clocks everywhere constructed thereafter in
this manner, but existing clocks were converted en masse. It is
hard to find a vertical clock of the earlier period that has not been
altered in this way.

This sequence of invention inevitably gave rise to controversy.
When Huygens' announcement of his new clock reached Italy,
there was an outcry of indignation and wounded *amour-propre*, and
Huygens was accused of plagiarizing Galileo. Among the plain-
tiffs were Leopold de Medici, Prince of Tuscany, Maecenas of
horologers and proud possessor of one of the earliest pendulum
clocks, built in 1658; and the Accademia del Cimento of Florence,
that short-lived (1657–1667) and much respected company of ex-
perimenters in the Galilean tradition.[5] Galileo was Florence's sa-
vant-saint—the more so as Florence had delivered him up to the
Inquisition—and Huygens' claim seemed to martyr his memory.

But it was not only the Italians who challenged Huygens.
There were skeptics north of the Alps as well: those jealous of
Huygens' success (and the world is always full of such people),

and then that multitude of clockmakers who did not want to pay royalties for using his design. Huygens obtained a *privileg*—we would say, a patent—in Holland, but his efforts to secure one in France were thrice refused, for just these reasons. As his agent, the astronomer Ismael Boulliau, reported: the Chancellor Séguier "has always replied that he did not want all the master clockmakers of Paris crying after him, and besides there was always the possibility that someone else had found this same way of [making] clocks."[6]

Huygens was vexed and hurt by these charges and spent much effort to clear himself. He asserted that he had not known of Galileo's design; and when he got a copy of it, he dismissed it as inferior to his own, even as impractical, first because Galileo had used a new kind of escapement, which Huygens found much more complicated (*embarasée* [*sic*]) than the traditional verge; and second, because Galileo, instead of suspending his pendulum freely from a cord or ribbon, had fixed it to a pivoting arbor. (See Figure 14.) That, he asserted, was probably why the Galileo clock had not worked: "I know from experience that because of that, motion becomes much more difficult and the clock is liable to stop."[7] Huygens was wrong about the first of these objections: at least, later horologists have been much taken with the originality of Galileo's escapement and its unrealized potential. And he was partially wrong about the second, because even pendulum clocks with fixed rods proved far superior to the older kind of clock with foliot or balance-wheel controller. Indeed, I am told that early fixed-pendulum bracket clocks can be rated within two minutes a week. That seems almost too good, explicable perhaps by the superiority of today's techniques and lubricants; but even allowing for that, we have here a gain in accuracy for the seventeenth century of at least an order of magnitude.

Yet Huygens was right in principle: to achieve truly high precision—say, variances of ten seconds a day or less—the pendulum must swing free, undisturbed by the irregularities of the train. In a letter of July 24, 1659 to Boulliau, who had expressed concern on this score, Huygens pointed out that his clocks gave only the tiniest impulse to the pendulum, no more than a breath, hence could not affect the rate of oscillation.[8]

Huygens also took note of the fact that Galileo had not succeeded in making a working pendulum clock. He had designed one; but his son had only begun to build it when the father died, and that was the end of it. "I accomplished," boasted Huygens, "what he was not able to achieve (*ce dont il n'a pas sçeu venir à bout*)"; and I did this without having from him or anyone else "any clue or lead to this invention." If anyone else can show the contrary, he wrote, you can call me "a rascal plagiarizer and anything else you want."[9] It was a fair boast; and some years later, when Huygens sent Leopold a copy of his *Horologium Oscillatorium* (1673), he had the satisfaction of receiving from the prince a disclaimer of these charges and innuendoes:[10] "In regard to the invention of the pendulum, I fully believe that it never came to your knowledge that Galileo had the idea of adapting a pendulum to a clock; because this was known to very few, and Galileo himself had never put into practice anything complete, as one sees from the little that was roughly made by his son, and I am certain that if you had known of it, you would have said so."

THE NEW TIMEKEEPER seems simple and obvious from the vantage of hindsight. At the time, though, the use of the pendulum posed some interesting difficulties. Two in particular are worth noting here, the first because it shows the paradoxical advantage of a little ignorance; the second, because it further exemplifies the law of error. The responses to both illustrate the technological version of the economist's derived demand: innovation begets innovation.

As to the first: Galileo thought that the oscillation of a pendulum is independent of the amplitude of swing; that only the length (the radius of the arc) matters. This is one reason why he thought the pendulum could serve as or in a *misuratore del tempo.* When he found that this was not true, he attributed the variance to air resistance. He was right about that: air resistance does make a difference. But it is a very small difference, and it was not to matter for another two hundred fifty years, when the ever increasing fineness of astronomical observations and the invention of better pumps made it worthwhile and feasible to put observatory regulators into vacuum chambers. The major source of the difficulty was the arc of oscillation: a pendulum weight swinging from a point describes a circle, and circular swings are not isochronous. To obtain isochronism, a *cycloidal arc* is required—that is, the path described by a point on the circumference of a circle rolling along a straight line. This is because the common cycloid (a very uncommon curve!) is a tautochrone (from the Greek *tautos*, the same, and *chronos*, time), which means that a ball or bead or pendulum bob starting *anywhere* along the curve and impelled only by the force of gravity will reach the bottom in the same time. This is not the kind of result that one would expect intuitively, but a little thought will explain why. The higher one starts along the curve, the greater the velocity of the falling object, and this increased speed makes up for the longer distance traversed.

This deviation of the circular pendulum from the isochronism of the cycloid is what the horologist calls *circular error*.[11] For very small arcs, where the two curves almost coincide, the difference is negligible, but it increases with the amplitude of oscillation. An increase of amplitude from one to two degrees of semiarc adds 4.95 seconds to a clock's daily rate, whereas an increase from five to six degrees adds 18.09 seconds.[12] Huygens, applying the pendulum to verge-escapement bracket clocks, very quickly became aware of this error, for a simple reason: verge clocks, in order to work, required an oscillation of the pallet arbor (the verge itself) of wide amplitude, yielding pendulum swings of twenty degrees and more.

How, then, to correct for circular error? (One must remember that Huygens initially had no knowledge of the tautochronism of the cycloid.) One way would be to build a clock whose pendulum would always describe equal arcs. That proved impossible. Huygens then tried for a steeper arc by placing slips (called *cheeks*) of metal of different shapes alongside the suspension cord of the pendulum to constrain the swing and thereby shorten the effective length of the pendulum—that is, make it move faster—as the arc increased. A little empirical testing showed that the shape that worked best resembled the cycloid, and from there it was a short, imaginative step to demonstrate geometrically that the proper arc was indeed a cycloid. Huygens was immensely proud of this result, which he published in his *Horologium* (1658)—prouder, perhaps, than of the pendulum clock itself. The latter, he wrote Pascal, embodied a certain amount of happy chance. (He was being modest.) But the former! That was the kind of thing that gave luster to mathematics and would be esteemed by his fellow savants.[13]

The demonstration was indeed an ingenious piece of reasoning, but cycloidal cheeks proved in the long run far less useful than Huygens anticipated. For one thing, they would work consistently only if the clock were perfectly fixed, and preferably vertical; any change in inclination changed the length of the pendulum. This might not have been a problem for wall clocks, but it would have seriously altered the performance of movable table and shelf clocks, to say nothing of a sea clock for use in navigation. Besides, there was an even simpler solution to the difficulty, as Huygens himself was aware, and that was to have the pendulum describe a very small arc, less than two degrees, for then the circular error is small enough to be negligible. But what about the verge, with its wide swings?[14] Huygens thought he could solve that problem by putting an intermediate wheel between pendulum and verge to gear down the swing. The idea was a good one in principle, but bad in practice: as we know, every additional moving part increases the friction and with it the irregularity of the wheel train.

Instead the solution took the form of a new *anchor escapement*, so-called because of the appearance of the arms that blocked and released the scape wheel. (See Figure A.2.) The anchor needed far less arc than the verge, and in combination with a seconds-beating pendulum (about a meter long in the latitude of Amsterdam or London) produced a rate that varied no more than ten seconds a day—a performance so splendid that the device was christened the royal pendulum. Astronomers, equipped with a finer instrument of time measurement than they had dared dream of, pursued the logic further and asked for even longer pendulums. The great clocks at the Greenwich observatory had pendulums thirteen feet long beating double seconds.[15] That was not the sort of instrument that would fit in everywhere.

The inventor of the anchor escapement remains a subject of dispute. Ernest Edwardes, in his *Story of the Pendulum Clock* (1977), says that no horological invention, with the possible exception of the pendulum itself, has given rise to so much controversy. Supposedly, the first clock to use it was one built in 1671 by William

Clement of London, though some would deny that, asserting that the clock's anchor escapement is a later conversion. But there is also a claim on behalf of Robert Hooke, that brilliant physicist and mechanician of suspicious and secretive temperament, who is said to have presented an anchor clock to the Royal Society soon after the Great Fire of London (1666). Hooke was fascinated by horology and devoted his talents to a wide range of clock problems, from escapements to regulators to tools. He was certainly inventive in the highest degree, but he liked to sit on his inventions until he could exploit them to his profit and satisfaction, which often meant filing them away for future attention; and then, when someone else came forth with a similar device, Hooke would furiously proclaim his priority. We shall see these bad habits at their extreme in the dispute with Huygens over the invention of the balance spring.[16]

The second problem was implicit in the very excellence of the new instrument: How could one verify the accuracy of so accurate a timekeeper? The new clock, after all, was not perfect. Good as it was, put two or more together in the same place, set them to the same time, and before long they would diverge. In July 1659 Charles Bellair wrote from France to Huygens to ask for help on this score: "Allow me to ask you whether in Holland, in those places where there are several pendulum clocks, they continue very long to sound the hours together; because I have had two of them converted and put a seconds pendulum in each [literally *un pendule de 3 pieds et quelques pouces*]. I have not yet been able to keep them going together four days in a row. Not that they're very far apart, and when one checks them with sundials, one cannot see a difference even after a week; but the precision of hearing is much more sensitive than that of sight."[17]

The reference to sundials further illustrates the way an advance in one standard implies or requires advances in others. The sundial had always been the time standard of reference and control, except for those special people such as astronomers who knew how to observe the transits of heavenly bodies. Yet the sundial is itself a less-than-fine instrument: atmospheric conditions, the angle of refraction, the character of the dial surface—these and other factors of variance made it hard to read a dial, even under the best conditions, to less than five minutes. "I don't know," went on

Bellair, "but maybe you have some idea [*quelque intention*] for getting a very sharp and neatly cut sunray or a well-cut shadow that would move as quickly as the minute hand of some of your clocks, so that we can adjust them quite precisely by the sun."[18] His plea was heard by instrument makers who, beginning around the turn of the seventeenth century, brought out sundials that would give readings to the minute.[19] They were too late: by then, a good pendulum clock with anchor escapement could keep time to within seconds a day.

Yet even an accurate reading by the sun would not have settled the matter. Sun time (French *temps vrai*) is not the same as *mean* (average) time—that is, the uniformly scaled dimension marked off by the mechanical clock. Sun time runs up to sixteen minutes ahead of mean time, and up to fourteen minutes behind. They coincide four times a year. These differences meant little or nothing when clocks were always way off and had to be reset to time, which meant sun time, whenever the opportunity presented. But now that there were clocks that kept good mean time over fairly long periods, one had to take account of the equation of time—that is, the ever changing correspondence of solar time to mean time—in order to verify and correct them. One way to do that was to consult one of the tables that now were published giving the equation for each day or week in the year; such a table was often pasted on the inside of the door of the clock case.

But that simply shifted the problem. Now the question was: Which time should one use? Should pendulum clocks be set to sun time, as the older clocks had been? To do so would mean resetting them at every opportunity, never a good thing to do to a precision mechanism. Besides, how accurate could such repeated adjustments be? The clock presumably stood inside, in a hall or chamber; the sundial was outside in the garden or on the wall. Two people, one to read the dial and call out the minutes, the other to adjust the clock, might succeed in transposing the time with reasonable accuracy; but any such reading was liable to be less exact than the converted mean time of a good pendulum clock allowed to run its course. Indeed, why convert at all? The availability of reliable, accurate timekeepers, working day and night, on cloudy days and clear, led people to order their lives more and more by mean time rather than the sun. Soon sundials were built with conversion scales for equation of time: instead of setting by the sun, people corrected the sun. Thus was taken one more step toward that emancipation from the natural diurnal rhythm that was begun with the substitution of equal hours for the varying temporal hours of ancient and medieval times.

Yet there were national differences in this regard—at least, this is what the "archaeological" record seems to tell us. The evidence takes the form of a special two-time clock, built to show both solar and mean time: the so-called *equation clock*. The technique, as ultimately developed, called either for a kidney-shaped cam, whose outline was defined by the curve of the polar coordinates of the equation of time over the course of the year, or for a highly complicated system of differential gearing. Here, as almost everywhere, we have the usual controversy over priority. The first such clock was apparently made by John Fromanteel, the clockmaker who brought the pendulum clock to England, to the specifications of the mathematician Nicholas Mercator, who presented it to the Royal Society in 1666 and then gave it to the king. The king neglected it, however, perhaps because his clockmakers did not understand it and could not make it work satisfactorily; and so it eventually found its way back to Fromanteel, and there we lose track of it. Some years later, Hooke showed Tompion how to make one, and he then built an equation clock for the mathematician Sir Jonas Moore, but the only testimony we have of its operation shows it well off the mark. One can understand, then, why Huygens felt that the highly accurate conversion tables were to be preferred.

Not until the turn of the century were these difficulties overcome and accurate equation clocks built by Tompion and Joseph Williamson, and, later still, by such continental makers as Julien Le Roy. These were marvelous machines, *tours de force* of mechanical art and ingenuity. But they were few and far between and, needless to say, frightfully expensive. At their best, moreover, they were never so accurate as the astronomical conversion tables, which cost a lot less. The very principle was flawed: these clocks did not provide an independent measure of sun time but simply derived it from their own mean-time beat. Thus, although they have always been a special test of clockmaking skill and a much-sought-after treasure for the museum or collector, they have had little or no influence on horological development.[20]

These clocks have, however, a special interest for the student of the social and cultural history of time measurement, for they mirrored the timekeeping habits and standards of their users. Equation clocks were built originally not so much to verify the mean time they kept as to convert this to the solar time people lived by—what the French called "civilian" time as against astronomical. The intention showed on the dial: the first such clocks linked the solar minute hand to the hour hand, so that the latter showed sun time rather than mean time; when striking, they rang the solar hours. The British were apparently the first to get away from this and follow the new precision timekeeping to its logical conclusion: with royal pendulum clocks, there was no point in using anything but mean time. Perhaps the cloudy climate had something to do with it. In any event, British makers came to build their equation clocks with the hour hand linked by motion work to the mean minute hand. Meanwhile the continental countries remained true to the older system (though the matter still bears investigation), which they continued to live by into the nineteenth century.[21]

T HE PRECISION of the pendulum clock led scientists and clockmakers alike to believe that the problem of the longitude was on the point of solution. All that was needed, apparently, was to adapt the new device to maritime use. A small corps of savants and mechanicians, Huygens among them, set themselves to the task. All efforts failed: the uneven motion of the vessel, communicating itself to the timekeeper, made isochronous regulation impossible.[22] For similar reasons, attempts to adapt the pendulum to pocket watches also failed. A few of these abortions have survived (there could not have been many of them to begin with), much cherished as mementos of one of horology's innumerable dead ends.[23]

The task, then, was to find a regulator that was more or less impervious to motion. This was the *balance spring,* also called the *hairspring* for its fineness—an elastic body attached to the balance and imparting to it its own periodic rhythm of compression and release, translated into oscillatory motion. With such a spring regulator converting potential energy into kinetic, the wheel train no longer had to drive a balance serving as an inertia brake. All that it had to do was provide enough impulse to restore to the balance spring the energy lost to the balance wheel and friction. A very small push sufficed. As a result, the balance could be made heavier, and hence more resistant to shocks and the effects of changes of position. Or the mainspring could be made lighter and less exigent: the stronger the drive, the less forgiving the irregularities in the wheel train.

It was Huygens again who effectively introduced the new technique. I use the word "introduced" advisedly: here again, as in so many other episodes of horological history, priority in invention was and is a matter of dispute. Certainly Huygens thought the balance spring was his. We have the sketch in his notebook; "Eureka," he wrote—I have found it. That was on January 23, 1675. A week later he sent the secret in cipher to the secretary of the Royal Society in London—a sealed claim on priority; and on February 20, he followed that with a letter giving the solution of the cipher: "The arbor of the moving ring [the balance wheel] is fixed at the center of an iron spiral." (This is a translation; Huygens sent his message in Latin.) In the meantime Huygens had had a balance-spring watch made for him by Paris watchmaker Isaac Thuret, which he showed to Colbert on January 31 and to the Académie des Sciences on the following day, with a view to obtaining a French patent. Here, however, as with the pendulum clock, he ran into counterclaims and enough opposition to make him eventually give up the idea. In particular, his technical collaborator Thuret seemed to think that the invention was his, or at least partially his. Thuret had even taken it upon himself to show the watch to Colbert without Huygens' knowledge, a week earlier than Huygens! The scientist found the clockmaker's pretension simply impertinent and made it clear to Thuret that he wanted a written apology and acknowledgment, or he would not let him make the new kind of watch. (At that time, Huygens still had hopes of getting his patent.) Thuret's letter gave him full satisfaction: "I am quite ready to disabuse those who think I am the author of the new watches, as you first communicated the invention and the first model was made to your instructions." This may have mollified Huygens, but not enough to persuade him to work with Thuret again.

The counterclaim of Robert Hooke was a more serious matter.

Hooke learned of Huygens' "spring watch" in February 1675 and immediately began showing about a spring-regulated watch of his own devising, presumably made for him by the great clock- and watchmaker Thomas Tompion. It was this watch or another made about this time—for Hooke, whipped up by Huygens' challenge, kept thinking up new springs and new arrangements—that Hooke caused to be engraved "R. Hooke invenit 1658, T. Tompion fecit 1675."[24]

Was that simply a piece of self-serving evidence, concocted ex post? In a way, yes; but also no. The fact was that Hooke had imagined and publicized the idea of a balance spring years earlier, and what's more, his thoughts on the subject had been communicated to Huygens via Robert Moray. Here is what Moray wrote to Huygens on August 1, 1665:

> I have never spoken to you of another thing which he [Hooke] has put forward in his lectures on mechanics (he gives one every Wednesday out of term). It is an altogether new invention, or rather twenty of them, for measuring time as exactly as your pendulum clocks, as well on sea as on land, for, according to him, they cannot in any way be disturbed by changes in position, or even the air. It is, in a word, to apply to the balance, instead of a pendulum, a spring, which can be done in a hundred different ways; and he even went so far as to tell us that he has undertaken to prove that one can so adjust the oscillations [*excursions*] that small and large will be isochronous. It would take too long to describe these in detail, and he claims to be publishing the whole thing in a little while; in the meantime you will no doubt understand a good part of what this is all about [*une bonne partie de ce qu'il y a à considérer*].[25]

From this it is clear that Hooke was not only working on this problem but that he had already envisaged a wide variety of solutions. It is probably not reasonable, then, to credit him only with the straight-line balance spring, as numerous writers have done (myself included), reserving the spiral spring to Huygens. Hooke almost surely had contemplated and publicly discussed these and other varieties of spring before Huygens turned his attention to the problem. He just had not taken the trouble to put one into a watch.

Why not? A. R. Hall suggests two reasons. In a little-noticed article of 1950 based on Hooke's papers, Hall notes that Hooke in

1660 was working on the solution of the longitude problem and that he was thinking of a balance spring as only one feature of a new kind of timekeeper that would also include a constant force device and a detent escapement.[26] Here was Hooke's mechanical genius at its highest. The remontoir went back to Bürgi, but a detent escapement! The very idea was supposedly not conceived until a century later, by Pierre Le Roy in Paris, who gave it the name of *échappement à détente*. (More on him later.) Yet Hooke had clearly worked out the principle and has left us a description of the device that, for all its ambiguity, leaves no doubt of the nature and feasibility of the mechanism. He even used the words "trigger" (noun) and "trigg" (verb), which correspond exactly to the French *détente*.[27]

Had Hooke carried out his design and built and perfected this timepiece, he might have changed the history of timekeeping and marine navigation. But Hooke characteristically was concerned to profit from his invention, and to this end he sought to assemble a group of backers who would not only finance its development but post a huge bond of £2,000 to ensure confidentiality. His friends refused, and Hooke, perhaps out of pique, simply buried the project. The idea of the balance spring was interred with it, maybe because Hooke was not ready to envisage it as an improvement to ordinary "civilian" watches and clocks, and hence separable from the marine chronometer. Once again, as Hall puts it, Hooke was "his own worst enemy."[28]

Meanwhile what was Huygens' reaction to Moray's letter? First a denial of priority, then skepticism regarding Hooke's results:

> In regard to the thought of Mr. Hook, which you were kind enough to tell me about, to apply a spring in clocks instead of a pendulum, I want to say that when I was in Paris in 1660, the Duke of Roanais [that is, Roancz] spoke to me of the same thing and even took me to the clockmaker to whom he and M. Pascal had communicated this invention, but under oath and promise before notary not to reveal it or attribute it. But I found their way of doing it no good at all, and I could see already that there were much better ways. But aside from the fact that this is not so easy a thing to do as pendulum clocks, I do not find it likely that these [balance-spring timekeepers] will be as accurate as these latter; for the movement of the vessel has to cause small irregularities in the motion of the

spring that would be hard to correct; and besides we don't know yet
that changes from hot to cold would not alter the vibrations in
some way. So that I think that Mr. Hook speaks a little too confidently of this invention of the Longitude in his preface [to *Micro-
graphia*], as with several other things.[29]

It is clear, then, that as of that date Huygens had not yet given
up on a pendulum marine chronometer; neither had he given serious thought to the possibilities of a spring regulator. Even so, if
his statement concerning his visit to Paris in 1660 is correct—and
we have no reason to doubt it—it seems fair to say, with Defossez,
that the balance spring was "in the air." Very possibly, people in
several countries were working on the problem. The fact remains,
though, that Huygens put forward his solution at the right time,
and as with the pendulum, the balance spring then caught hold
and swept all before it.

Instant obsolescence: overnight, as it were, the world's stock of
watches had become essentially useless. We shall have occasion
later to consider the effects of a similar revolution in time measurement—the shift from mechanical to quartz crystal regulators.
This has happened in our own day, and we cannot but be impressed by the performance of these new instruments, which are
accurate to within less than a minute a year—as against, say, a
minute a day. But this gain of more than two orders of magnitude
is less radical in its social and cultural implications than the earlier one. The reason is simple. Watches good within a minute or
even two per day are, in conjunction with radio or telephone time
signals, more than accurate enough to enable the wearer to meet
the temporal requirements of our civilization. Anything more
precise is something of a psychic luxury, like hi-fi equipment that
faithfully transmits sounds inaudible to the human ear. But the
introduction of the balance spring revolutionized time discipline.
The gain in accuracy—from a half-hour or more per day to, say,
five minutes—brought watch performance very close to the standard we now employ in ordering life and work.[30] The balance
spring thus laid the material basis for what even we would consider punctuality.

It now paid to show minutes. Yet it was not immediately obvious to watchmakers how to furnish the information. Clocks were

already showing hours and minutes by hands working off the same arbor (what the horologist calls *motion work*), but watchmakers of the late seventeenth century experimented with a variety of displays—among them picturesque *wandering-hour, sun-and-moon* (for daytime and nighttime readings), and *differential* dials (see Figures 15–17)—before settling on the concentric hour and minute hands that we are familiar with. These lively layouts were not conducive to good timekeeping, but they made for colorful, attractive faces, and there always have been, and always will be, people for whom a watch is a bauble or ornament rather than an instrument. As a result, such watches continued to be made long after the issue was settled. Certain makers specialized in them, and other makers were only too happy to buy them and sign them with their own name. The purists scorned them as antifunctional. The English expatriate watchmaker Henry Sully, writing in Paris in 1717, denounced these and other novelty watches as the epitome of horological foolishness, and Savary's *Dictionnaire du commerce* thought it useful to repeat this condemnation as late as 1761.[31] Collectors today love them and pay a premium for their distinctiveness.

Even when watchmakers settled on what seems to us the natural form, they found it hard to accept the economy of a single numbering for two scales (minutes and hours). Users, they feared, would be confused, so they provided separate indications for hours (roman numerals) and minutes (arabic). Almost invariably dials numbered the minutes at five-minute intervals. Only in the third quarter of the eighteenth century was it thought possible to make do with fifteen-minute indications, and not until the end of the century, to dispense with minute numeration entirely. Changes in dials, in other words, describe the growing familiarity of the public with time measurement, a process completed in the twentieth century, when we have for the first time watch dials without any markings; the position of the hands is enough to tell us the time.

Minute hands were followed by second hands, beginning in the 1690s. The first watches to show seconds are sometimes called doctors' watches, on the assumption that they were made for pulse-taking. This was surely true in some cases, but these were certainly bought by others than doctors and one has the impression that the seconds dial served primarily to tell the user that the watch was going. (This is what it still does today.) The point is that there was no technical advance at this time comparable to the invention of the hairspring and thus yielding a new level of accuracy that warranted this further indication. Rather, it was simply the layout of the verge movement, with its contrate wheel turning once every sixty seconds, that made it easy to insert a seconds bit.

Such second hands were useful primarily over short intervals—say, fractions of a minute—and not only for doctors. Astronomers needed them along with the more accurate pendulum clocks, which could not be moved about; and navigators could use them to time multiple sightings. The growing importance of small units of time led to the invention in the 1740s of center-seconds watches—that is, watches whose second hands ran off the center arbor and tracked the circumference of the dial; the larger the arc, of course, the more visible and divisible the count of seconds. These watches were customarily fitted with a stop lever, which allowed the user to freeze the result, the better to make his count. The only trouble was that the whole movement stopped. That played havoc with time measurement, in the sense of minutes and hours, but the very arrangement indicates a special set of priorities. The difficulty was not solved until 1776, when Jean-Moyse Pouzait, a young Swiss maker, invented the independent seconds train.[32]

By 1770 the logic of this pursuit of ever finer time measurement led to the appearance of the first center-seconds watches with fractions of seconds marked on the dial; the earliest I have seen show fifths. Who cared about fifths of seconds in those days? There were, to begin with, the astronomers and navigators mentioned above; but also another constituency—namely, owners and trainers of race horses. In no society was horse racing more cultivated and cherished than in Britain, and these fractional center-seconds watches were characteristically British; they do not appear elsewhere for another generation or more. When made with the new cylinder escapement (see below) by makers of the top caliber, they represent the highest degree of precision achieved by the watch before the invention of the pocket chronometer in the 1780s.

The new controller gave a sharp boost to the demand for time-pieces. To begin with, old users wanted to replace their watches, now obsolete. Some of them thought to save money by having their old watches converted to balance-spring regulation, but the result was almost invariably inferior to the new standard of performance; sooner rather than later, they bought a new watch. Meanwhile new users were won over by what was a vastly superior and hence more useful article. We have no statistics on watch output during these years, but the changes in technique and industrial organization tell the story. It was this period that saw the first use of special-purpose tools in watch manufacture, the introduction of standard calibers and batch production, and a strong tendency toward specialization and division of labor. (More on this below.)

All these tendencies were reinforced by concomitant changes in attitudes toward time and time measurement. Watches were now accurate and dependable enough to be credible. Accordingly, people used and relied on them more than ever. Recall again John Suckling's quip: everyone's watch goes differently, so each looks for an "authentic" timekeeper to set his own by. No one, in other words, trusted his own watch. Not quite a century later, in 1711, Alexander Pope said it differently: " 'Tis with our judgments as our watches, none / Go just alike, yet each believes his own."[33] So much was this so that the Savary—the standard dictionary of commerce of the mid-eighteenth century—warned readers against excessive horological confidence and puffery: "One should put little credence in the common notion most people have of the accuracy of their watches."[34]

8   Approaching the Asymptote

E VEN A FEW MINUTES A DAY, however, are too much for the purpose of calculating the longitude. Watchmakers now had to reduce that already small variance—a twentieth or less of what it had been when watches were invented. Here they ran into another law: *the last gains are the hardest.* (Compare the exponentially increasing energy consumption of a prime mover—say, an automobile engine—as it approaches the limits of its power.) The curve of technological improvement is typically asymptotic, approaching ever closer to some limit, but by ever smaller increments.

So with time measurement. The remaining sources of error were subtle, sometimes slow to show, and relatively weak; the remedies not simple; the results so marginal as to seem not worth the effort. Yet the goal of an accurate marine chronometer remained an irresistible lure, and the best minds and hands in the profession were ready to drop everything else to concentrate on this task.

Two sources of error now came to the fore. The first was change in temperature; the second, friction. Neither was immediately obvious. Temperature changes, for example, are irregular in pattern and, under most circumstances of personal habit and domestic management, limited in range. Both cause and effect, moreover, are neither visible nor audible, and over a period of days and weeks can more or less average out. In the presence of other sources of error, then, temperature effects are not easy to detect, segregate, and measure. Yet they could make a significant difference: to the rate of a pendulum by expanding and contracting the rod, hence changing the radius of arc; and to the period of a balance, by changing its moment of inertia and the force of the hairspring. These changes could be very small, but no instrument is so unforgiving of even small error as a timekeeper: seconds add up to minutes, and minutes are too much.

Surprisingly enough, it was not easy for the scientists of the day to grasp the nature of the problem. For one thing, they were not always ready to accept the fact of temperature effects on solids. Thus Huygens, who had a strong personal commitment to the well-behaved pendulum, cites in *Horologium* (1658) the observation of the Belgian astronomer Godefroy Wendelin that a given pendulum oscillated faster in winter than summer, but only to refute it: "But as he confessed in that examination to have used only sandglasses, together with sundials and common mechanical clocks himself, perhaps not too much attention should be given to his account; many will doubt that he has made a correct observation. To me, certainly, it was not given to observe anything of this kind."[1] For that matter, Wendelin himself never imagined that it was the change in temperature that was responsible for the change in period; instead, he attributed it to a hypothetical change in the speed of the earth's rotation, supposedly faster in winter. As late as 1690 Huygens contested Denis Papin's opinion that pendulum rods expand when going from cooler to warmer climates. He was supported in his skepticism by other careful observers, who took their finely engraved brass rules with them to tropic lands and unsurprisingly found that brass pendulum rods measured the same as before.[2]

Craftsmen, unhampered by theory, knew better. A number of branches of industry were based on the empirical observation that heated metal contracts on cooling. Iron cannon, for example, had been made for centuries by placing heavy rods around a core and wrapping them with hoops of white-hot metal that bound them ever tighter as they cooled. Now that timekeepers were so much more accurate, clockmakers and watchmakers quickly became aware that temperature made a difference. It took them another fifty years and more to find a way to compensate for it: in clocks by means of John Harrison's gridiron pendulum (1720s), George Graham's mercury pendulum (1726), and John Ellicott's self-adjusting pendulum bob (1730s); and in watches by means of bi-metallic or other compensation curbs on the hairspring (John Harrison, 1730s; Ferdinand Berthoud and John Arnold, 1760s); or better yet, by a self-compensating balance (Pierre Le Roy, 1765; John Arnold, 1773; Thomas Earnshaw, 1784). Such refinements, however, which represented one of the highest expressions of the watchmaker's art, were long reserved to chronometers and similar precision timekeepers, and it took another hundred years before they became a standard feature of even a good "civilian" pocket watch.[3]

A second source of error which now came to the fore was friction—more precisely, the residual friction that remained after centuries of improvement in the cutting and articulation of the train. This gave rise to deviation not only over the course of an hour or a day, but over longer periods—and not always in the same direction. Friction varied, for example, with the position of a watch, which would keep a different rate on the night table than in the pocket, and with temperature, because the oils behaved differently in warm and cold. Friction would also change with use and wear, so that the rate of the clock or watch would diverge more and more from that established by the latest cleaning. "I find," wrote a correspondent of Huygens in 1659, "that when one oils [these clocks], that makes the pendulum swings wider and the clock slows; and on the contrary, that when the wheels are very dry and the amplitude of the pendulum is smaller, it runs fast."[4]

All of these errors now became important, and all engaged the interest and talent of savants and craftsmen alike. The attack on friction took place at three points: the meshing of wheel teeth and pinion leaves, the holes in which the pivots turned, and the escapement.

Friction at the mesh point. The shapes of teeth and leaves were a function of design and craft skill. Not until the late seventeenth century do we have a mathematically derived theory of the most efficient profiles of meshing gears. Interest in the question was generated not only by horology but by other applications of mechanics, and here, as in so many similar circumstances of simultaneous inquiry, the story of discovery is a battle of priorities, confused by the usual mingling of theory and application. The French have their claimants: the architect-geometrician Gérard (alias Gaspard) Desargues, author in the 1640s of a privately circulated manuscript treatise on conic sections, and Philippe de La Hire, whose *Traité des épicycloïdes et de leur usage dans les méchaniques* appeared a half-century later in 1694. The Danes have the astronomer Olaf Roemer, who in 1674 presented to the French Academy of Sciences a paper "on the form to give to the teeth of wheels so that they mesh as well as possible without jumps or friction." Leibniz and Huygens both credited Roemer with the contribution. La Hire, writing twenty years later, managed not to mention him at all. Meanwhile a Swiss mathematician named Nicolas Facio or Fatio (we shall meet him again) sent his thoughts on the subject to Huygens, and in England both Newton and Hooke were looking into the matter. The answer, as stipulated in the title of La Hire's treatise, is that the most efficient profile for gear teeth is an *epicycloid* that is, the curve described by a point in a circle rolling on the circumference of another circle.[5]

This was not the kind of result that a clock- or watchmaker would arrive at intuitively. Nor was it the kind of information that made possible in itself a drastic improvement in gear trains, at least not for small watchwork: there is a limit to the precision of hand-guided tools. Still, the recommended profile was not far from those already achieved by centuries of trial and error, and the possibilities of more accurate work were substantially enhanced by the invention of wheel-cutting engines. Polymechanician Hooke was the inventor of one of these (c. 1672), and he was followed by many others, for once the principle was understood, any first-class watchmaker could make his own.

This was only the beginning of a growing trend toward special-purpose machines and tools, designed to simplify and ensure the regularity of repetitive tasks. The English took the lead here, as John Carte, a Coventry watchmaker working in London, was pleased to boast: "The English . . . have invented that Curious Engine for the Cutting the Teeth of a Wheel, whereby that part of the work is done with an exactness which farr exceeds what can be performed by hand: Then there is an Engine for equalling the Ballance wheel: Likewise the Engine for cutting the turnes of the Fusie: And lastly the instrument for drawing of the steel pinion wier: All which ingenious inventions were first conceived and made at *Leverpool* in *Lancashire* in *England.*"[6] But makers in other countries followed fast on their heels, and such works as Anthoine Thiout's *Traité de l'horlogerie, méchanique et pratique* (1741) show clearly that these were then familiar tools in a major center such as Paris.[7]

The implications were enormous. The new devices made possible the employment of less skilled labor; they also lowered costs, helped widen the market, and laid the basis of a finer division of labor. In the long run, they helped turn an art into an industry and provided the models for eventual mechanization and interchangeable parts.

Friction at the pivot holes. The steel pivots of the arbors of the wheel train and escapement turned in brass—brass plates or bushings. The friction generated was eased by lubricant oils, which were generally organic in origin and changed character with time and the seasons. The oil made the pivot holes collecting points for dust, which fixed itself in the softer brass and made the surface abrasive. With use the pivot holes changed from comfortable, well-lubricated bearings to tiny laps, rubbing merrily away at the pivots themselves. Meanwhile the timepiece would be choking in its own waste. Thus, Tompion's long-case clock for the new observatory at Greenwich, built in 1675, started gaining seriously in a matter of months as the pendulum swings grew ever shorter and finally stopped in November. Astronomer-royal John Flamsteed blamed the oil: "The dust mixed with the oil in the pivot holes was macerated into swarf which clogged the wheels and took off a great part of the force which the weight ought to have on the wheel which forces the pallets that move the pendulum."[8]

The solution lay in harder bearings that would not provide a bed for abrasive particles, and in better oils. The first of these improvements came in the early years of the eighteenth century, when Geneva mathematician Nicolas Facio, working in London with two French refugee watchmakers named Pierre and Jacob

Debaufre, learned to pierce rubies and used them as bearings and end stones for the balance staff. Facio was typical of that marvelous age of scientific revolution when savants could know almost everything and work at anything, when the great minds of Europe were in constant communication, sharing their results, posing challenges and problems to one another, competing in the rapidity of their solutions. As a young man, Facio had measured the height of Mont Blanc from the terrace of his family's chateau and had drafted a map of Lake Geneva. Later on, like every astronomer worth his salt, he had ground his own telescope lenses and found better ways to do it. His calculation of the distance of the earth from the sun, his observations of Saturn's rings, and other precocious work might have won him election in his early twenties to the French Academy of Sciences; but he was a Protestant, and this was the time of the Revocation—an end to tolerance. Defossez regretfully (rhetorically?) remarks: "Oh, if only Fatio had had more flexibility and less conviction!"[9] He might better have addressed his reproach to the French crown.

In any event, France's loss was England's gain. Facio went to London, where he became a friend and intimate of the greatest scientists of the day, was elected to the Royal Society at the age of twenty-four, and joined the informal competitions in mathematical problem solving. One such competition where he made his mark involved a question posed by Newton: What was the form of least resistance for a solid? The idea was to establish a mathematical basis for the keels of ships. Facio sent in a correct, but very involved, solution. Guillaume de l'Hospital and Jean Bernoulli found simpler ones, which led Facio to take up the question again and publish an elegant answer in 1701.

This was the man who turned his attention to the problem of pivot bearings. We do not know what experiments he and the Debaufres undertook before arriving at their solution. Presumably they tried the hardest stones available, found the diamond too hard to pierce and too easily cleft by use, and settled for the next-hardest materials: rubies, later sapphires and agates.[10] They submitted their request for a patent in 1704, and it was initially granted but then withdrawn. It was successfully opposed by the Clockmakers' Company, which argued that such a monopoly would reduce the British watch industry to servitude. The patent, after all, could not be enforced abroad, so that makers on the continent would soon be making jeweled watches; and "if such Watches as these Persons pretend to make should come into use, all other Watches will be undervalued, and consequently few or none of them will be made, and all the Workmen now imploy'd therein must become Servants or Tributary to these *French* Patentees or go into Foreign Parts to exercise their Trades."[11] Never before, the Clockmakers asserted, had any horological improvement pretended to patent protection; at least, no Englishman "ever yet had assurance enough to sollicit the *Legislature* to Establish a *Monopoly.*"[12]

But these were Frenchmen (Genevans were seen more as French than as Swiss in those days), and—the Clockmakers here assumed a wounded tone—"though we retain our Compassion towards the persecuted Protestants of *France,* We could never expect, That they would attempt to take away our Livelihoods."

To clinch their argument, the Clockmakers subsequently came up with an earlier watch by Ignatius Huggeford with what appeared to be a ruby end stone on the balance staff. This watch was clearly never examined, for when it was taken apart some 250 years later, the ruby proved to be a mere decoration. This revelation came too late for Facio and the Debaufres. In the meantime, an increasing number of British makers learned the technique, until by the third and fourth decades of the eighteenth century, jeweled bearings and at least top end stone for the balance staff became standard features of any first-class watch. The use of stones for the other pivot holes came later, though I have seen a watch by George Graham, made in the 1720s, with almost every hole jeweled.

The British watchmakers need not have worried: the home industry enjoyed a monopoly of watch jeweling for a hundred years. I can think of no comparable advantage so long enduring. Not until the very end of the century do we see the first jewels appear in continental watches, and then only in those of the most demanding and expensive makers. Instead, continental makers contented themselves with placing a small steel end piece (the so-called *coqueret,* or little cock) as butt to the balance staff; the pivots continued to turn in brass as before. When, after the middle of the century, a few innovative makers experimented with precision timepieces for marine use, they adopted the costly and cumbersome solution of antifriction rollers. It was Abraham-Louis Breguet who finally made the breakthrough. He began in the 1790s with predrilled stones that he bought in England, probably from John Arnold. Then he persuaded "at great danger and expense" an English jewel maker to come and work for him in Paris; in those days English artisans were forbidden by law to emigrate, and the penalties for seduction were severe. The Englishman then trained French workers to do the job.[13]

Why this long English monopoly? The usual explanation is that the English succeeded in keeping the art secret.[14] Perhaps; but I find that reasoning unpersuasive, if only because so many British craftsmen learned to pierce jewels once the Clockmakers' Company succeeded in breaking the Facio patent. Surely the nations across the Channel, with their long experience in jewelry and precious stones, did not lack for skilled hands to learn this kind of work—as indeed they demonstrated when jeweled bearings came into demand in the first decades of the nineteenth century. My own sense is that the continental makers did not use rubies because they did not believe them necessary, and this for two reasons: they had other, more important sources of error to worry about and they were less concerned for precision than the British.

In all fairness, it should be noted that not every British clock- and watchmaker was convinced of the superiority of jeweled bearings. Paul Philip Barraud (active 1780–1815), one of the leading artists in the trade, felt that oil in brass holes could stand up better to long use and made his regulators and chronometers accordingly.[15] Here he touched upon what had by then become the bane of horology: bigger sources of error had been largely eliminated, but lubrication remained a mystery. Give me the perfect oil, the great Breguet quipped, and I'll give you the perfect watch. Every maker had his own ideas on the subject, his own private recipe, and horological craft manuals simply ignored the topic. In the seventeenth century just about everyone used animal oils, produced by rendering fat and skimming off the clear liquid that floated to the surface. A few preferred vegetable oils—the first pressing of the olive, for example. It is not clear that any of these lubricants was significantly better than any other, nor did the seventeenth and eighteenth centuries see any marked improvement in this regard. Such gains as were made took the form of improvements in application. Oil has a nasty way of spreading away from the points that need lubrication to places where it simply attracts dust and makes trouble. Watchmakers such as Henry Sully and Julien Le Roy worked out in the 1720s and 1730s appropriate shapes for oil sinks, designed to keep turning pivots bathed while preventing oil travel. Yet the best cure for wear was frequent cleaning—just the kind of thing that most watch owners were inclined to postpone or forget.

Friction at the escapement. New escapements yielded the greatest gains in minimizing the perturbations due to friction and irregularities of the train. The late seventeenth and eighteenth centuries were a period of fertile experiment in this regard, but we need concern ourselves here only with the four most successful devices, which shaped the character of precision timekeeping for the next two centuries. All four aimed at neutralizing the train, but they did it in very different ways. Two of them did it by blocking the movement and holding it while the balance wheel accomplished its swing; we call them *dead-beat escapements* (French: *échappements à repos*). And two of them did it by separating balance wheel from train almost completely; only for that tiny moment when the balance released the scape wheel and in turn received impulse from the train was there any contact. We call these *detached* or *free escapements*. All four eliminated the recoil characteristic of the older verge and anchor escapements—that is, the backward movement of the train on each beat as the pallet blocked the scape wheel and the balance wheel or pendulum made its swing. (You can see what I mean if you look at the second hand of a very old grandfather clock.) By so doing they did away with a major source of friction, stress, and loss of energy.

The new dead-beat escapements were both the work of George Graham (1673?–1751). The first, which is in fact known as the dead-beat escapement, was a modification of the anchor for use in pendulum clocks. The name comes from the fact that the pallet stones are so positioned as to fall dead on the teeth of the scape wheel: no recoil and minimum friction and perturbation. (See Figure A.2.) Graham first used the dead-beat in 1715. (There is some evidence that Thomas Tompion attempted something like it as early as 1675; but if so, that was forty years earlier. The domination of the anchor escapement in the interval would indicate that Tompion's invention was not a success, for a true dead-beat would have found wide use.)[16] In combination with a compensated seconds-beating pendulum (you will recall that the mercury pendulum was also Graham's invention), the new escapement made possible a rate varying as little as a fraction of a second a day—in other words, an improvement in precision of between one and two orders of magnitude. It was this instrument that became the preferred timekeeper of scientific observation and remained so, with only ancillary improvements, for two hundred years. It is hard to think of any mechanical device in the annals of modern technology that has reigned supreme so long.

Graham's other dead-beat escapement, for use in watches, was the cylinder, the first examples of which date from 1727–28. In this escapement each tooth of the scape wheel butts first the outside, then the inside, of a pivoting, cut-away cylinder. It is held there a moment, and then another moment, while the balance swings, and during those pauses the train simply cannot transmit its irregularities to the controller.[17] (See Figure A.3.)

Was the cylinder watch an improvement over the verge? The opinions of today's experts differ, as did those of Graham's contemporaries. British makers took well to the new escapement, and for about fifty years the best of them (Graham himself, John Ellicott, Thomas Mudge, et al.) used it almost exclusively in their finest watches. The only important exceptions, such as the verge watches that Mudge made for the king of Spain, were timepieces made for export to countries whose watchmakers did not know the escapement and could not be expected to maintain it.

The French disagreed. Their watch industry was making a comeback after some fifty years of technical conservatism and commercial stagnation, and their *chefs de file,* foremost among them the brothers Pierre and Julien Le Roy, made opposition to the cylinder escapement a matter almost of national honor. They had some valid technical objections, in particular to the wear and tear occasioned by the constant rubbing of brass scape wheel teeth on steel cylinder, which over a period of months could change a good rate to a very bad one. (As always, the steel wore first, for the softer brass, with particles of dust and dirt imbedded, could be as abrasive as a file.) To counter this, the cylinder escapement needed lots of oil; and oil, we have seen, was as much enemy as friend. (The verge escapement did not need oil.)

Yet these problems should have been more challenge than impediment. The answer, after all, lay simply in good maintenance and better materials: steel scape wheels and, for a while, stone-tipped or full-stone (ruby) cylinders. The British worked out these improvements and then, ironically, abandoned the cylinder in the early nineteenth century for what they felt were better escapements. The French and other continental makers, on the other hand, learned to make a proper cylinder only toward the end of the eighteenth century and then, abandoning their objections, produced tens of millions of cylinder watches over the course of the nineteenth. These became the staple of the industry, the typical inexpensive watch, and the verge was finally abandoned.[18]

The British preferred the detached (free) escapements—in particular, the so-called *lever,* known as the *anchor* on the continent. (See Figures A.4 and A.5.) (The British chose to emphasize the nature of the pallet that blocked and released the scape wheel; their side-action version looked like a lever. The French and Swiss chose to stress the appearance of the pallet arms; they looked like an anchor.) The inventor was Thomas Mudge, Beethoven among watchmakers. Mudge made his first lever escapement for a clock in 1754; it may have been only a model. He made his first lever watch in about 1769 for King George III, who gave it to Queen Charlotte, and it is known to this day as Queen Charlotte's watch. At the time he built it, it was almost surely the most precise watch in the world: if Mudge's tests are to be believed, the cumulative variation averaged less than a second a day.[19] Yet Mudge made only one other, giving up this promising line of development to devote himself exclusively to the quest for an effective marine chronometer.

Mudge's escapement found few imitators, all of whom looked on the new device as a highly difficult, experimental form. Not until another British maker, Peter Litherland of Liverpool, invented (or rather, reinvented) the *rack lever* in 1791, did a form of lever escapement enter into commercial production.[20] (See Figure A.4.) To be sure, the rack lever, so called because a pinion on the balance staff meshed with a rack on the end of the lever and pushed it back and forth with each oscillation, was not a detached escapement. Pinion and rack remained in constant contact, and the irregularities in the train were passed right on to the controller. But the rack lever was easily converted to a detached lever by dropping all but one of the pinion leaves on the staff and using that one to kick horns (projections) on the end of the lever back and forth. This was first done by Edward Massey around 1814, and in that form the lever found ever wider acceptance.[21] By the 1820s it dominated British manufacture. It came slower to the continent, which was in the process of shifting from verge to cylinder; and when, in the 1840s, the Swiss took up production of lever watches on a commercial scale, it was as much for export to Britain as for European markets.[22]

Once watchmakers eliminated the bugs and worked out the angles and ratios, the lever proved the most reliable precision escapement for day-to-day wear. If complemented by a compensation balance and properly adjusted, it would keep time within ten or fifteen seconds a day; it was also uncommonly resistant to sharp movements, so that it almost never stopped. Small wonder that from the middle of the nineteenth century it came to dominate all other forms, especially in better watches. Thrifty or impecunious buyers still had to settle for cylinder watches, but even these yielded by the end of the century to a form of lever watch, the *pin lever*, so called because the carefully formed jeweled pallets were replaced with far less costly steel pins. It was the pin lever that made possible the Roskopf watch (Swiss, 1866), "la montre du prolétaire"—forerunner of the Ingersoll dollar watch, the Mickey Mouse watch of the 1930s and 1940s, and the Timex of the 1950s. Thomas Mudge could never have imagined that his experimental detached escapement would one day be found in just about every mechanical watch made, in the tens of millions every year.

The other major detached escapement was the detent, from the French *détente,* meaning trigger or release. The derivation reflects the French priority in this regard: the first timepiece with a detent escapement was built in 1748 by Pierre Le Roy (1717–1785), son of the "celebrated" Julien and himself described by Baillie as "the most eminent horologist of France."[23] Le Roy's version of the detent escapement proved impractical; we shall see that he himself abandoned it for something closer to a lever escapement. But the principles on which it was based found application in simpler forms that offered the highest precision a balance-regulated time-keeper was capable of. (See Figure A.6.) The essence of the escapement lies in its reduction of contact between train and regulator to the barest minimum: an instant when a pallet on the pivoting balance arbor pushes the locking spring aside to allow the scape wheel to advance; another instant when a tooth on the scape wheel catches up with the balance arbor and imparts enough impulse to keep it going. Most of this takes place at speeds too high for the eye to follow. Like the cinema wagon wheels that seem to be standing still because the spokes advance at the same rate as the film, the scape wheel of a detent escapement seems immobile as it flashes from one position to the next.

Thanks to this separation from the wheel train, the balance regulator in such timekeepers was essentially free to keep its own beat. With balance wheel properly compensated for changes in temperature and with an isochronous hairspring, such a mechanism could keep time within a second or less per day. This represented a gain in accuracy of at least two orders of magnitude by comparison with the best watches of the early eighteenth century. It was the detent escapement, then, that made possible in balance-regulated timepieces a standard of performance approaching that of seconds-beating pendulum clocks and thereby realized Huygens' dream of a practical sea clock. From the 1780s on, the detent was a standard feature of the marine chronometer, a position it was to hold for almost the next two hundred years—until the invention of the quartz regulator. Nomenclature reflected the link: with time and use, the *detent escapement* became known as the *chronometer escapement*.

Which brings us back to our story of time measurement and the longitude.

9   The Man Who Stayed to Dinner

W E GO BACK to the late seventeenth century. Recall that after the invention of the pendulum regulator, Huygens and others attempted to adapt it for use at sea, without success. The focus of innovation then passed to timekeepers with balance regulation, but these at their best could not perform so well as pendulum clocks, to the intense disappointment of those savants who thought for a moment that a good marine chronometer was at last within reach.

All this time the flow of impractical, even crackpot, suggestions continued. One of the most grotesque, surely, was a proposal of 1687 to substitute a living clock for a mechanical chronometer. The scheme rested on an adaptation of Sir Kenelm Digby's "powder of sympathy." Digby was a quack of mercurial temperament—in religion and politics he shifted direction like a weather vane. He claimed to have discovered a curative powder that worked not by application to wounds, but to something connected with but separate from them—used bandages, say, or the knife that had made the wound. Digby pretended to have made patients jump merely by dipping their dressings into his powder. Do you see the possibilities now? Our inventor—not Digby, I hasten to add—proposed that all ships set sail with a wounded dog

aboard; meanwhile, back on land, someone would dip some powdered bandages from the dog's wound every hour on the hour in a
bowl of water; the dog would then yelp. However far away the
poor animal was from London, it would yelp the hour with an accuracy that no ship's clock could match. Then all the captain had
to do was to compare local time to yelping-dog time, and he had
his longitude.[1]

Needless to say, this proposal, like dozens of others, never found
application. More to the point, the greatest minds of the age gave
up on clocks and decided that the solution to the longitude problem would be found elsewhere. Sir Isaac Newton, for example,
writing in 1721 in regard to a proposed new longitude watch, argued that longitude "is not to be found at sea by any method by
which it cannot be found at land." Now in those days the longitude was found on land to one-half of a degree (Newton says one-
fourth of a degree) by astronomical observation—that is, by comparing the time of a given celestial event at place of sight with the
time of such event at another place of known longitude. The
trouble was, there was no way as yet to make comparably accurate observations from the deck of a rolling ship so that "by the
same method the longitude might be found at sea without erring
above two or three degrees." What is more, each degree of longitude at the equator is sixty-nine miles; an error of two or three
degrees could easily spell disaster. Newton recognized that "astronomy is not yet exact enough," but it had been much improved "of late," and some further improvement would make it
"exact enough for the sea: & this improvement must be made at
land, not by Watchmakers or teachers of Navigation or people
that know not how to find the Longitude at land, but by the
ablest Astronomers."[2]

The clock- and watchmakers disagreed, so much so that a number of them, including the very best, were prepared to give up
everything else, even the most lucrative practices of "civilian" horology, to devote themselves exclusively to this quest.

One of these was the darkest of dark horses, John Harrison,
carpenter and son of a carpenter, self-taught clockmaker, who
came down from a small, out-of-the-way village near Hull to
London, announced his intention to compete for the Great Prize
offered by Parliament in 1713–1714 (12 Anne c. 15), and then won

it. He is, as much as anyone in history, a symbol of what raw talent can do if married to tenacity and self-confidence. Soviet historians of technology love him, because he represents for them the intellectual triumph of the working man.

John Harrison presumably learned about clocks by fixing them; that is the way in places too small to enjoy the advantages (?) of specialization and division of labor. In a village like Barrow-on-Humber, the smith and the carpenter had to do a little of everything. From repairing, Harrison went to building. His first clocks were conventional, but after the announcement of the Great Prize, news of which reached even to Barrow, he set about with his brother James to build clocks of a higher degree of precision, clocks that would be a testing ground for ideas that might later be incorporated in a ship's timekeeper. He was guided in this by common sense and feel; but it should not be thought that Harrison was ignorant of what we would call scientific principles. Someone, presumably a visiting minister, lent him a copy of Nicholas Saunderson's lectures on natural philosophy at Cambridge University, and Harrison found these so valuable that he copied text and diagrams *in extenso* for his own use.

He had, then, some general knowledge of mechanics; also some mathematical skills. But the characteristic of his work that most impresses is his technical ingenuity and imagination. He could not afford to work in brass, but he did know how to cut and shape wood, and he made a virtue of what any London or Paris clockmaker would have seen as a totally inappropriate material. His plates and wheels were made of oak, carefully cut and glued so that it would not warp; the arbors and pinions were fashioned of boxwood; the bushings were originally of brass but later of lignum vitae, a superhard wood imported from the tropics. Thanks to the oils in the wood, no further lubrication was required, the movements stayed clean, and friction and wear were negligible. Further to reduce friction, he used antifriction rollers and built his scape wheel with roller pinions (something like lantern pinions: but the leaves were spindles that turned as they meshed with the teeth of the train wheels). The escapement itself was a new one, of his invention, which has come to be known from the lifting motion of the pallets as the *grasshopper escapement*. And to compensate for hot and cold, Harrison made use of the different coefficients of expansion of brass and steel to build a composite pendulum rod that would maintain its length at different temperatures. The result was a clock (he actually made two) so accurate that Harrison had to check it by stellar transits, which he managed to track without any of the equipment of an observatory; and had to use special tables of equation of time that allowed for small changes within the leap-year cycle. Harrison claimed that his clocks varied by no more than a second a month, which would make them, if his measures were correct, the most accurate timekeepers known until Siegmund Riefler's regulator of 1889.[3]

Harrison completed his two revolutionary clocks with the aid of his brother James, though he alone signed the instruments; wrote up his methods and results; then went down to London to seek support for his next step, the building of a marine chronometer that would meet the conditions for the Great Prize. One can well imagine what a testing experience it must have been for this young man, gifted and self-confident as he was, to go from a tiny backwater town to the great metropolis and seek audiences with prominent and busy men. He saw Edmund Halley, of comet fame, who sent him to George Graham, the nation's leading clockmaker. Harrison was reluctant: Graham might steal his ideas. But Halley more or less told him that if Graham was not interested, the Board of Longitude would not be. And he gave him some good advice: "Speak to the point." Harrison, in his nervousness, may have given an impression of loquacity.

Harrison got to Graham's about ten in the morning. The first moments of interview were difficult. One can understand that Graham may have been skeptical, even suspicious, at the start. But once he realized that Harrison knew what he was talking about, indeed, that in making a proper gridiron pendulum he had accomplished something that Graham and his fellow clockmakers had essayed in vain, Graham gave him all the time he needed. The two men talked all day and into the evening. In the end, Graham gave Harrison funds without security or interest and put him in touch with others who were also ready to help. The episode says a great deal for "Honest George" Graham as man and craftsman; but it also says something for the openness of British science and technology. The clockmakers of Paris would have been less generous.[4]

Harrison went back to Barrow and spent six years building his first marine timekeeper, which can be seen working today at the National Maritime Museum in Greenwich. It is something to see, especially if one goes there by water, down the Thames from Westminster, under Blackfriars Bridge and past St. Paul's and the Tower, under the Tower Bridge that millions have confused with the London Bridge of the children's song, then left along the London docks, and right again past the Isle of Dogs to Greenwich Reach to land hard by the Royal Naval College. I know of no better way to feel the immensity of old London and the greatness of Britain's maritime and commercial achievements, which owe much to Harrison's inventions.

Harrison's No. 1, along with Nos. 2, 3, and 4, is to be found in a special room of the museum devoted to navigation and timekeeping. It is a large (three feet high, wide, and deep, weighing seventy-one pounds) and, to the camera eye, a most ungainly device. The architecture is all wrong: it is an unshapely mass of rods and springs and wheels, with balls and knobs poking out above and below. (See Figure 18.) It surely looked better in its wooden case, without its entrails exposed like the museum at Beaubourg in Paris. But if it were in a case, we could not see it at work, and we would be much the poorer for that; for it is a wondrous device, made with the awkwardness of a first approximation, Rube Goldbergian in its complexity, but fairly breathing creative genius. Its silent beat is majestically regular, and even an inexperienced, uninformed observer cannot but wonder at the care and skill that went into every part. The wheels are of hard oak (except for the scape wheel, which is brass), with the teeth, also wood, mortised into a groove around the rim. The pinions are of lantern construction, but with rollers rather than fixed rods. The escapement is Harrison's silent, almost frictionless grasshopper. The regulator consists of two large straight-bar balances, mounted on antifriction wheels and connected by cross-wires. They swing, therefore, as if geared together (this is what Bürgi had done), but with much less friction. (As the reader will have inferred by now, one of the salient features of the mechanism is the pains Harrison took to minimize friction.) The force of the two mainsprings is equalized by a central fusee. The four helical balance springs are attached to levers mounted on bimetallic gridirons to compensate for changes in temperature. This was the first such compensation device ever installed in a balance-regulated timekeeper.

H.1 was tried at sea only once, and then on a voyage from London to Lisbon and return—hardly a real test, for there is little change in longitude on such a north-south course. Still, the results were most encouraging, and Harrison was able to correct the master's reckoning of the return landfall by some ninety miles. The Board of Longitude gave him a small subsidy to persevere, and he went home to build a better and smaller machine.

H.2 took four years to build, and H.3 seventeen. H.2 was bigger and heavier than H.1; H.3 hardly smaller. Neither of these was tried at sea, but each was an opportunity for Harrison to try new solutions to the old problems of irregularity, friction, and temperature change.[5] With the presentation of H.3 in 1757, moreover, he announced that he would now work on something much smaller. Clearly he had already begun to think along different lines. When H.4 was completed in 1759, it proved to be no larger than a coach watch: fitted in its outer case, it measured only 5.2 inches in diameter and could be held easily in the hand. It was this watch, which Harrison had thought to build as an auxiliary to No. 3—what we would call a deck watch—that finally met the standards set by Parliament in 1714 and won the Great Prize: it was, in the words of Rupert Gould, "the most famous timekeeper which ever has been or ever will be made."[6] "I think I may make bold to say," wrote Harrison, "that there is neither any other Mechanical or Mathematical thing in the World that is more beautiful or curious in texture than this my watch or Timekeeper for the Longitude . . . and I heartily thank Almighty God that I have lived so long, as in some measure to complete it."[7]

I shall not go into the detail of this mechanism, which interested readers can learn about by consulting Gould's classic history. Let me say only that Harrison could not use his grasshopper in so small a device, so he went back to the age-old verge, but a verge such as had never been made before—almost dead-beat, that is, with hardly any recoil—to the point where Gould treated it as a different escapement entirely.[8] Regularity was maintained by the use of a remontoir; friction was minimized by high-numbered pinions and extensive jeweling (the pallets of the verge were diamond rather than steel, their backs shaped to a cycloidal curve); and the effects of temperature change were compensated by a bimetallic curb on the balance spring.

Again, one cannot but wonder at the creativity and imagination of this man. He was getting on in years now; he was sixty-six when he completed H.4, and sixty-eight when this model apparently met the conditions of 12 Anne 15 on a voyage from London to Jamaica (November 1761–March 1762). One can hardly doubt that, had he been younger, he would have come up with still other and better solutions to the longitude problem. But he was old, and he got caught up at that point in a running battle with the Board of Longitude, which chose to be very sticky about awarding Harrison the prize money promised by law. Harrison did build (or cause to be built) one more chronometer, H.5 (very much like H.4), which he completed at age seventy-seven! In effect, though, he spent the rest of his life fighting for the credit and prize money he felt he had earned.

The quarrel between Harrison and the board has come down in history as the classic battle between David and Goliath, between the lonely, small individual and the massive power of vested interests and nitpicking bureaucracy. Harrison certainly chose to see it this way. He became convinced in particular that the future (from 1765) astronomer-royal Nevil Maskelyne was prejudiced against him and his work, and that he exerted a nefarious, malevolent influence on the board in favor of a technique that Maskelyne himself had helped develop, that of *lunar distances,* or lunars as they were called.

The principle of lunar distances was simple enough. The moon moves faster across the sky than any other visible celestial body—an average of thirty-three seconds of arc for every minute of time. This rapid movement lets the moon serve as the hand of an astronomical clock whose dial is the celestial dome and points are the stars. If one could only read at a known local time the angular distance of the moon from some fixed star and compare this with the predicted time of the same observation for some place of known longitude, the difference in time could then be converted into a difference in space.

As the above statement makes clear, the method would require four things: (1) a means of observing and measuring lunar distances; (2) a timekeeper good enough to give a reasonably accurate measure of local time, in other words good enough to keep time from a solar reading during the daytime to the reading of the lunar distance, usually at night—say, over a period of six to eighteen hours; (3) a table of predicted lunar distances and associated times at a place of known longitude; and (4) a method of "clearing" the *observed* distances from the distorting effects of refraction and parallax—the latter especially important because of the nearness of the moon relative to the star it was measured from—and of reducing them to a common base with the *predicted* distances in the tables.[9]

The first of these was not available until John Hadley's invention in 1731 of the octant that bears his name. The earlier backstaff, or Davis quadrant, simply did not give readings close enough for this purpose, as the best it could do was come within five or six minutes of arc (at the equator, one minute of arc is approximately equal to one nautical mile). The octant meanwhile was accurate within one minute of arc, a major improvement, but could do this only under favorable conditions. Reading points of light in the sky on a rolling deck took a steady hand, for the finer the instrument the more sensitive it was to shake.[10] Even then, an error of one minute of arc translated into thirty minutes or so by the time one finished the complicated calculations required to "clear the distance."

The second requirement was a good deck watch. The balance spring made this possible from the last quarter of the seventeenth century, and subsequent improvements in watch construction (jeweling, more accurate cutting of moving parts, better lubrication, harder steels) realized this potential. Here is Isaac Newton describing best equipment and best observational practice in 1725; he was almost surely thinking of the work of such a maker as George Graham: "A good jewel-watch kept from the air in a proper case, & examined every fair morning & fair evening by the rising & setting Sun, & kept in an eaven heat, may be sufficient for knowing the time of an observation at sea till better methods can be found out."[11] A generation later, around the middle of the century, the best watches available were the center-seconds cylinder watches of such makers as Ellicott and Mudge, though any large-format balance spring watch (what contemporaries initially called pendulum watches) by a first-class maker would keep time within a minute or two a day—if well maintained.

The third requirement, a table of predicted lunar distances, was no easy matter, for the moon's motion, like that of any moving body caught in the gravitational fields of two or more larger bodies, is the resolution of compound and changing forces. No one can measure the labor that went into this research. "More arithmetic and algebra have been devoted to it than to any other question of astronomy or mathematical physics," wrote Sir Frank Dyson, then astronomer-royal, in 1922 in his foreword to Gould's treatise on *The Marine Chronometer*. Not until Newton's contributions to celestial mechanics was it possible to attack the problem correctly; and even then the results were not good enough. Newton's lunar tables of 1713 reduced the error regarding the moon's position from 10–15 minutes to 5–6; but the latter still yielded longitude errors of 2.5–3 degrees. The grand old man (he was seventy-one) confessed that predicting the motion of the moon "is difficult, and it is the only problem that ever made my head ache."[12] It took thousands of observations of the moon by Flamsteed, Halley, Pierre-Charles Lemonnier, Jacques Cassini, and others, plus further mathematical analysis by Alexis Clairaut and Leonhard Euler,[13] before a little-known, self-taught German mathematician-astronomer named Tobias Mayer finally produced the tables that made lunars a feasible method for finding the longitude at sea.

Strangely enough, Mayer had not worked out these tables with a view to solving the marine longitude problem. He felt that the moving platform of a ship would not permit observations of sufficient accuracy. But he was a correspondent of Euler's, whose solar and lunar tables and planetary equations had taught him how to analyze the moon's motion, and it was Euler who urged him to submit for the Great Prize.[14] Mayer had already communicated his materials in 1752 to the Royal Society in Göttingen, where he was professor at the university and taught mathematics. Now in 1755 he sent them to the Board of Longitude, which put them to the test in 1757–1759 off the coast of Brittany. They produced readings accurate within thirty-seven minutes—not good enough to win the largest prize, but good enough for second best. To be sure, the circumstances did not yet satisfy the specific conditions for the trials, which required that the longitude be found after a long voyage to the West Indies, but there was no reason to believe that lunars would not serve as well there as off the coast of Africa.

Imagine now the sense of accomplishment that was felt in astronomical circles. (Remember the pleasure that Huygens felt at working out the theory of the cycloidal pendulum.)

Maskelyne was one of the believers. In 1761 he was sent by the Royal Society to St. Helena (off Africa) to observe a transit of Venus. He took a set of Mayer's tables, calculated the longitude by lunars, and reported in the *Philosophical Transactions* in 1762 that he had come within 1.5 degrees of true. He was not alone: the French, who had also been working along these lines for years (Lemonnier, Clairaut, the Abbé de la Caille), were quick to present Mayer's materials in a form suitable for marine navigation. From 1761 on, the *Connoissance des temps ou des mouvements célestes,* a yearbook of astronomical ephemerides that went back to 1679, gave positions of the moon at twelve-hour intervals; and the 1761 edition gave a specimen table of positions at four-hour intervals especially arranged for use by seamen. The French were thus the first to act upon the new observational possibilities. But they never carried through, and it was Maskelyne who earned the title of "Father of Lunar Observation."[15] (We shall see a similar failure of French execution in the development and exploitation of the marine chronometer.) In 1763 he brought out *The British Mariner's Guide,* which introduced the lunar method to seamen; and in 1765, when he was named astronomer-royal, he published the first annual *Nautical Almanac and Astronomical Ephemeris,* with lunar tables for the year 1767.[16]

Just at the point, then, when Maskelyne saw astronomy triumphant, there was that spoilsport John Harrison, coming with his secret boxes and usurping the fruits of decades of patient observation and brilliant mathematical analysis.

Yet in the end it was Harrison who was right and Maskelyne wrong. Why? Because of the practical difficulty of the lunar method. I have already mentioned the observational problem: the truest predictions of lunar positions and best trigonometric techniques could not compensate for a bad reading. The best one could do was multiply the readings so as to avoid the errant result. Good practice called for four observers working simultaneously, each confining himself to one of the data required, and making four or five sets of observations within a period of six to eight minutes. The results would then be averaged, ideally while taking into account the lapse of time between the different sets.

But this was only the beginning. The measures, as made from the deck of the ship, then had to be reduced to what they would have been for an observer standing at the center of the earth. (This was the fourth of the requirements.) There were a number of ways to do this, but all were extraordinarily complex and time consuming, even with the help of logarithms (without logs, none of this would be thinkable), and all had to be done the hard way—there were no pocket calculators in those days. As a result, even so experienced a reckoner as Maskelyne, who had the advantage of understanding the procedure (which is more than most navigators did), needed some four hours to work out a longitude by lunars. I am not prepared to say how long someone less instructed might take or what the risks of error were in so lengthy a procedure.[17]

None of this discouraged Maskelyne. Harrison was right: the astronomer was partial to lunars. But Harrison was almost surely wrong about motive. Maskelyne's original support for lunars was based on intellectual conviction and congeniality, not on personal animus. The animus came later, on both sides, and embittered a difference of judgment. Maskelyne came to dislike Harrison and took every opportunity to be difficult, perhaps even unfair.[18] Harrison replied in kind. Each saw the other as representative of a larger group of adversaries and pests: Harrison inveighed against "priests and professors" (Maskelyne was a minister), while Maskelyne proclaimed his contempt for "mechanics."[19] The two men were at each other like cat and dog, and when the workmen whom Maskelyne brought to Harrison's dwelling to take delivery of his timekeepers managed under Maskelyne's eyes to drop his model No. 1 and damage it severely, Harrison could not be convinced that the mishap was accidental.

Meanwhile the Board of Longitude had all the advantage of legal ground. The original act of Parliament was unfortunately vague about the conditions for awarding the prize, in part because no one at the time had any clear notion where the solution would lie; least of all did the experts and savants expect it to be horological. Now, confronted by Harrison's claim, the board chose to construe the act strictly, and wisely insisted on the submission not of a method that worked once but of one that would work again and again and be accessible to all.

Harrison's clock had not yet met these conditions in 1762–63.

There was some lingering doubt as to its accuracy, not unjustified; Harrison had not provided a rate for his clock until after the voyage was over. (An ex post rate is like an ex post prediction: it is hard to be wrong.) Even if one accepted his figures, moreover, the clock had proved itself at best twice (going and coming). Was this a happy accident? Harrison and the board agreed on the desirability of a second trial, which was conducted in 1764. Once again H.4 performed admirably, with a mean error of estimate of the time difference between Portsmouth and Barbados of only thirty-nine seconds (equals 9.75 minutes of longitude or 11.2 miles at the equator, well within the best limit of thirty minutes of longitude set by the Act of Queen Anne). In five months the clock had lost only fifteen seconds, a tenth of a second a day![20]

That seemed to dispose of the accident objection. But it was not enough. The position of the board, which now included the newly appointed astronomer-royal, Maskelyne, was that this was not yet a solution to the problem of the longitude—indeed that it could not be, so long as the mechanism remained a secret, took years to build, and could not be reproduced by others. (One can see their point of view.) In 1765, therefore, Parliament passed 5 Geo. III c. 20, which required Harrison to explain the principles and working of his chronometer and to turn over his earlier machines, and which withheld the last half of the prize money until "other timekeepers of the same kind shall be made" and proved capable of determining the longitude within thirty minutes. This last provision, which was to be a source of continuing dispute and acrimony, was interpreted as requiring that these timekeepers be made by someone other than Harrison.

This is not the place to follow this controversy in detail; the story is well told by Humphrey Quill, who does justice, I think, to both sides. Harrison eventually did get his reward, but only because King George, himself (like Louis XVI of France) an amateur horologist, intervened on his behalf: "By God, Harrison, I'll see you righted!" It was Parliament that voted him the money, and even then the Board of Longitude managed to "bilk" him—the word is Gould's—of some £1,250 that they had paid for H.2 and H.3 and now deducted from the sum due him.[21] It was a bittersweet end for a story of extraordinary achievement.

With all this, it will come as a surprise to learn that Harrison's sea clocks, for all their art and precision, are not the ancestors of the modern marine chronometer. Of Harrison's many ingenious inventions, only the maintaining power (the spring that keeps the clocks beating time while they are being wound) has been retained in later mechanisms. The rest were superseded by simpler solutions. His contribution, then, lay not in his devices but in the simple fact of their effectiveness: he demonstrated that the job could be done. He proved that a clock that could eliminate those last significant sources of error (friction, expansion and contraction due to temperature changes, and the influence of irregularities in the train) would indeed serve to tell the longitude. Ironically his forte proved to be not so much his ingenuity and craftsmanship, great as these were, but his grasp of the principles of horology, which well exceeded that of more learned contemporaries and served as an example to them.

Harrison's splendid long-case clocks, with their grasshopper escapement, gridiron pendulum, and almost frictionless roller pinions, suffered the same fate. They performed better than any others of that day; one of them did not vary a minute in ten years. Yet it was George Graham's dead-beat regulator with mercury pendulum that became the standard instrument for astronomical observation. Here, as with the marine chronometer, it was the combination of simplicity with high performance that decided. Harrison spent his last years crying out for recognition of his machines and inveighing against those who preferred the Graham model. This put him in an awkward position: Graham, after all, had been the first to help him, and Graham was long since dead. For Harrison, though, the issue was one of fact and transcended personal ties and obligations of gratitude. He had never, he affirmed, said or written anything on this matter that he had not said earlier to Graham himself. We may believe him.[22]

Harrison died in 1776. The *Gentleman's Magazine,* which had devoted one and a half columns to George Graham's obituary in 1751, had only a few words to say about the most extraordinary autodidact in the history of arts and crafts, a man whose inventions laid the basis for accurate navigation, fostered world trade, and saved untold lives: "March 24th, Mr. John Harrison, in the 84th year of his age. He was a most ingenious mechanic, and received the £20,000 reward for the discovery of the longitude."

10   The French Connection

THE PROOF that a marine chronometer could be built bore fruit at this point not so much in England as in France. The two nations were great competitors and often adversaries. From the late Middle Ages on, their rivalry was the leitmotiv of European diplomatics and war, obscured sometimes by the rise and pretensions of a third power such as Spain in the sixteenth century or Holland in the seventeenth, but always there, because England and France were the prime candidates for European hegemony. There was also a long-standing disagreement about English claims to the throne of France. These went back to the fourteenth century and the Hundred Years' War. They were just memories by the eighteenth—memories livened, however, by tales of the Black Prince and Prince Hal and *la Pucelle d'Orléans*, of Crécy, Agincourt, and a burning stake at Rouen. And what is more tenacious than a tale heard at a father's knee or on the Shakespearean stage? The rulers of England continued to include among their dignities the crown of France. They were not to give up the claim until the Peace of Amiens in 1802, and it took another fifteen years to erase the *fleur de lys* from the British coinage.

Before the Industrial Revolution, France was the more populous country and richer in aggregate, hence the stronger land power. But England had the advantage in ships and seamen and exploited this superiority over the course of the eighteenth century to strip France of most of its colonial possessions. Not without difficulty: France had its own sources of naval strength and, in spite of England's intimidating record (the Spanish and Dutch could offer testimony on that score), was not prepared to concede the seas without a battle. This maritime struggle was linked to commercial rivalry. For both countries the eighteenth century was a period of rapid growth of trade and competition in what were known as colonial wares: sugar, coffee, tea, tobacco—what I like to call Europe's "big fix." These commodities were the source of fabulous fortunes and windfall revenues, and their profitability depended directly on the efficacy and productivity of the merchant marine.

Each therefore had good reason to try to beat out the other in the race for the discovery of the longitude and then use its advantage as a military and commercial weapon. Today, certainly, the whole subject would be treated as a state secret. But the sense of what we call national security was less developed then and was outweighed in this instance by the scientific character of the problem. The search for the longitude was perceived from the beginning as a project of intellectual and humanitarian concern transcending national interests and boundaries. This is not to say that the spirit of competition was absent. On the contrary, the participating scientists and mechanics were as sensitive as ever to questions of priority, while their rulers and countrymen drew honor from their achievements. But such prizes are not the rewards of secrecy.

In both countries the research effort required and called forth new and costly modes of investigation. The late seventeenth and eighteenth centuries, for all their wars and colonial rivalries, were a golden age of collective observation and verification, much of it undertaken by interdisciplinary teams conceived, recruited, and guided by newly sprung learned societies, especially the academies of Paris and London. Science was entering an entrepreneurial mode, whose highest expression was a new vehicle of experiment and inquiry—the expedition.[1] This was a logical derivative of a technique long since worked out by the military:

one or more vessels, or a wagon train, suitably equipped and furnished with logistic support, dispatched to distant places to operate for months if need be, with instructions to kill (figuratively speaking) as many intellectual birds as possible. Such expeditions might have a principal, opportunistic objective—the desire, for example, to observe a particular celestial event such as an eclipse. But typically they also took the occasion to carry out geodetic, hydrographic, and cartographic surveys; to catalogue and collect flora and fauna; to make ethnographic observations; and to gather such other information as an intelligent curiosity might discover.

These projects were not without political interest. Among the earliest, for example, were Russian missions to Siberia, the Aleutians, and Alaska in 1720–1741. There the Dane Vitus Behring was the precursor of empire, as James Cook was later to be in Australia and the South Pacific. For the scientific community, however, such considerations were incidental. They had their uses as an argument for government support. But what mattered to savants was the intoxicating prospect of a new world of knowledge and understanding.

This magic casement went back to the first oceanic discoveries of the fifteenth and sixteenth centuries, to the visions nursed by Henry of Portugal, surnamed the Navigator, and his team of mariners, instrument makers, and astronomers in their land's-end command post at Sagres. These first contacts with new places and new peoples had the most profound impact on European culture and on the European psyche. Their scientific effect was initially limited, however, in that European science was not yet ready to make much of the opportunity offered. For that, a revolution in theory and method was required, which was largely the work of the seventeenth century. This scientific revolution in turn was closely linked to the availability of new instruments—telescope, microscope, thermometer, barometer, pendulum—that made possible observations finer than any before and posed issues never suspected.

One of the first voyages to take advantage of these new opportunities was that of Jean Richer, sent to Cayenne (Guyana) in 1672–1673 by the French Académie des Sciences to establish at site the length of a seconds-beating pendulum. This was on the face a horological problem, but its interest lay in its ramifications into the physics of heat and mechanics. Richer found a figure of 990 millimeters, as against 994 at Paris, and Huygens, using Newton's law of gravitation, deduced therefrom that the earth was fatter at the equator and flattened at the poles. Newton agreed. Jean-Dominique Cassini, Italo-French astronomer and organizer of the Paris Observatoire, disagreed, and with him a number of outstanding savants who found the very idea grotesque. The debate ran on for decades. The highest officers of state got involved and persuaded the French king to send out two further geodetic-geographic expeditions by way of testing (among other things) the Huygens-Newton hypothesis. These only confirmed the outrageous notion. They were also the occasion for testy quarrels among the participating scientists that provided weeks of good gossip for the salons of Paris. Science in those days was *à la mode,* and there was not yet that gulf separating it from the humanities that has given rise in our time to C. P. Snow's thesis of the Two Cultures.

The problem of the longitude was ideally suited to this mood and mode of discovery. The inquiry demanded extensive observation and analysis, required the combined efforts of savants in several disciplines, and was tightly linked to the systematic testing of the newest instruments, among them a whole series of experimental timekeepers. It was in this context that the French carefully followed Harrison's work while pursuing their own horological efforts; and that British scientists and officials made it a point to share Harrison's achievement, as well as their progress in the use of lunar distances, with the world at large. In 1763, after conclusion of the Treaty of Paris, the French ambassador to London was encouraged to bring horologists over to examine Harrison's machine. Nothing came of the matter because Harrison, who by this time felt persecuted, would not show his instrument except on payment of a large sum. A few years later, however, Ferdinand Berthoud, watchmaker in Paris and a leader in the French effort to produce an accurate sea clock, came over again and, after negotiations with Harrison failed (the French did not offer enough), learned what there was to be learned of Harrison's No. 4 from Thomas Mudge, who was a member of the committee appointed by the Board of Longitude to examine Harrison's clock and oversee its reproduction.[2] Harrison, needless to say, was furious at what he felt was an unpardonable breach of confidence by one clockmaker at the expense of another. In fairness to Mudge, he had no intimation that Harrison's inventions were to be kept secret. The ultimate responsibility was the board's, which showed itself throughout more conscientious in its defense of the government's money and interests than of Harrison's rights. In fairness to the board, though, the commissioners felt that Harrison, by submitting for the prize money, had surrendered his machines to the public domain; furthermore, that the whole purpose of the competition was to propagate the new technique as widely as possible. That meant not only Britain but the world.

Berthoud, as it turned out, could do little with what he had learned. In a letter from London he pointed out to the French government that the construction of the watch was simple enough, but its execution very difficult. Certain parts could not even be made in France, notably the ruby bearings.[3] What Berthoud could get from Harrison, he had already taken—namely, the bimetallic grill for temperature compensation. He had used this in the very first of his marine chronometers in 1760 and had continued to use it thereafter. But Berthoud did not copy Harrison's escapement; instead he tried an array of dead-beat and frictional-rest mechanisms, mostly variants of the anchor and the Graham cylinder.[4]

Meanwhile Berthoud's rival Pierre Le Roy, son of the leading French watchmaker Julien Le Roy, was working along different lines. Pierre was not the craftsman Berthoud was; but like Harrison he had a good grasp of the principles of accurate timekeeping, and he was put on the right track by his critical approach to George Graham's cylinder escapement. As noted earlier, French watchmakers in general disliked the cylinder, for right and wrong reasons; and none disliked it so much as the Le Roys, in particular Pierre's uncle, also named Pierre. Now the younger Pierre, seeking a remedy for the friction of the cylinder, designed in 1748 the first detached escapement for balance-controlled timekeeper: the controller (the balance) was free of contact with the scape wheel through most of its oscillation.[5] Le Roy achieved this result by employing a spring-loaded detent to block and release the scape wheel—*the first true detent escapement.* Le Roy put his guiding principle metaphorically: like a doctor who lets nature work for him in healing a patient and makes it a point not to interfere (a

very prudent strategy, given the medicine of the time), he preferred not to cure the defects of the cylinder but to avoid them
entirely by letting the balance swing free.[6] It was the right therapy, but the patient died: Le Roy was not satisfied with his new
device and sought a new form of detached escapement.[7] (See Figure 19.)

He took some years finding one that he thought acceptable and
incorporated it in 1766 in a sea clock that he submitted the following year for a newly announced prize of the Académie des Sciences. Unfortunately, the projected tests at sea could not be carried out and the prize was not awarded, although the clock was
followed on land by the astronomers Lemonnier and César
François Cassini de Thury and did well enough to merit "much
praise." Then a private Maecenas stepped forward, François-
César Le Tellier, marquis de Courtanvaux, captain-colonel of the
king's Swiss guard, grandee of Spain, and member of the
Académie des Sciences—a most impressive and varied bouquet of
titles and distinctions. Le Tellier had a small ship built to serve as
a floating chronometric laboratory, and Le Roy submitted two
clocks for testing, the original and a copy. In charge of the tests
was Alexandre Pingré, cleric and astronomer, veteran of the oceanic expedition of 1761 to observe the transit of Venus across the
sun. For some three months Le Tellier's vessel, the *Aurore,* cruised
the North Sea, touching at Dunkerque, Amsterdam, Rotterdam,
and other ports. This was not the same as an ocean voyage, although the ship did encounter some stormy seas. At the end, the
judges found that one of the clocks had gained too much, but the
other had been "effectively isochronous." In fact both did very
well, and had allowance been made for the underlying rate, either
might have competed for and won the British prize. The better
clock, we are told, was accurate within two minutes of longitude.[8]

Yet there was still the academy to be satisfied, and that required a test comparable to the ones imposed on Harrison, specifically an Atlantic crossing. In 1769 the academy renewed the
challenge, doubling the prize, and again Le Roy submitted. This
time it was the French naval frigate *Enjouée* that served as laboratory, taking the clocks from France to Newfoundland and the
valley of the St. Lawrence, thence to the west coast of Africa and
north along the Spanish coast back to port. The man in charge of

the tests was the astronomer Jacques-Dominique Cassini (the young son of the astronomer Cassini de Thury), who found that Le Roy deserved the double prize, not so much because his clocks were everything one could ask for—they did less well than Harrison's—but because they were a step in the right direction. The better clock's rate, Cassini wrote, "seemed generally regular enough to justify this reward, whose primary aim is to give encouragement to further research."[9] The French, then, were more generous in their response than the British had been with Harrison; but the money stakes were considerably smaller (thousands of livres rather than thousands of pounds), and the Board of Longitude had in fact given Harrison a number of partial rewards by way of encouragement.

By that standard, Le Roy had earned his prize, for he was definitely on the right track. Ask a sample of horologists to name the inventor of the modern marine chronometer, and the answer will be Pierre Le Roy, primarily because that clock of 1766 is thought to have had the salient feature of just about all later chronometers—namely, a chronometer (detent) escapement, in this case with *pivoted detent*.[10] A pivoted detent? That's what it says in the display case in the Musée de la Technique in Paris. More to the point, that's what Gould said, and generations of horologists have taken this to mean the kind of spring-loaded escapement used in subsequent marine and pocket chronometers and generally known by that name.[11]

In fact, the escapement of Le Roy's improved sea clock of 1766 was no longer a detent escapement in the commonly accepted sense. I say this even though Le Roy dubbed it an *échappement à détente*.[12] In mechanical terms it was a lever; indeed, the catalogue of the collection of the Conservatoire des Arts et Métiers describes it as such, as "a sort of anchor [equals English *lever*]"[13] It blocks and releases the scape wheel by means of two locking arms (hence the image of an anchor) on a turning arbor; these in turn are thrown in and out by two disengaging arms that are kicked back and forth by projections on the rim of the balance (see Figure 19). Because this rim moves very rapidly, the motions of the locking piece are abrupt, too swift to follow—almost as fast as those of spring-loaded detents. But a quick lever is still a lever.

Le Roy's sea clock, then, was not the prototype of the marine chronometer, at least not in execution. It was, however, its inspiration. It did have a *detached* escapement, hence a freely oscillating controller. It was also the first timepiece to compensate for temperature change not by curbing the balance spring but by adjusting the moment of inertia of the balance. And it was the first to use an isochronized balance spring. To quote Gould: "All three of these devices are to be found in every chronometer made today—all three are present in Le Roy's machine—and the three are not to be found, combined or even separately, in any timekeeper of earlier date. Accompanying the machine was a memoir describing, in most lucid style, the principles of its construction and the calculations on which its proportions were based. On the joint evidence of machine and memoir, I regard Le Roy as the greatest horological genius who ever lived."[14]

All of these devices were drastically modified in subsequent marine instruments. Le Roy's compensation balance, which made use of the differential expansion and contraction of alcohol and mercury, was never taken up: an easier solution was eventually found by Thomas Earnshaw, who laminated brass and steel to produce the classic bimetallic circle.[15] Similarly Le Roy's pair of spiral balance springs, copied on those used in watches, was superseded by John Arnold's cylindrical (helical) coil from the mid-1770s on. And so also for his escapement: the lever had a great future in "civilian" watches, but it derived from Mudge's work and not Le Roy's. As for chronometric precision, the future lay with one or another form of spring-loaded detent, the sort of thing that Le Roy had tried and abandoned.

For all that, Le Roy showed the way—in two senses. First, he understood, as no one before, the physical and horological principles on which an accurate timekeeper could and should be built. This was no accident. Aside from his own intelligence, he had learned much from his father and uncle, and he was a faithful reader of the mathematical and scientific literature of his time. Second, he described these principles and the machines that embodied them with such detail and clarity as to enable others to copy and eventually improve on his work. His books were eagerly read and studied in England, and there is reason to believe that at least one sea clock was built in London to the specifications of his chronometer of 1766.[16] According to the physicist-astronomer

John Bevis, moreover, this copy was the work not of a professional watchmaker but of a skilled mechanic—testimony to the reproducibility of Le Roy's design. So far as I know, this clock has since disappeared, which is most unfortunate because it represents a kind of missing link between Le Roy's inventions and the development of the modern chronometer by English horologers.[17]

Le Roy himself was very much aware of the significance of his design, so different in spirit and execution from Harrison's almost inimitable masterwork. In his *Exposé succinct* he noted that his clock had far fewer mobiles than Harrison's, fewer teeth per wheel, fewer leaves per pinion, no ruby bearings or shaped diamond pallets. "Nothing," he wrote, "can be too simple in a machine for sea purposes, and to be transported into all places and climates." As a result, he felt, the clock could be easily copied and widely used: "All this I think equally favorable to [my watches'] solidity and their easy execution; two things of great moment in making up these instruments, which in order to come into general use should be executed and set to rights, if not by every workman, at least however by the generality of such, whose talents are a little above the common sort."[18] The passage of years confirmed him in the conviction that what really mattered was the possibility of industrial production: "It is not enough for the complete solution of this great problem of the longitude [simply] to make a watch that keeps time at sea with the accuracy required. The task of making others like it must not be so difficult that one cannot always be sure of their supply; if not, this invention, instead of being of universal use, would serve only a small number of Navigators lucky enough to have one."[19]

For these reasons his clock, he asserted, was better than any that had preceded it, including that of Harrison, "and I am prepared to affirm [*j'ose assurer*] that all those that will be used in the future will be copied very closely on this model."[20] A proud boast; but it takes a proud man to drive himself to accomplish such a feat, to sacrifice on the altar of science the comfortable, easy earnings of the civilian horology that Pierre had inherited from his father.[21] To put his triumph in perspective, Le Roy cited the seventeenth-century astronomer-astrologer Jean-Baptiste Morin on the chances of measuring longitude by means of a clock. "It would be folly," wrote Morin, "to undertake it . . . I don't know if the Devil himself could do it."[22] Look at me, Le Roy was saying—I have done it.

Patriotism was a spur. Like his father before him, Pierre le Roy had all the self-esteem and *suffisance* that characterize the more gifted representatives of French genius; and nothing has more power to stir French genius than the prospect of beating the English. "The honor of having given my fatherland an instrument so useful and so much desired, of having sustained in this regard the reputation of our Arts among Foreigners, especially those of a nation that has always been our competitor and rival; this honor, I say, which Frenchmen always hunger for, joined to that of two double academic palms, was a sufficiently flattering recompense for twenty-five years of work and sacrifice of all kinds undertaken to achieve this discovery."[23]

The point of such self-gratulation, of course, was that the recompense had not been enough. Le Roy had every reason to expect fame and the gratitude of his king and country. He was disappointed.

The disappointment came from the favor shown his rival Ferdinand Berthoud. Berthoud (1727–1799) was Neuchâtelois by birth, had learned watchmaking from his brother, then at eighteen had gone from the mountains to Paris to learn more about the art and seek his fortune. He remained in France the rest of his life, but for Pierre Le Roy he would always be a foreigner, just as Sully had been for his father Julien. Berthoud was everything that Le Roy was not—and vice versa. He did not have Le Roy's understanding of principles, so that his work proceeded by trial and error; but he was a marvelous craftsman and worked hard and fast, so that he could turn out one timepiece after another while Le Roy was following his principles to their conclusion. His first eight sea clocks, finished between 1760 and 1767, employed an extraordinary diversity of drives, controllers, escapements, and compensation mechanisms, many of them borrowed from the work of his predecessors without sufficient thought of their suitability to marine use. All of them, moreover, went through a number of conversions: No. 3, for example, started with a clock-type, partially dead-beat escapement, was changed to a double virgule (one of the rarest and most difficult escapements known), and finally ended in 1771 as a pivoted detent.[24]

To Le Roy, Berthoud's fumblings and tergiversations in the choice and construction of escapements were absurd. All that copying: What did it show but incompetence? "I do not like to follow in other people's footsteps," wrote Le Roy.[25] All those doubts: in a sarcastic passage, Le Roy cites Berthoud as being pro the ruby cylinder escapement, then con; pro the pivoted detent, then con—all within the pages of a single book. At that point, the detached (free) escapement seemed to intimidate Berthoud because it moved so fast: "The promptness of its effects frightens the imagination."[26] One can almost feel Le Roy's contempt: the free escapement had never given him any trouble. Ferdinand Berthoud, he assured his reader, could also learn to work with it, but he must make it as Le Roy did and not multiply the parts and springs unnecessarily. But that, of course, is the besetting temptation of cunning hands: to solve each difficulty by adding a corrective mobile or two, rather than by building on the right principles.

For Le Roy, Berthoud had no principles—of chronometry, that is. How else could he assert "the perfect conformity of principle between my [Berthoud's] clock No. 1, completed before 1761, and all those that I have built since?"[27] More honest, Le Roy felt, was Berthoud's unconscious admission that "unfortunately I allowed myself to be led against my own principles, and I have used up in divers experiments time that would have been better employed following only my own ideas."[28] That, after all, was Le Roy's way: "I was intimately persuaded that my principles absolutely had to lead me to an excellent model of marine chronometer, without my relying on anyone else's inspiration."[29] To which he adds a touch of candor that implies again an invidious comparison with the foreigner: "May I say it, that I was animated by a sentiment wherein one will perhaps recognize some of that noble pride that is natural to a Frenchman."[30] In the meantime, none of Berthoud's judgments could be definitive: "I am prepared to assert," wrote Le Roy, "that that's not the author's last word, and that he will change his mind a third time."[31]

Le Roy's ferocity on the subject was triggered by the publication in 1773 of Berthoud's *Traité des horloges marines,* which managed to treat the history of marine chronometers without mentioning the work of Pierre Le Roy, or of anyone else for that matter. "I shall not permit myself," wrote Berthoud, "to speak of the attempts made by other Artists since my first trials and the publication of my researches. It is not for me to judge my contemporaries."[32] What trials? asked Le Roy. Berthoud's machines had not been ready for the sea test of 1768, which had proved the quality of Le Roy's clocks and won him the prize of the Académie des Sciences. What publications? The man's own work testified to his ignorance of the principles of chronometry.

Berthoud, for Le Roy, knew only how to publish his own merits: the *Traité* of 1773 had been preceded by a series of brags in the gazettes and other popular prints. It was bad enough that this foreigner, "born a subject of the King of Prussia,"[33] had succeeded in persuading the French government to appoint him Horloger-Mécanicien de Sa Majesté et de la Marine at an annual stipend of three thousand livres—a thousand days' pay for skilled labor in the Paris watch trade, or about $75–100,000 in today's money.[34] (I don't mind, said Le Roy. He minded terribly.) Or that the French government bought and distributed dozens of copies of Berthoud's *Traité* and awarded him for that monument of puffery an additional pension of fifteen hundred livres per year. But to appropriate Le Roy's merit and credit—that was too much! "After thirty years of work," Le Roy wrote, "I thought I could enjoy in tranquillity the fruit of my sleepless nights—that is, the little reputation that they had earned me."[35] Instead, he had fallen prey to a master of advertising and public relations: "I concede that M. B[erthoud] has known how to get infinitely more out of the discovery of sea clocks than I; I will even confess all my ineptitude in that respect." That was his fault: he had worked rather than talked. He had invited victimization. While inventors, he wrote, are busy inventing, others, "full of outside activity, move heaven and earth [*mettent tout en œuvre*] to profit by their work and more often than not succeed."[36]

History has not been kind to Ferdinand Berthoud. Le Roy's contemptuous strictures have left their mark, and subsequent generations of horologists have tended to dismiss Berthoud's work as bumbling, uninspired, and irrelevant. "A talented plodder" was what Gould called him.[37]

That is unfair. Berthoud was not the mechanician Le Roy was. He did not truly understand first principles; he may not even have understood his own work. But he was a superb craftsman, he had an open mind (too open?), and once he found the right direction, he made machines that worked well. He also was as tenacious as Le Roy was not, so that the course of chronometric development in France went through his clocks and not those of Le Roy. After producing this astonishing array of experimental pieces, all of them a disappointment, Berthoud hit pay dirt with his No. 9 of 1771, which began with a ruby cylinder but finished after three alterations with a pivoted detent—his first successful use of a free escapement. It was, as Le Roy pointed out, a cumbersome, needlessly complex machine, but it marked a turning point in Berthoud's work: from then on he more or less stayed with detents, purifying as he went. (See Figure 20.)

What was needed at that point was some emulation, a multiplication of sources of innovation and improvement, especially in the direction of simplicity and reproducibility. With Le Roy gone, however, Berthoud had no rivals, perhaps because of his privileged position with the authorities. (The French were to retain this reliance on a chosen purveyor to the navy, on an *horloger de la Marine,* until the retirement of Henri Motel in the mid-nineteenth century.) Ferdinand trained his nephew Louis Berthoud (1754–1813) to take his place, and from the mid-1780s on Louis was producing some of the most exquisitely finished chronometers ever made, cleaner and better than those of his uncle. Yet given these standards, there was a limit to what these two men and their helpers could turn out: about eighty pieces by the uncle, one hundred fifty by the nephew.[38] French chronometers remained rare and unconscionably expensive.

11 Fame Is the Spur

IN THE LAST QUARTER of the eighteenth century, the British turned the marine chronometer into an object of industrial manufacture and commercial use. The pioneer here was John Arnold (1736–1799), the rising star of British horology. Arnold was born the son of a watchmaker in the country town of Bodmin. He quarreled with his father, perhaps because his father was trying to teach him the trade, and went off to Holland, where he learned watchmaking and German. The latter was to prove almost as useful as the former, because Arnold later made his way in London by impressing king and court. George III spoke English (unlike his great-grandfather, the first of the Hanoverian dynasty), but Queen Charlotte was from Mecklenburg, the ties to Germany were still strong, and conversation in German not infrequent.

King George was also a devoted amateur horologist, who loved watches and liked nothing better than to take them apart and put them together again. He also liked good watchmakers. We have seen how he took up John Harrison's cause and got him his prize money. John Arnold won George's heart by presenting him in 1764 with a minute repeater (in itself a great rarity) so tiny as to fit into a ring. This triumph of miniaturization was an instant success and brought Arnold international fame. The Tsar, on hearing of it, offered him a small fortune—£1,000—for another, but Arnold refused rather than diminish in any way his homage to his sovereign.

Very few of Arnold's early "civilian" watches have survived. One that has, marked John Arnold No. 1 (but Arnold later admitted making more than a dozen watches with that number), is to be seen in the Musée International d'Horlogerie at La Chaux-de-Fonds. It, too, is an exquisite miniature, a small minute repeater in a blue enamel chinoiserie case. It was sold at auction in London in the early 1950s for the trifling sum of about a hundred pounds. How the British let it out of the country is hard to understand, but museums were slower then to defend the national patrimony. Indeed, the price indicates that no one was much aware of the watch's quality and importance—except perhaps the fortunate buyer.

On the strength of that kind of work, with or without royal favor, Arnold could easily have made his fortune by catering to the needs and whims of the wealthy. Instead, like John Harrison, Pierre Le Roy, and Thomas Mudge before him, he chose to devote himself exclusively to the construction of high-precision sea clocks. From the start, his machines marked a sharp break with the complicated instruments of his predecessors. Beginning in the 1770s he produced a series of relatively simple timekeepers using (1) a pivoted detent escapement, (2) a helical, isochronous balance spring, and (3) a compensation balance of his own invention. Then, some time thereafter in the early 1780s, he shifted from pivoted to spring detent: the lubrication of the pivot was the source of difficulties and irregularities in use.[1] (See Figure A.6.)

Here he ran into competition. From about 1781, a young journeyman watchmaker named Thomas Earnshaw (1749–1829) had been making chronometers with his own spring detent, and this, against considerable prejudice and resistance, drove Arnold's version out in the space of a generation. This victory was based, not on influence or wealth—Earnshaw was vastly inferior in both respects—but on cost and performance. His design was simpler; and he needed no oil on the teeth of his scape wheel (an inestimable advantage in ensuring long, trouble-free service). In the words of the Board of Longitude, his machines were "incomparably bet-

ter" than those of his rivals.[2] (See Figure A.6.) Earnshaw refined his mechanism and layout by the end of the eighteenth century, and these, once fixed, became the general model for marine
timekeepers for the next century and a half—dominance matched
only by that of Graham's dead-beat pendulum regulators. As
Commander Gould put it in his lecture of 1935: "If I were to show
you, side by side, an Earnshaw chronometer of 1795 and one
made this year, you would be puzzled, I think, to discover any
material difference either in their appearance or their mechanism—a very remarkable tribute to the mechanical sagacity of a
man who died more than a century ago."[3] (See Figure 21.)

The coincidence in time between Arnold's introduction of his
spring detent and Earnshaw's of his gave rise to the usual dispute
over priority. In this instance the quarrel was the more acrimoni-
ous because Earnshaw accused Arnold of plagiarizing his invention. As he tells the story, he had made a number of watches with
his new escapement and shown or spoken of them to other
makers. This readiness to publish his discovery was not moti-
vated, as with Le Roy, by the hope of fame or by a desire to be a
benefactor to mankind. Rather, Earnshaw was moved by economic necessity. He was one of those journeymen who could not
afford to set up on his own and hence worked for others; it was his
hope that his new device would bring him work. In addition, as
his account shows, he took an irrepressible pride in his achievement, to the point of bragging even to someone who he had every
reason to believe would pirate his design. To be sure, he might
have patented it, but a patent then cost about £100 in fees and
Earnshaw did not have that kind of money. Instead, he thought
to interest some London masters in the new escapement, so that
they would contribute to the cost of filing in return for a share in
royalties. One of these was Thomas Wright, watchmaker to the
king.

The watch trade is a small world, and it was not long before the
report of Earnshaw's new escapement reached Arnold, who had
been having trouble with his pivoted detents. The oil tended to
deteriorate, and once the pivot stopped turning freely, the detent
would not return fast enough to block the scape wheel. So, as
Earnshaw tells it, Arnold contrived to get a look at one of his
pieces by telling Wright that it would not really matter if he

showed it, since he (Arnold) had already applied for a patent on his own design of a spring detent escapement.

When Earnshaw heard of this, he was furious. Wright defended himself by saying that he had not given Arnold permission to open the watch and had protested when he did: "Mr. Arnold, I will not have the watch opened." To which Arnold had haughtily replied by asking if anyone in Wright's shop knew how to make a watch anyway and then answered his own question by saying that "so far from being able to make a watch, none of them knew what o'clock it was." This insolence was enough to provoke even a theeing-thying Quaker. "Mr. Arnold," said Mr. Wright, "it does not signify whether I can make a watch or not, I don't fear getting plenty of employ at mending thine, and if the watchmakers do not know what o'clock it is they can know by going to Greenwich for it as thee does."

Is this what happened? Earnshaw says that this exchange was confirmed to him by Robert Best, Wright's foreman. It does have the ring of truth. In any event, Earnshaw, on learning this, hied himself off to the Registrar's office, where he found that Arnold had not yet submitted for a patent. He then promised a small reward to anyone who would notify him if and when Arnold did apply. Arnold applied a week later and received a patent covering, among other things, a new escapement (No. 1328 of 1782).[4]

That Arnold saw Earnshaw's escapement before making this submission is probably true; at least Arnold never denied it, and Earnshaw fairly shouted his charge of plagiarism from the rooftops.[5] But we know that Arnold's spring detent was significantly different from Earnshaw's: Could he have seen the latter, modified it substantially, and prepared a patent application, all in the space of a week? Vaudrey Mercer, Arnold's biographer, says no: "Almost impossible." No doubt; but there is no evidence that Arnold had in fact perfected his own version of the spring detent escapement at the time of his submission. The patent application is so summary, both in diagram and description, that it is a wonder that any patent was awarded. Indeed, if someone had attacked it, it would have been hard for Arnold to prove what he had invented.[6] (See Figure 22.) Fortunately for Arnold, or unfortunately, there was no need to attack it, since the Earnshaw alternative was better. Mercer's judgment, which strikes me as plausible, is that Arnold had already been working on his own version of the spring detent, and the sight of Earnshaw's watch spurred him to get in a patent application as quickly as possible. That it was a hasty submission is obvious. Meanwhile Earnshaw, who could not afford the patent fees, got Thomas Wright to apply for him (No. 1354 of 1783).[7]

Earnshaw never got over what he felt was Arnold's duplicity. In subsequent years he launched a number of public attacks on Arnold, whose commercial success (no one charged more for his work) and personal prestige exasperated him. These continued after Arnold's death and bear an uncanny similarity to Le Roy's disparagement of Berthoud: they fairly drip with sarcasm and with disdain for Arnold's alleged shortcomings as a craftsman and technician.[8] Earnshaw denounced, for example, the large size of Arnold's marine and some of his pocket chronometers—pieces so prized today by collectors as "of the best kind"—as a blunder copied only by fools. Oil, he said, was the source of Arnold's trouble and "made him run his Time-keepers to the absurd size he has done, merely for the purpose of having a Large and Powerful Main Spring to drag the Works on through all Impediments of oil and dirt." The inevitable result was heavy wear: "witness the cutting and tearing of his Machines." His own chronometers, by way of contrast, because they did not require oil on the scape wheel or detent, were free of this: "They will not find those cuttings and tearing in any part or hardly be able to discover in the oldest of them the least trace or place mark'd on which the escapement has acted."[9]

To Earnshaw's criticisms Arnold (and later, his son) returned a decorous silence. He could afford to, and besides, discretion was the better part of valor. Note that John Arnold was not a man to hide from controversy or turn the other cheek. One of the two prints we have from him during and after the contretemps with Earnshaw is a fierce "Answer . . . to an Anonymous Letter on the Longitude" (London 1782), in which he heaps scorn on his non-professional critic who cannot put his work where his mouth is. Arnold concludes his drubbing with a Popean shaft: "I wage no war with *Bedlam.*" The other publication (1791) is a collection of testimonials to the excellence of his timekeepers, prefaced by a very brief encomium of his contribution to navigation. This makes an invidious comparison between the hundred Arnold watches that were then working for "the preservation of the shipping of these kingdoms" and the one or two machines—"were they ever so perfect when completed"—submitted by some people for the longitude prize over an entire lifetime. The reference is presumably to Harrison, and possibly to Mudge. But it carefully avoids any reference to Earnshaw, whose model was even more susceptible of industrial manufacture than Arnold's.[10]

Earnshaw's strictures and sneers rang rudely in a culture given to understatement. Although his was the injury, he came across as a tactless troublemaker; and unlike Harrison, he had no royal protector to cover him for sins of accent and tone. Also unlike Harrison, he had Maskelyne on his side: after forty years the astronomer-royal was still the dominant figure on the Board of Longitude, and Earnshaw's clocks, we have seen, had the best record of performance on the board's tests. But Earnshaw had his own Maskelyne in the awesome personage of Sir Joseph Banks, baronet, privy council, Officer of the British Empire, president of the Royal Society, and member ex officio of the Board of Longitude.

Banks and Earnshaw came from different worlds. Banks had been born rich and well connected; had been educated at Harrow, Eton, and Oxford; enjoyed a personal income from inherited estates of £6,000 a year (equals almost a million 1982 dollars). He had always been able to do what he pleased. What he pleased was to collect botanical and zoological specimens, and in 1766, at the age of only twenty-three, he was elected to the Royal Society in recognition of a rich haul of plants and insects brought back from an expedition to Newfoundland and Labrador. Two years later he gave £10,000 (well over a million of today's dollars) to outfit James Cook's first voyage to the South Pacific. Banks went along, accompanied by a sizable retinue of humans and animals—a trained botanist, two artists (no photography then), a draftsman, a landscapist, a secretary, four servants (two white, two black), and two greyhounds, to say nothing of a mountain of baggage and creature comforts. He brought back from this trip a treasure of specimens and drawings, plus sweet memories of Tahitian innocence and hospitality so unconstrained that the island came to symbolize the New Cythera. When Cook prepared his second voyage, Banks once again offered to go along and help finance the expedition. This time, though, he wanted to bring a retinue of fifteen. When it proved impossible to accommodate so large a number in appropriate comfort without compromising the seaworthiness of the vessel, Banks stamped his foot on the dock and appealed over Cook's head to the admiralty; and when that did not work, he withdrew in a huff and settled for a voyage to Iceland. Cold comfort indeed.[11]

Banks became president of the Royal Society in 1778, when he was all of thirty-five. He had held this office, then, for over twenty years, brooking no contradiction, when he took it upon himself to combat Earnshaw's claims. He was, Gould tells us, "an enthusiastic patron of the Arnolds, both father and son"; now he "set himself . . . to oppose, tooth and nail, anything tending to exalt or reward another chronometer maker to the detriment of their reputation."[12] Great Cham of British science, friend and colleague of anyone who mattered, dispenser of research funds, patron and arbiter of careers, Banks was the kind of adversary who could turn even the Board of Longitude around.

In this campaign pro Arnold, Banks had the help of (was urged on by?) an old friend, Alexander Dalrymple, the influential chief of the Hydrographic Office of the Royal Navy and a fellow of the Royal Society since 1771. Dalrymple had long been a vigorous advocate of the chronometer as the key to knowing longitude. In accord with this conviction, he had bought and used some of John Arnold's earliest timepieces, which he liked so well as to publish in 1788 a short laudatory pamphlet: "Instructions Concerning Arnold's Chronometers or Timekeepers." Dalrymple was surely well aware of the long-standing rivalry between Arnold and Earnshaw, and we need have no doubt which of the two he favored. What role he played in 1803–4, when Banks made his attack on Earnshaw's claim, we do not know for certain, though we may well infer it from a letter he wrote the *Times* in 1806 criticizing an earlier advertisement by Earnshaw, a letter that he then worked up into a book.[13] Earnshaw himself felt that it was not so much Banks as Dalrymple who was his real adversary, for it is surely to the hydrographer that he refers in the introduction to his *Longitude: An Appeal,* where he expresses his respect for Banks and his regret that he had never won his patronage: "I am persuaded that the opposition I have met with originated not in his mind, but was biassed and stimulated by some person who pretended great knowledge in such matters, and who thought to assist his friend by keeping me down."[14]

On March 3, 1803, the Board of Longitude resolved that Earnshaw's watches had "gone better than any others that have been submitted to trial at the Royal Observatory," and that Parliament should give him a reward at least equal to that received by Mudge (£3,000), provided that he disclosed the construction of the machines in such a way as to make possible their reproduction.[15] Before Earnshaw could satisfy that condition, Banks issued (1804) a pamphlet, "Sir Joseph Banks's Protest," denouncing the decision and praising the work of Arnold and the Swiss expatriate Josiah Emery. This in effect reopened the whole proceeding, the more so as rumors began to reach the board that the improvements they had credited to Earnshaw were really John Arnold's. To settle the dispute, the board then asked Earnshaw and John Roger Arnold (acting for his deceased father) to submit diagrams and explanations of their respective devices.

At this point Banks, acting on his own authority, solicited comments from leading London chronometer makers concerning the merits of the two men. The answers were largely unfavorable to Earnshaw, who had managed to alienate most of his fellow makers over the years by shouting far and wide that their signed work was often his; a watch finisher and contractor should have the good taste and discretion to maintain anonymity. Some of their opinions are clearly derivative: when P. P. Barraud writes that Arnold had obtained a patent for a spring detent before Wright (acting for Earnshaw), it is clear that he never saw the specifications or diagrams of the Arnold patent. And some of their opinions are gratuitously misleading, as when the estimable Robert Pennington wrote: "These two escapements as they are here called, are certainly the same with only a little difference in the construction, and as to any advantage that either has over the other if any, is very small indeed."[16] So small that Pennington, along with his fellow chronometer makers, was then turning to the Earnshaw version. Commenting on these unfriendly witnesses, Gould allows himself a rare moment of sarcasm: "watch and chronometer makers," he calls them, "who indulged in a good deal of hard swearing, and obscured the issue with all manner of envy, hatred, malice and other Christian vices."[17]

One exception was the watch jeweler William Frodsham, who though dependent on Earnshaw's adversary John Brockbank for work, would not deny his own recollection. When asked by the Board of Longitude when he had seen Earnshaw make his improvements in the detached escapement, he answered: late 1780 and early 1781. Had he heard of any such improvement by Arnold at that date? "Not till after Mr. Earnshaw's was out. He showed it to Mr. Vulliamy and my father [who worked for Vulliamy], and Mr. Dwerryhouse."[18] Decidedly, Earnshaw was free with his favors; but that is to his credit and was not without its rewards. Some eight years earlier he had shown Frodsham a better way to make jewels—bread upon the waters.

By this time, though, almost everyone was behaving badly. Not John Roger Arnold, who kept his calm and submitted a sober, factual description of his father's work. But Earnshaw was in a rage that recalls Harrison's. Banks must have seemed to him some kind of evil genie, a *deus* come down *ex machina* not to bring a happy ending but to complicate the drama and thwart its resolution. His complaints became intemperate, even vituperative, and when the board in an effort to appease both sides offered Earnshaw and Arnold equal prizes of £3,000 each (less sums already received, of course), Earnshaw was in no way appeased. He waited until he had his money in hand, then inserted a long advertisement in the press recounting his triumph and denouncing Banks for his partisan interference. Banks tried to get the board to respond by prosecuting Earnshaw for libel, but they had had enough. Banks thought a moment of initiating his own suit, but on reflection, perhaps on counsel, gave up the idea. "From his own point of view, this was probably wise, but he deprived posterity of a great deal of amusement. Banks in the witness box against Earnshaw would have been a sight for the gods."[19]

There is no need here to pursue the dispute further, for it did go on, pamphlet upon pamphlet, charge upon charge. It was the kind of fight that no one could win. Earnshaw did have the consolation of seeing his escapement triumph, but he never got the financial reward he coveted, though he continued until his death in 1830 to make watches and chronometers for both marine and civilian use, some of them superb, some of them of current quality. He never could afford, as Arnold could, to confine himself to the carriage trade.

With the death of the principals, the debate subsided; but it never died, or rather it has been revived in our own time, with passions astonishingly unabated. There are Arnold partisans and Earnshaw partisans, and they are still exchanging invidious comments on their respective champions. Whether these loyalties reflect more the characteristic infatuation, even fanaticism, of collectors or the financial interests at stake (these watches and clocks are worth a small fortune) is hard to say. Surely something of both, as illustrated by a recent *cause célèbre,* which might well provide the matter for a Sherlock Holmesian "Case of the Cryptic Clock."

This was a previously unknown table or bench chronometer, signed Arnold, discovered in 1980 and brought to Christie's auction house in London for sale. The machine itself was neither numbered nor dated, and the signature by itself did not reveal whether the clock had been made by father or son. Presumably it was some kind of experimental model, not intended for sale, and the manner of its construction, with the escapement concealed behind the dial plate and a sleeve, was unique—an arrangement not found in any other Arnold clock, or for that matter in any standard marine chronometer. Christie's, suspecting that they had come upon something very special, had the machine taken down and examined into its farthest recesses; then, on the basis of internal evidence (type of box, balance, enamel dial, wheel crossings, and so on) decided that they had found the earliest known Earnshaw-type spring detent chronometer. They put the date at about 1775.

This was indeed a momentous discovery, one that bid fair to put an end once and for all to the Arnold-Earnshaw controversy. If Christie's were right, the clock was evidence that some seven years before Earnshaw ever built such a device, Arnold had already made one. Could Earnshaw have invented his own escapement independently? Maybe; but the auction house reminded potential buyers that Thomas Wright, the man who took out Earnshaw's patent for him, was maker to the king, and suggested that in that capacity he might have learned of Arnold's ideas and "put up Earnshaw to execute them." Wright, Christie's went on to say, "an overtly religious man, showed himself to be devious and was ejected by the Quakers." Such a man was obviously capable of anything.[20]

When this discovery was published—and Christie's trumpeted their good fortune far and wide—it caused an immense stir in the otherwise small world of horology. I myself was asked by an enthusiastic employee whether I had heard about the great clock coming up for sale—"the most valuable clock in the world." Friends called me from New York and London and wrote from Geneva and Zurich. The London *Times* (April 18, 1981) quoted the head of Christie's clocks and watches department as saying that his find was equivalent in its way to the discovery of evidence that Shakespeare's plays were indeed written by Francis Bacon. "A Plain Jane exterior deliberately hides a mechanism which is six to eight years earlier than students of horology have always been led to believe. For its maker, John Arnold, it means a reappraisal of his role as the Father of the English Marine Clock." As for Earnshaw, he came out of this looking very much a thief. The *Times* said as much, and so did the *Tribune de Genève,* and no doubt other newspapers as well.

That was Christie's peak moment. Almost immediately the objections began to pour in, not only from those who were known to be Earnshaw partisans, but from conscientious Arnoldites. Vaudrey Mercer, for example, Arnold's biographer, dated the machine at about 1811, largely on the basis of internal evidence: the dial had an up-down indicator, something that Arnold *père* had never used; the compensation curb was of "the type used by John Roger in 1811 when he first began to try the Earnshaw detent in his marine chronometers."[21] Others pointed to the similarity between the bimetallic compensation curb and those used by Breguet around the turn of the century, which only led Christie's to insert a fanciful addendum suggesting that the young Breguet might have visited England after completion of his apprenticeship and influenced young Arnold. Still others noted that the spring detent with its separate passing spring was of an advanced type, and that the earliest form, as patented by both Arnold and Earnshaw, used a single spring.

Even more decisive was the illogic of the tale: the whole thing made no sense. It is simply not to be believed that Arnold invented such a device around 1775 and then stood still and silent for decades under Earnshaw's accusations and aspersions. Arnold, we saw, was not a man to suffer obloquy or criticism lightly. Nor can we believe that Arnold would have permitted Wright to patent his (Arnold's) escapement without opposing it as unoriginal or suing him for infringement. All he had to do was put this clock forward, and the matter would have ended for Earnshaw-Wright as it had almost a century earlier for Facio and Debaufre. In other words the story, if true, made out Arnold to be a fool or a saint, and he was neither. Even more serious, it painted him as a horological incompetent, for why would he invent a superior escapement and then abandon it to others, while persisting in the production of timepieces that were harder to make and performed less well?

Under this hail of objections, Christie's retreated, discreetly disguising their retraction as a typographic error: a loose insert in the catalogue warned readers that the date in the catalogue should read "circa 1775–1790" rather than 1775. The rest of the text remained, though it no longer had much point—unconscious testimony to the tenacity of Arnoldite loyalties and the cost of changing page proof. By the time of the sale, a perfectly good and interesting experimental chronometer was so tainted that no one was ready to bid. The house had to buy it in.

That was not the end of the story. Christie's still had the clock. Normally if a lot does not sell, the object is returned to the owner, who then pays a commission to the house for its good efforts. (Salt in the wound.) In this instance, though, Christie's had apparently advanced a very large sum upon the clock, on the expectation that it would realize a small fortune, and now they were left holding the security. It is not for those of us on the outside to know what then ensued, but after some months had passed, Christie's put the clock up for sale again, this time with less fanfare. Even so, the long catalogue description matched the earlier for detail, and the house let it be understood that those who preferred to believe the first conjectures would accept this as a clock of revolutionary historical importance. (Which reminds me of the Duke of Wellington's reply to a gentleman who mistook him for a Mr. Smith: "If you believe that, Sir, you'll believe anything.")

Apparently no one did, for the clock sold at £4,600—no small sum, to be sure, but far less than hoped for (and apparently far less than Christie's advance) and probably less than it was worth as a unique bench chronometer from the shop of John Roger Arnold.

Christie's comparison of their discovery with the Shakespeare-Bacon controversy was apter than they had realized. Just as many people persisted for centuries in refusing to credit ordinary Will Shakespeare with such lofty work (only a great gentleman could do as much), so plebeian Earnshaw has seemed a poor candidate for greatness. In this case, of course, both rivals did wondrous work, and history has room for both. But it has made the second first. *Honi soit qui mal y pense.*

T HIS RETURN of technical primacy to Britain caused the proud Pierre Le Roy much regret. In a letter of 1783 (two years before his death) to the controleur-général des finances, he spoke of twelve years of inertia: "This is the more unfortunate in that the reputation of our watch and clock industry abroad may be affected by it. If England alone is engaged in manufacturing marine chronometers, if we have nothing to oppose to theirs, other countries will conclude that the English are better than we in all horological work." He went on to write that he should have raised the alarm sooner, that he should have been training apprentices. The war had intervened, and besides, he thought it difficult to find people capable of following out his ideas. The master watchmakers, he said, are too busy making money by making for and selling to the general public. "Besides, the few workers who are presently in Paris, even granting them the intelligence required, would not be readily available to do a new apprenticeship; they would need substantial indemnities to devote themselves to the task."[22]

That the workers in Paris were capable of such difficult work, Louis Berthoud and Abraham-Louis Breguet were soon to show. But it would have taken money, enterprise, and tenacity, and Pierre Le Roy unfortunately had none of these. Before he died, his countrymen recognized his contribution; the Academy of Sciences had awarded him a double prize of four thousand livres for each of his two marine timekeepers; and from 1777 on he enjoyed a pension of twelve hundred livres (say, $30–40,000 of our money) "for the invention of the marine chronometer." But he died, I fear, a bitter man.[23]

From the 1780s Arnold, first, and then other British makers began to produce chronometers as they would any other established type of timekeeper. By the early years of the nineteenth century the mechanism was more or less standardized and had taken on the appearance it was to keep from then on. The work was divided among a large number of specialists, scattered in shops in London and Lancashire. Output during this first generation—say, by 1815—was on the order of four to five thousand instruments. Of these, about half came out with the signatures of the big three: Arnold, Earnshaw, and Barraud. As we have seen, French output could be numbered in the hundreds: eighty from Ferdinand Berthoud, one hundred fifty from Louis, less than a hundred from Breguet. The differences reflected not the superiority of British production—although Arnold and Earnshaw enjoyed the highest reputations at home and abroad—but the size of British demand and purchasing power. (A good chronometer cost anywhere from twenty-five to one hundred pounds—from half a year's to two years' salary for a skilled worker [$10,000 to $40,000]. French machines cost even more.) They also reflected, though, the willingness of the British to sacrifice art to utility: the British machines were good enough. Louis Berthoud, whose instruments are the summum of detail and finish, thought even Arnold was too ready to trade quality for economy. To be sure, many British chronometers were superb, setting astonishing records of precision and reliability. To this French makers retorted that such results were less the work of intention than of chance: so large a number of chronometers was bound to produce a few pieces three or four standard deviations above the mean.[24]

The availability of good timekeepers accurate to a second or so a day did not immediately render the lunar method obsolete. On the contrary, the partisans of astronomical navigation worked steadily at improving and simplifying the technique, so that the time required was cut in half by the second part of the nineteenth century, and then in half again in the twentieth. The effort, however, was disproportionate to the results achieved. Only in that first generation or so, when chronometers were still in their costly infancy and had not yet established their reliability, were lunars preferred. Thus in 1809 the navigation handbook issued by the Naval Council in Amsterdam devoted only two of six hundred sixty pages to the chronometer: there were some very good time-keepers, the author admitted, but they were too dear and delicate for use by most sailors.[25] At that date, then, so far as the textbooks were concerned, the first and best method was still that of lunar distances.

So far as the textbooks were concerned. . . I do not believe that the procedure was ever so popular with seamen as these manuals seem to suggest, or that the chronometer made its way so slowly as a number of historians have asserted.[26] It took a well-trained, sharp-eyed, intelligent, eager-beaver navigator to make these observations and then do these tedious calculations day in and day out, and the conjunction of those four virtues can hardly have been common. They were found, if anywhere, in the navy, and it is no accident that the lunar technique persisted longest there. But in the merchant marine, the quick, easy recourse to a clock or watch gave it the preference early, even where no chronometer was available.

This is a crucial point. Students of the subject have allowed themselves to infer the adoption of the chronometer from the number of instruments made or in service. But the boxed chronometer was not the only timekeeper that would do the trick. Makers such as Arnold and Earnshaw were turning out an even larger number of so-called pocket chronometers or deck watches, which masters and mates were buying for themselves at perhaps half the price of a boxed instrument. These would not normally perform so well as a gimbaled marine chronometer. Worn in the pocket, though, they suffered less temperature variation, and this compensated for much of their intrinsic inferiority. A good one would keep time to a second or so a day. And in a pinch, even a good civilian pocket watch would serve.

Indeed, there is reason to believe that most captains preferred dead reckoning to lunars. Gould cites the example of an American vessel seized during the Napoleonic Wars at Christiania (now Oslo) because it had neither chart nor sextant aboard: the authorities assumed that it must have come from the British Isles in violation of the blockade. To this all the American captains in port made protest, saying that many of them made transatlantic voyages without these instruments, "and we are fully persuaded that every seaman with common nautical knowledge can do the same." Gould offers the story as evidence of the late adoption of the chronometer; but without a sextant, no lunars either.[27]

There were now readily available both clocks and watches that could keep time within a fraction of a second a day—one part in 10^5 or better. These devices, moreover, with good care, would provide accurate service indefinitely. A pocket chronometer by Arnold or Earnshaw from the 1790s will keep just as good time today, two hundred years later—a far cry from the ten-year life expectancy of the medieval clock. Indeed, so good was the model that it killed off its makers: the British chronometer industry eventually shrank to one firm because, once the needs even of an expanding merchant marine were satisfied, the demand for replacements was very low. The makers wrought too well.[28]

I SHALL NOT pursue this story of the quest for precision. It went on. The solution of the longitude problem did not end the effort to measure time ever more accurately and in ever smaller units. By the late nineteenth century the use of electric current and vacuum containers made possible mechanical controllers entirely independent of the train and cut variation to a hundredth or even a thousandth of a second a day—one part in 8.6×10^6 or 10^7. At this point man's clocks were more accurate than nature's, and it was possible for the first time to detect, measure, and explain irregularities in the earth's rotation. Then the introduction of the quartz crystal reduced error by another order of magnitude to one part in 8.6×10^8; and then (though surely not finally), the use of atomic vibrations as controller (the first cesium resonator, at Teddington, England, was built in 1955) yielded an accuracy of one part in 30 billion (3×10^{10})—that is, one second in a thousand years. Subsequent gains have averaged an order of magnitude every five years. As a result of this heightened accuracy, the old definition of the second as one 86,400th part of a solar day was no longer suitable; the solar day was simply not uniform. After an attempt to redefine the second as a fraction of a given solar year was rendered obsolete in its turn, the General Conference on Weights and Measures decided in 1967 to detach the second entirely from celestial clocks: it was defined as "9,192,631,770 periods of the radiation corresponding to the transition between the two hyperfine levels of the ground state of the cesium-133 atom."[29]

No one knows what time is; certainly no one knows how to define and explain it to the general satisfaction. But we sure know how to measure it.

III   MAKING TIME

If we date the beginning of mechanical timekeeping from the late thirteenth century, we have a span of seven hundred years of manufacture—from the heavy turret clocks of medieval Europe to the modern wristwatch. Today mechanical timepieces are, for their price, better than ever. They are the most efficient machines in the world: the energy liberated by a cubic centimeter of gasoline—a thousandth of a liter—will keep a watch running for 585 years![1] And the most reliable: that same watch, if well maintained, should still keep time within ten or fifteen seconds a day—an error of 1.7 parts in ten thousand (10^4)—at the end, as at the start, of that long period. Yet in the face of a revolution in the technique of time measurement, mechanical clocks and watches are obsolescent. Not obsolete: they still sell in the tens of millions every year. But the new product is superior, both technically and commercially, and we are witnessing the end of one era, the beginning of another.

Seven hundred years, then: they have constituted an epic of trial and error, systematic experiment, scientific inquiry, technological achievement. They have also witnessed continuing change in the mode and methods of production, in what the economist calls process as opposed to product. This change has lain as much in the relations of manufacture as in the tools employed, in the software as much as in the hardware. I can think of no industry that so well illustrates the stages of manufacture: from ad hoc, itinerant teams of craftsmen to the factory, from complete versatility to utmost specialization and division of labor, from one-of-a-kind works of art to uniform assemblages of interchangeable parts.

This sequence of forms, which was conditioned by changes in the size of the market and in both product and equipment, was accompanied by major shifts in the locus of manufacture. The mode of production was intimately linked to place—to local supplies of labor and knowledge, to entrepreneurial values, to commercial institutions. In the course of these seven hundred years the center of clock and then watch manufacture shifted from Italy to Germany, from Germany to France, from France to England and Geneva, from England and Geneva to the Swiss Jura. There it stayed, in the face of a strong American challenge, for over a hundred years—until the introduction of the quartz watch signaled the birth of a new industry. The giants of the new solid-state technology were the United States and then, even more, Japan. The Swiss are belatedly trying to catch up, while at their heels is a growing pack of pretenders whose greatest asset is their low wages. In this respect the new industry is very different from the old: design is everything and can be exported past frontiers and tariff barriers where commodities often cannot go. Skill counts for less. Assembly can be anywhere, for almost anyone can case these integrated circuits. This is a brand-new ball game, and it would be a rash historian indeed who would predict the ultimate winners. There is one safe bet, though: the long Swiss hegemony, a domination without parallel in the annals of manufacturing industry, will not be repeated.

12 Clocks in the Belfry

THE EARLIEST TURRET CLOCKS were big and so crudely cut that they needed very heavy weights for driving. Their construction was just that: a process essentially comparable to the building of a structure. The building was done by teams, which were brought to the site and did most of the work, including the forge work, right there. These jobs took months, for it was often necessary to build not only the clock but the housing, and the remuneration of the team usually took the form of payments in kind (lodging and subsistence) as well as money. The core of such a team was the master clockmaker and those assistants who had some experience of these matters. In the beginning these people were drawn from among blacksmiths, braziers, and other workers in metal; there was also a pool of craftsmen familiar with machines and wheelwork, for the water mill was already a major feature of the European industrial economy.[1] Probably the largest body of mechanics was to be found in the church, particularly among those monastic houses actively engaged in mining and metallurgy. These made extensive use of water power and played a major role in diffusing knowledge of best engineering practice throughout Europe.

Clock manufacture in its earliest stage was based on locally available talent: teams of craftsmen and laborers were assembled ad hoc and then disbanded. With time and experience, however, certain clockmakers acquired a reputation for excellence and were sought after far and wide. Some of them did little else: when not out in the field, they were busy in the shop filling orders for smaller, domestic timepieces and maintaining and repairing the existing stock. The Jewish clock- and instrument-makers of four-teenth-century Aragon are examples of expert horologists, as are Jacopo and Giovanni de' Dondi of Padua, father and son, and Henry de Vic (Wick), brought in from Germany (Lorraine?) to build the palace clock on the Ile de la Cité in Paris. We know that Dondi *père* was given, as a reward for his services, the right to bear the surname "del Orologio" (of the clock)—an honor much cherished by his descendants. This was a sign of the still excep-tional character of such services: the making of a complicated clock was an act of technical prowess; but this title was also the announcement of a new profession.

Sometimes a clockmaker came, as it were, with his clock. These machines needed care and attention: the *service après vente* was as important as the sale itself. Many a first-class clockmaker, then, having built a costly machine for an important personage, a king or church dignitary, found himself permanently engaged. This did not mean that he did not make other clocks. On the contrary, given the rapid wear and obsolescence of these devices, those who could afford them were constantly ordering new ones, both for themselves and as gifts for others, while stipulating new require-ments. Service as clockmaker to a king entailed supplying a stream of timepieces, as well as going out occasionally on de-tached service. One could make a gift of one's clockmaker—on loan, of course.

We have the story of one such loan in the mid-fourteenth cen-tury. We owe the story to Ramon Sans (alias Remo or Ramundus Sanç), notary (that is, part attorney, part steward) at the court of King Peter (Pere) IV of Aragon. Like all good notaries, Sans was a meticulously careful man, and when his master instructed him to put together a team of craftsmen to build a clock and install it in his palace at Perpignan, Sans kept a detailed record of every expenditure down to the last sou and denier (shilling and penny). This listing was then neatly copied and bound up in a book of accounts, consisting of seventy-seven folio sheets, which has been preserved in the Archives of Aragon in Barcelona. It constitutes the earliest and by far the most detailed record we have of the building and installation of an early clock; as such it is one of the great documents of medieval economic history.

A first report on this material was made by L. Camós i Cabruja of the Historical Archives of Barcelona and appeared in a *Festschrift* published in that city in 1936. Those were hard times in Spain, and most people had their minds on other things than medieval clocks. Furthermore, the report was written in Catalan, not a familiar language to most scholars, and, as legions of disappointed authors can testify, there is no better way to consign something to oblivion than to put it in a *Festschrift*. In those circumstances, the very word "publish" loses its meaning. So it was with Camós i Cabruja's essay.

It was not until a generation later, in the 1960s, that the Museum of the History of Science in Oxford obtained a microfilm of this account book and C. F. C. Beeson, a specialist in the early history of the mechanical clock, undertook its analysis. He needed help with the Old Catalan; and he called on the Iron and Steel Institute, the British Steel Corporation, and the Centre de Recherches de l'Histoire de la Sidérurgie in Nancy for technical advice on the metallurgical techniques of the fourteenth century. His report on the subject to the Antiquarian Horological Society was published in its journal as "Perpignan 1356 and the Earliest Clocks." It is one of the richest, most informative pieces I know of in the history of medieval technology, or any technology. This, however, was only an appetizer. Beeson, along with Francis Maddison of the Museum of the History of Science, was determined to put out an edited edition of the full manuscript and persuaded the Antiquarian Horological Society to sponsor publication. The work took years, and Beeson died too soon to see the book come out. But with Maddison's help and encouragement he did complete the manuscript, which was long delayed in publication; it idled in galley for years. It has now come out and will take its place alongside *Heavenly Clockwork* as a fundamental reference work on early horology.[2]

Beeson deserves well of History; so does Ramon Sans, that punctilious steward who wrote it all down in the first place.

Sans received his commission on January 15, 1356: to assemble the workers and materials required, provide board and lodging for the outsiders, furnish and equip the work space, pay all bills, wages, and salaries, and keep a record of disbursements. The work took nine months and cost 1,667 livres, net of monies realized by auction of unused materials and materiel after completion. Equating top wage then with top wage now (four sous per day, equals $60–$75), we find that the clock would have cost half a million or more 1980 dollars. No small sum.[3]

Of this amount, over a sixth went to pay the contractor-master clockmaker. This was Anthony Bovell (possibly Bonelli), *plomberius* to Pope Innocent VI at Avignon, who was released to King Peter for this project. Bovell was typical of the versatile technicians of his day: plumber by title, he was engineer, mechanic, architect, and builder all rolled into one, and he needed all of these skills to do this job. He brought ten clocksmiths and other specialists along with him from Avignon, whom he paid personally, along with an array of special-purpose tools.

The clock as he planned it would be a large one—about two tons. The bell would weigh twice as much. The tower where it would be placed had to be reinforced and extensively modified. To do all this, Bovell turned a corner of the castle into a workyard. Two large rooms were cleared out and converted into a forge and machine shop. Bovell had three furnaces built, installed anvils and grindstones, had carpenters make bellows on the site. The work force, which numbered nearly a hundred, included masons, bricklayers, smiths, founders, plasterers, ropers, carters, and others. Local shops and merchants supplied a steady stream of raw materials and equipment, some of it drawn from outside the kingdom. The raw iron came from nearby smiths in bars or blooms; it then had to be heated and hammered and reheated and hammered again until it had lost much of its carbon and attained the consistency and malleability of merchant iron. This then had to be beaten into the forms required. Bovell had the carpenters prepare wooden patterns for this purpose, and iron and patterns were sent out to a water-powered tilt hammer. By the time the pieces came back, over nineteen hundredweight of them, they cost thirty-four livres eleven sous. That came to thirty-six sous per hundredweight, almost $700 of our money.

Then the machine had to be assembled, fitted, and adjusted, and the clock and bell raised to the top of the tower. This in itself was a major project: Bovell devised a mobile derrick to bring these huge objects to the tower and used a crane and windlass to hoist them up. By comparison, the driving weights were easy: cast in lead, they weighed only four hundred sixty kilograms altogether.

One can well imagine the lively interest of the people of Perpignan in these goings-on. So eager were they to crowd and watch the work that Bovell had to divert the carpenters a moment to building a spectator barrier. The orders for food and supplies and the enhanced demand for labor must have brought a flush of prosperity to the town. When the huge bell was successfully cast—no small feat—the royal family and the whole work force celebrated with a feast on the spot. The rest was in a way anticlimactic: it took two months to build the bell frame, lay out the course of the ropes, place the bell hammer properly, and time the clock and the striking. There seems to be a law governing any complicated task, whether construction or writing books: the finishing always takes longer.

As often happens, the making was the best part of the clock's history. From 1379 at least—the clock would then have been about twenty-three years old—it gave repeated trouble; hardly surprising, for this was a fair lifetime for a fourteenth-century turret clock. In 1387 the new king, John I, gave up and decided to eke out the machinery with some human assistance. He wanted two men, not necessarily clockmakers—they could be tailors, shoemakers, or furriers, he said—to be hired at a penny a day to ring the hours at the appropriate times. At the same time he endowed the cathedral chapter in his capital of Barcelona with funds for the same purpose: two men would live in the cathedral spire, track the time with a sand glass (King John was not going to waste his good money on clocks), and take turns ringing the bell.

Yet people are not necessarily more reliable than machinery. That, after all, was why the clock was invented in the first place. The ringers of Perpignan did not do their job. The people murmured and grumbled; their plaint reached to the court in Barcelona; and in 1399 the then king Martin ordered the governor of the province to look into the matter. What then happened, we do not know.[4] The fifteenth century was a time of wars for Aragon, and timekeeping may have been forgotten. When the French occupied Perpignan in the sixteenth century, they converted the palace into an arsenal. By that time, the clock had long fallen into disuse and disappeared. Beeson thought that the bell had been broken up and the metal used to cast cannon; I would conjecture, the clock too. Iron, and bell metal even more, were too precious to toss aside on the rubbish heap.

The Perpignan clock was a big, royal project. Other jobs were much smaller, like the tower clock that was built in the fourteenth century for the castle of a noble lady of Nieppe, in northernmost France. We have the contract between the lady's agents and the clockmaker, who lived about fifty kilometers away:

The year 1379, the first of January, was ordered of Pierre Daimleville, maker of clocks residing at L'Ille [today's Lille] to make a clock for my most respected lady [*très redoutée dame*], Madame the Countess of Bar, Lady of Cassel, and to put same and install it in her castle of Nieppe; the clock to weigh, all wrought, 301 pounds of iron, which iron he is to furnish for said work; and in case he thought said work would not be solid enough and he put in extra iron, wherever it was needed to strengthen the work, and if it were properly used, my said lady will pay for all the iron in the work over and above 301 pounds; and for doing said work loyally and well in the opinion of workers and people who know and are expert in this kind of work, the said Pierre will have and take with him the sum of 11 francs in gold or the equivalent, that is, 38 gros of Flanders per franc, for making said clock and for the 301 [pounds of iron]; my said Lady will have it given him and the surplus weight paid, as said.

Item: The said Pierre will place and install the said clock in the steeple where the other clock is at present, and for such time as he takes to put it in, he shall have his expenses paid by my said Lady without other wages.

Item: If said clock should be defective and if it were not installed in the proper form and manner, he would be required to correct the trouble at his own expense, cost, and charge, as stipulated by good workers expert and knowledgeable in the matter.

Item: There is to be lent and delivered by Caspar Molinet, for and on behalf of my said Lady, all sorts of cut and wrought timber, and this installed and placed where he shall order it placed for installing and placing the said clock.

Item: The said Pierre is to have and hold for regulating said clock every year a tunic made of the officers' cloth every time Madame does her livery, and it will be at the expense of my said Lady every time he comes to look at the said clock and that it needs anything; and he must come there whenever asked to; which work the said Pierre is to deliver complete and installed in the said steeple by next Easter; all of which things have been done and ordered by Colard Lévesque and Jehan de Chastillion, clerks and secretaries of my said Lady.[5]

The contract is of interest for many reasons, both for what it says and what it does not say. What it does not say is the nature of the clock to be built. This was presumably agreed upon orally between my Lady of Cassel (or her agents) and her clockmakers; but the fact that the specifications are not given in the contract would seem to indicate that Daimleville was going to make a standard tower clock, without complications and possibly without bell. This standard, moreover, was manifestly current in the trade. Else how could one refer to "proper form and manner" and provide for arbitration of differences by "good workers expert and knowledgeable in the matter"?

What the contract does talk about is iron. Three times it mentions the amount of iron to be supplied, which is some indication of its value. To be sure, it does not mention a price; presumably Daimleville would be reimbursed for costs. Nor does it mention what is to happen to the old clock. Almost surely the clockmaker expected to take it away—without charge to my Lady, of course.

One thing is clear: clocks were already familiar features of the urban and sometimes even the rural landscape. We do not know when the earlier clock was installed or if it was the first at Nieppe; but it probably takes us back another decade or two, and this in an out-of-the-way village. All these references to expert opinion, moreover, show that Daimleville was anything but unique; this "maker of clocks" was member of a craft.

IF ONE TRIES to understand clockmaking as a craft and branch of manufacture, one has to keep in mind that almost as much time and effort went into maintenance and repair as went into the original construction of timekeeping devices. Our records on this score are necessarily spotty: Who could possibly keep track of the endless round of adjustments and changes, to say nothing of retouches to decoration and appearance, needed to keep any clock in good shape from day to day? Still, some of the stories that have come down to us allow us to imagine what was required: if the king of Aragon finally gave up on a tower clock in , Perpignan and had human bell-ringers hired to sound the hours, one may be sure that he had exhausted the mechanical alternatives and had spent a large sum in the effort.

Such accounts as have survived, then, are necessarily incomplete. They touch normally only on the important events, omitting the routine and the trivial; and even the big jobs are so laconically described that it is not always possible to reconstruct the story. Still, there is much to be squeezed out of these records. Take those of the cathedral chapter at Cambrai in northern France. These concern, among other things, an astronomical show clock that was the wonder of the region and a subject of mention and myth in all the local chronicles. One story had it that the clock was built around the end of the fourteenth century by a shepherd, no less, whose eyes were then put out to prevent him from doing the same thing elsewhere. The story sounds so absurd as to be incredible, but one must remember that such a wonder clock was a lure to pilgrims (the tourists of the Middle Ages) and hence a source of income—a new substitute for or complement to those religious relics that were still the greatest asset any church could have. Churchmen were known to anathematize one another over the authenticity of relics; a little mayhem to protect a similar monopoly may not have seemed implausible to the people of that day. So far as we know, though—and we may take comfort in the thought—the Cambrai clock was not built by a shepherd, or by a bishop for that matter, but by a clockmaker; the maker was not blinded; and far from being a creation of the 1380s and 1390s, the cathedral clock went back at least to the beginning of the century.[6] So early an installation may simply reflect the wealth of the chapter—one of the richest in the area, as the accounts show. But Cambrai was also a textile center (from which our word "cambric" is derived), and hence a place of market bells and work bells. The church may well have been catering to secular as well as liturgical requirements.

The Cambrai accounts are unfortunately incomplete for our period.[7] They do not really begin on a sequential basis until 1332, and for the century thereafter more than one year in three is missing, so that it is not easy to reconstruct the history of the clock(s). (There is good reason to believe that the original machine was replaced at least once in the course of these years.) The earliest reference is an isolated item from the year 1318: seventy-five sous paid to Symon and Sausset *pro horologio custodiendo et retinendo,* for taking care of the clock and maintaining(?) it.[8] Then we have nothing for some years, until 1348, when there is a burst of entries indicating either a major overhaul or the construction of a new timepiece. Some sixty écus remained to be paid for work done the previous year—a substantial sum when one remembers that one écu (given in the accounts as equal to twenty-seven sous, or one livre seven sous) represented more than ten days' wages of a skilled laborer. During the new year the cathedral chapter paid further room, board, and wages to master clockmaker Johannes Biekes, to a specialist for the calendar (who, exceptionally, received his due in gold florins), and to at least three painters, plus a small fortune for carpentry, masonry, ironwork, and lead work. The total expended was about three hundred livres, of which almost half went to the master of the work. Clerics were invited from places around to see the wondrous machine and celebrate with the chapter. A show clock, after all, is for showing.

Almost half a century later, in 1396, the clock had to be rebuilt. The conclusion of the contract with Master Mahieu of Soingnies, a maker of some repute, was the occasion for a banquet at which eight persons managed to put away some fifteen livres' worth of drink and viands—the value of ninety-six days' pay! The work ran on into 1398, some of it on the site, some of it in Mahieu's shop in Valenciennes. A scribe was hired to paint the numbers and zodiacal signs to specification, a painter to do a background picture around the clock, and an organ builder from St. Quentin to repair the trumpet of the angel that stood atop and sounded the triumphant hour to the four points of the compass; two craftsmen in Lille were engaged to work with the master of the clock on the bellows for the angel's trumpet and on such other matters as he required. Total cost: 143 livres, five sous, three deniers—over a thousand days of skilled labor.

Three years later the master clockmaker had to be called in again, this time designated as Mathieu of Valenciennes but the same man. Mathieu in turn engaged a Master Reginald to make a new zodiacal circle, Pierre the painter to paint the zodiac in astrolabe style, and a certain Hideux (who could bear such a name?) to fix and replace some of the figures of the procession of the hours. These accessory projects cost less than two livres in all. Mathieu received more than sixteen: good clockmakers were harder to come by than painters. He was also appointed clockmaker to the cathedral and placed on retainer: one new robe a year of the same quality as the one given the bailiff.

In 1408 a new craftsman makes his appearance: Jehan or Jean Warnier, locksmith and worker in iron. Warnier received twelve livres that year for making a new detent (the release mechanism for triggering the bell) and a wheel to ring the *conditor,* turn the angel, and drive the procession of the hours of the cross; also for making a new pulley for the rope holding the weight that powered all these motions. But Mathieu was still in charge, receiving twenty-eight livres ten sous for unspecified work on the clock.

And so on. The years passed, and the accounts tell us of continuing maintenance. Even so, 1435 brings the mention that the clock "was in great ruin." The chapter sent for a smith of Lille, Jacques Yolens, "expert in this science" and builder of the belfry clock in the city hall there, and paid him five livres seventeen sous six deniers for advice—surely one of the earliest instances of technical consultation in the historical record. Yolens visited the Cambrai machine along with Colard Lefevre, clockmaker to the cathedral and himself a smith, if one may judge by his name; and then Lefevre did a major overhaul that cost one hundred twenty livres, in addition to the usual outlays for painting, sculpture, and the like.

THERE IS NO POINT in continuing the saga further: it is clear that who built a clock gave a hostage to fortune; also that clockmaking was a diverse and complicated art, characterized from the start by collaboration among the crafts and division of labor.

13 The Good Old Days
That Never Were

D IVISION OF LABOR is something that we have come to
see as a law of collective production: as soon as two or
more people work together (or separately) to make something,
they apportion the tasks differentially. Such discrimination in as-
signment has for two centuries now, ever since Adam Smith, been
perceived as natural. Corresponding as it does to differences in
skill, talent, and experience, it enhances the quality and quantity
of the collective product. Since division of labor is almost invariably associated with a hierarchy of tasks and status and hence
with an unequal apportionment of the income earned by this
labor, it is impossible *a priori* to know for any given shop or enterprise whether a bigger and better product will yield a bigger me-
dian share—that is, whether a bigger pie means bigger pieces for
most of the workers. But it has been generally assumed that specialization works to the advantage of the individual worker as
well as that of the collective enterprise—and this for at least two
reasons. For one thing, every worker can do some things better
than he can do others and will maximize his own return by con-
centrating on those he does best: the doctrine of comparative advantage holds for individuals as well as nations. (To be sure, one

may make a good case for diversification into less familiar tasks by way of personal development; the analogy is to the infant industry argument for tariff protection. But such deviations from today's optimum for the sake of better tomorrows are not so much a denial of the division of labor as a refinement of it.) For another thing, more productive enterprises make for lower costs, bigger sales, and (other things equal) more jobs, and more jobs should translate into higher wages.

This conventional and perhaps complacent view of the benefits of division of labor has been challenged in recent years by iconoclastic economists seeking to justify and legitimate an egalitarian mode of collective production. To this end they have felt it important to show that division of labor and its accompanying hierarchy are anything but intrinsically more efficient and therefore natural; rather, that they have been imposed by exploitative employers who have perceived specialization as a device for limiting the worker's skill and hence his autonomy, as a means of enforcing his subordination.[1]

There is some truth to this contention, as we shall see at a number of points in our history of the clock and watch industries. But insofar as masters or employers sought to introduce division of labor (really further division) as a means of enhancing their control, they did so not *a priori* but on the basis of previous experience of specialization and its advantages. The division of labor, in other words, which goes back as far as the record takes us, was father, not daughter, to the thought.

Analytically and historically, we can distinguish three levels of specialization: among crafts, within a craft, and in the yard or shop.

Specialization among crafts is the oldest, no doubt, and the least problematical. Long before the mechanical clock was invented, going back to ancient times, manufacture had divided and subdivided into a wide range of specialties, associated usually with the raw materials employed. Metal workers (smiths) formed one trade, leather workers another, textile workers another, millers still another. No one can be surprised by this: the skills required for these divers crafts are so different that it would be foolish for any one person to try to do them all. The only exceptions would be those persons and communities cut off from trade and hence obliged to rely on themselves for just about everything. In clock-making, as we have seen, every machine was the collective product of a team. The master of the project bought his iron and lead from forges and foundries, contracted out for special parts such as dials, hired painters for numbers and decoration, carpenters and masons for structural work, sculptors and casters for automated figures. Sometimes these collaborators were available locally, but as the history of the Cambrai clock shows, the radius of recruitment and transport swept a sizable region. Roger de Stoke, builder of the astronomical clock of Norwich cathedral (1321–1325), had to have his complicated dial done in London—one hundred fourteen miles away as the crow flies. The job took three tries, bankrupting at least one engraver, and it was not until Roger himself went to London to supervise the work that he got what he wanted. The heavy dial then had to be hauled to site in Norwich, presumably by oxcart. In those days, that may well have taken a fortnight.[2]

Such collaboration among crafts became even more marked once the clock had been successfully reduced to a watch. The watch was from the start an ornament as well as an instrument, an *objet d'art* as well as a machine. Of necessity, then, watchmakers turned to outsiders for case and dial, for with the best of will they could not match the burin or brush of the specialized artist. Watch cases and dials, moreover, are light and valuable enough to travel far, hence quickly commanded an international market. The very existence of such a market promoted in turn comparison and emulation and gave rise to standards and expectations that few could meet—further encouragement to specialization. In the 1520s David Ramsay, later first master of the Clockmakers' Company of London, found the beautiful star-shaped case of one of his watches in Blois, France, where it had been engraved by an internationally famed artist named Gérard de Heck, himself an expatriate from Holland established near the French court to cater to the richest clientele for luxury watches in Europe.[3] Some decades later we find the enamels painted by Toutin and his successors, again in Blois, adorning watches signed from London, Amsterdam, Strasbourg, and other centers. And still later, at the end of the seventeenth century and beginning of the eighteenth, the Huaud family were turning out polychrome watch cases by the score, to be fitted to batch-made movements of standard size all across Europe.

The second level of specialization, *fission within the craft*, had also gone far by the time the clock made its appearance. Textile manufacture had long divided among such specialists as spinners, weavers, fullers, dyers, and shearers; and so simple and routinized had some of these tasks become that the work force had been reduced to mere hands. The fullers and dyers in particular had become a proletariat, centuries before the Industrial Revolution. The same fragmentation was observable in the metal trades, where iron workers (blacksmiths), tin- or whitesmiths, coppersmiths (braziers), and of course gold- and silversmiths all constituted separate crafts.

In the same way, clockmaking ramified as the industry grew. The craft began as an offshoot of older trades: most of the early makers were smiths, either blacksmiths or braziers. (We hear more of the former, because the big turret clocks were made of iron, and those are the ones memorialized in the record. But as noted above, chamber clocks were just as old if not older, and they must have used brass very early.) Some of these smiths "cut their teeth" on locks and similar assembly work; later on, with the invention of small arms, we find clockmakers emerging from that milieu as well. Also from big arms: clockmaking and cannon founding made an incongruous but complementary pair. Witness the career of Jacques Cutriffin, alias Jacques de Fribourg, *magister horologiorum et bombardorum regalium* (master of the royal clocks and cannon) to Charles V of France: he was builder and custodian of the tower clock of Grenoble and went on to be comrade-in-arms of Joan of Arc.[4] Clearly, in the context of that technology, a gifted metalworker or mechanic could move about from one device to another: what counted was a feel for the material, a sense of wheelwork and moving parts, and a good pair of hands. Kaspar Brunner, for example, famous as locksmith and cannon founder (there's a combination!), worked more than three years (1527–1530) to put the tower clock of Bern right. Was he learning as he went?[5] The best craftsmen soon found themselves busy enough with clocks to leave all else, but no clockmaker ever lost his skill for making just about any iron piece. Indeed this versatility is a salient characteristic of the profession: anyone who can make so complicated and finely adjusted mechanism as a time-keeper can generally make any part thereof or anything else that uses the same materials.

The one exception, perhaps, was the spring that drove the smaller clocks from the fifteenth century on, and later the balance spring (French *spiral réglant*). Although in theory a clockmaker could make his own springs, it made no sense to try to do so, for good spring steel was the arcane specialty of certain smiths and no amateur could pretend to do as well. The very making of steel of any quality was a long and difficult process, an endless repetition of heating, hammering, folding, and hammering again. Good steel, moreover, was only the raw material. It then had to be heated, drawn, and tempered, and the tempering in particular was a secret process, with each blade- and springmaker using his own special recipe. I have read somewhere that African smiths used urine for the purpose (lion's urine for the blades of brave warriors), and I have no doubt that European steelmakers resorted to their own esoteric, semimagical quenching agents. Obviously we know little about these things, though one French refugee in London in the late seventeenth century is reported to have made tougher springs by tempering them in talc (perhaps for slow cooling). In the eighteenth century the first instruction books appeared, providing cookbook recipes for aspiring springmakers.[6] But no written formula could supply the ingredients of touch and sight that came only with experience, and no real artist would tell the world what he used to quench his hot metal—sometimes not even his own assistants. The best of watchmakers, then, have always found it advisable to leave this part to the experts, who in the eighteenth and nineteenth centuries sometimes signed the springs as the watchmaker his watch.[7]

Multicompetence and diversification were the exception from the seventeenth century on. Where once they were symptoms of insufficient demand, and hence a need to find and meet custom where it was, they later became the hallmark of the exceptional artist who worked largely to order, made few pieces but those exceptionally fine and costly, and could thus permit himself some idiosyncratic diversion. We find Daniel Quare of London, inventor of the repeater watch, making some of the best stick barometers of the early eighteenth century. A hundred years later, the greatest of Danish makers, Urban Jürgensen of Copenhagen, designed and sold superb pocket thermometers in watch form; these are still working today and are still much admired. And in the 1830s and 1840s, the house of Breguet in Paris produced to order high-speed rotating optical instruments for experiments in the physical sciences.[8]

All of these, however, were sidelines. Clock- and watchmaking was all-devouring. In a growing market, the good makers had all they could do to keep up with demand, especially for run-of-the-shop products, and the press of work in combination with the routinization of technique made versatility costly and uncongenial. As a result, the industry as such split over time into two branches, big work and little (*gros volume* and *petit volume*), tower clocks and domestic timepieces; and then split again, between clockmaking and watchmaking. This is not to say that the great artists confined themselves to one or the other category. Their customers would not have allowed them to: someone who bought watches at Tompion's or Graham's expected to be able to buy clocks there as well. To satisfy such clients, one sold both, though often the only contributions of the maker of record were the signature on the backplate and, by implication, the warrant of performance. This kind of ostensible diversification was possible only on the basis of subcontracting with outside specialists—that is, it rested on an even finer division of labor within the horological profession than the split between big and small. When one of Quare's customers wanted an equation clock, Quare contracted with Joseph Williamson to do the equation work—just as Brockbank and other London watchmakers bought pocket chronometers from Thomas Earnshaw, Breguet imported tourbillon *ébauches* from Jacques-Frédéric Houriet in Switzerland, and all the leading British watchmakers of the nineteenth century bought marine chronometers from the specialists in London's Clerkenwell.

The same divergence between principle and practice, appearance and reality, characterized *division of labor within the shop*, the third level of specialization. The earliest clockmakers inherited, like other medieval artisans, the ideal of the "compleat" craftsman: a good worker should be able to make a good product in its entirety, from start to finish. This ideal, however, was potential rather than actual; translated into real terms, it said that each master should know and be able to direct every aspect of manufacture. In the shop itself, no one was expected to do everything. Instead, the work was divided and assigned according to the skill required and the agreeableness of the task. There was plenty of rough, menial work to do—stoking the fires, beating the plates, cleaning the metal, and so on—and this was the responsibility of the beginners (apprentices) who could not yet be relied on for finer jobs. At the other extreme every shop had to have at least one master craftsman, to design the pieces, plan the layout and articulation of the parts, and then—after the components had been made—file them to fit and adjust their performance. Often this was what the master did and no one else; or it was entrusted to or shared with the most gifted worker, the man who had everything required to open his own shop except the capital. This was the person most likely to marry the master's daughter or niece, if he had one—or the master's widow, who was often much younger than her deceased husband and needed help to carry on the trade.[9]

This hierarchical organization of the shop, based as it was on experience, was the instrument for the transmission of knowledge and skill from one generation to the next. Practice made perfect, and the simplification and repetition of tasks made for much practice. It was precisely the possibility of so defining and assigning the work that made the hiring of unskilled trainees possible and ultimately profitable. In a trade such as clockmaking, which took some years to learn, apprenticeship would have been far too costly had it been necessary to treat each newcomer as an equal and let him try his hand from the beginning on all kinds of work. The loss in botched projects would have been intolerable. Besides, rank (that is, experience and capital) has its privileges. Bosses assign dirty work to beginners not because they want to assert their dominance but because they have better things to do with their time and the time of their more skilled workmen. The aim is not hierarchy for its own sake, but efficiency and the maximization of return.

The clockmaking craft broke down, then, as did all urban crafts, into three categories: the masters, who were owners of the shops and equipment, employers, and teachers of labor; the student apprentices; and those journeymen (*compagnons*) who had completed their apprenticeship but were not yet in a position to open their own shop. This hierarchy existed before the creation of guilds or *corporations*. It was implicit in the learning process and the economics of production. The guilds, when they came, merely institutionalized it and fixed it in law.

C LOCKMAKERS' GUILDS CAME LATE, three hundred years after the invention of the mechanical clock, a good two centuries after clockmaking existed as a defined trade. When they came, they appeared as offshoots of earlier, more comprehensive groups of metal workers or locksmiths for whom clockmaking was one of several appropriate activities.[10] The secession of clockmakers' companies was a recognition of the growth of the industry as an independent branch to the point where its professional concerns and commercial interests were no longer roughly congruent with those of its fellows. This divergence of interests was especially marked for those making the smaller pieces, domestic clocks and watches. The makers of turret clocks (*Grossuhren*) usually preferred to stay behind in order to preserve their right to engage in other kinds of iron work. As might be expected, the advantage of secession depended on the size of the market. In a small city such as Montbéliard in the Franche-Comté (now in eastern France), clockmaking and locksmithing never separated out. The normal occupational designation was *serrurier-horlogeur* or *serrurier et horlogeur,* and both groups were together in St.-Eloy's guild. When in 1737 Jean Reichert was admitted to the company, it was on the express condition that he not be able to stop the old master locksmiths from "working at clockmaking with their own hands."[11]

The primary aim of these guilds or *corporations* was to restrict competition. Their ideology derived from the zero-sum outlook of the village community, from which most of these urban craftsmen were separated by only a generation or two. In the village, where land was limited, growth by one peasant could only be at the expense of others. Fortunes were not equal, but an ethic of equality prevailed that found expression in a large number of shared rights. In such a community the laws of the marketplace were seen as unjust and socially subversive: one did not allow one's neighbor to go under; one did not charge him interest in his time of need; one did not exploit his misfortune to take from him his land and independence. On the other hand, charity ended at the village line. One did not admit outsiders to the community and share with them its common resources.

The same for guilds: they were fraternities organized to help one another by defending their "turf" against trespassers. This was not the way the masters would have put it. They spoke instead of the quality and reputation of their manufacture, their duty to their customers and their art. To this end, they restricted entry to those who had spent long years learning the trade and could prove their skill by the production of a difficult masterpiece. They required each shop to ensure the quality of what it sold by selling only pieces of its own manufacture. Above all, they prohibited the practitioners of other crafts from making and selling what had been reserved to them, and forbade even their own members to sell imports from outside, however well made. In general, theirs was an anticommercial ethic. The criterion of performance was not the ability to compete in the marketplace and undersell one's fellows, but to work well and conscientiously according to the standards of the craft. A man who worked well, however inefficiently, was entitled to sustenance by the community he served, and the guild was there to make sure he got it.

From the technological and artistic points of view, the system was not without virtues. It did ensure the transmission of an extraordinarily high level of skill. (Compare in this regard the quality of leather work or jewelry done today by Italian workers, who still spend years as apprentices learning the secrets of the craft, and that of the self-taught practitioners who sell their wares on the fringes of American college campuses.) And it did make possible the production of labor-intensive products of the highest complexity and artistry, the kind of work that took months and even years to complete.

On the other hand, the effort to prevent competition and gross inequalities of status and fortune led to technological conservatism and constraints on growth, both of the industry as a whole and of the shops within it. Most guild members were opposed to new things and new ways of making them, especially those ways that entailed capital expenditures beyond their means. They were

joined in this by their workmen, who saw in short cuts and ingenious devices a threat to employment. Even more feared by journeymen were division of labor and its attendant specialization of
function: first, because such division devalued the craftsman's
skills and diminished his status; and, second, because it jeopar-
dized his access to all-around knowledge and training and hence
his future. Division of labor, moreover, so inexpensive to begin
with and hence especially attractive to thrifty (tight-fisted?) masters, was almost inevitably a stimulus to the invention of labor-
saving devices: the simpler and more repetitive the operation, the
more it lent itself to mechanization. In these respects, therefore,
the workers were more conservative than their employers. Meanwhile, since most guilds defended the interest of their weakest and
most timorous members (anything less would have entailed their
dissolution), they were compelled to wage a ceaseless struggle
against the forces of change.

The guilds were also a standing barrier to personal ambition
and advancement. All of them put ceilings on the number of apprentices that could be trained in one shop at one time, of journeymen that could be employed, of applications that would be
accepted for admission to mastership. These limitations were
aimed ostensibly at the preservation of equality among the masters; but their effect was also to hold the young journeyman in a
position of prolonged subordination and exploitation, for until he
became a master, he could not sign his work or sell it to the public. In Augsburg, where journeymen in the mid-sixteenth century
were paid some twenty to twenty-four kreuzer a week plus room
and board (say, about a gulden a week altogether) and where a
run-of-the-shop timepiece that took four to eight man-weeks to
make was worth thirty gulden, it obviously paid the masters to
keep the journeyman in a dependent state: there was no surer
source of profit.[12] In light of the above, we can understand better
these limitations on the number of workers per master: they were
needed to ration supply in a market where employers were able to
pay labor less than its marginal product. Naturally, the clockmakers' guild at Augsburg also forbade journeymen to accept
out-of-town employment: this cheap labor was too good to share.

Why did workers put up with this? The answer is that many
did not. The records of apprenticeship are dotted with stories of
flight and pursuit. For the first two years, an apprentice cost more

than he was worth, for if we are to believe the masters, he spoiled almost everything he touched. But after that he was an asset and could find more remunerative employment elsewhere if he was not happy with his master and was prepared to jump contract.[13] By the same token, journeymen often found ways to sell their work to the highest bidder. In Blois and Augsburg, craftsmen working in their own rooms were the principal source of supply for those masters who could not keep up with demand. Unfortunately for the guild, they were equally ready to sell their work to interlopers from out of town or from other trades.

Other journeymen were ready to serve their terms and wait their turn. They were more patient by temperament; their masters won their loyalty and affection; or they believed in the values and advantages (not negligible) of community and fellowship. Besides, there were short cuts to advancement within the system. For example, guild constraints were invariably relaxed for relatives by blood and marriage; indeed, the quickest way to secure membership in a craft was to marry the daughter or widow of one of the masters. This, too, was a way of protecting territory against outsiders. Such tolerances, however, could at best mitigate a fundamentally restrictive strategy, not cancel it.

Fortunately these constraints, which were honored more in the breach than in the observance, were of limited effect. For one thing, they constituted at best a holding action against the forces of business enterprise and commercial change. The very establishment of guilds was a response to the competition inspired by widening of the market and growth of demand. The guilds were attempts to restore the *status quo ante*, the comfortable, semimythical ideal of local autarky, and as such were always behind-hand and on the defensive. Their records are filled from the start with accounts of alleged violations. Sometimes the guild was able to enforce its regulations with the aid of the civil authorities, which cherished these companies as sources of revenue and instruments of social control. Sometimes the guilds *were* the civil authority: there were town councils that consisted largely of representatives of the class of master craftsmen, who held rights of citizenship (*Bürgerrecht*) from which journeymen were excluded. Such councils could be depended on to defend the interests of their constituencies.

Even so, the writ of the town council extended only to the town line, and many an enterprising clockmaker found it possible to work just outside, where air and industry were freer. Blois was surrounded by a veritable belt of village centers of manufacture: Chouzy, Ouchamps, Vineuil, Saint-Gervais, and others. Augsburg had its suburb of Friedberg; Nuremberg had Fürth; Hamburg, Altona.[14] To be sure, the guilds made every effort to enforce their regulations outside the town; especially was this true in France, where the *corporations* were authorized by the royal government in Paris. But even where the central authority was of a mind to sustain the guilds in their pretensions, it often took with one hand what it gave with the other. The French crown, for example, felt itself to be above and outside the scope of guild regulations. It was free to bring in and employ the finest artists available, regardless of credentials, and it created to this end a privileged enclave in the Louvre that constituted a major breach in the guild monopoly. It also sold patents of exemption from taxes and from corporate rules and searches to tradesmen and craftsmen *suivant la cour*—that is, attached to the court. The French crown always needed money and had no scruples about collecting for monopoly privilege with one hand and for exceptions with the other. The whole process of give and take had a touch of fiscal genius. These special appointments—between three and four hundred of them in the seventeenth century—were much sought after, and some of them sold for thousands of livres.[15] Similar exceptions were made in Augsburg, where the emperor intervened on a number of occasions on behalf of clockmakers working in his service.

Some sense of the problems and resorts of an enterprising outsider may be gained from the story of Georg Roll, clockmaker and merchant of Augsburg. Roll was born in Liegnitz (Silesia) in 1546, learned the clockmaker's trade from his parents, and then wandered around Europe picking up jobs where he found them. In 1565 or 1566 he came to Friedberg in Bavaria, just outside imperial Augsburg, opened a shop, hired workers, and began making and selling clocks. This contravened the rules, for even in Friedberg, which lay beyond the jurisdiction of the Augsburg guild, only a married man could open a shop and train apprentices, and Roll was a bachelor. (How could one entrust young apprentices to the care of an unmarried man?) But his enterprise was tolerated because he bought the work of the masters in the town and sold it for them. He also made it a point to solicit the custom of the Bavarian court in Munich. This brought in profits and held the promise of protection if needed.

Protection was needed. The clockmakers of Augsburg resented the success of this interloper, the more so as Roll tried to sell his wares in Augsburg as well. Even worse, he hired their own journeymen away, to which they responded by banning such renegades from ever working in Augsburg again. Furious with Roll, they incited their apprentices and employees against him, harassed him with insults, even drove him from the city when he attended the wedding of the daughter of the locksmith Thomas Geiger. Nothing was sacred enough to shield him from their wrath.

Roll turned to Friedberg town for help. Friedberg asked Augsburg to desist. Augsburg pointed out that Roll was a bachelor and expressed concern that such a person was not a suitable tutor to young apprentices and would inevitably disturb the peace—that is, offend public morals. Letters went out to Ulm, Nuremberg, and Breslau to ascertain their practice regarding bachelors. We have Ulm's reply: bachelors were allowed to open shops there. This was not enough to mollify Augsburg, which insisted that Roll was an "uproarious rogue, botcher, and troublemaker." Friedberg turned now to Duke Albrecht of Bavaria, who granted Roll his special protection because he was working for the court; and when Augsburg still resisted, Friedberg reminded the bigger city that Friedberg was part of Bavaria and that Augsburg had nothing to say about industry there.

The issue was finally settled by an agreement that Roll would confine himself to working for the Bavarian court, a promise that he more or less kept for some years. But then, in 1575–76, we find him traveling in Latin Europe, going as far as Rome and Naples with a chest full of clocks and watches to the value of five thousand gulden. On his return to Friedberg, he learned that the Augsburg authorities had refused to admit some of his former employees to practice. Roll entered a suit on their behalf. Even more, he decided to attack the citadel from within. In 1578 he asked to be made a burgher of Augsburg and expressed the intention of buying the right to open a shop—not a clock shop, but a *Kramerei*, or general store. Duke Albrecht and the city of Friedberg supported the petition; the clockmakers of Augsburg accepted it only on condition that Roll stop selling timepieces. So the request was granted, and Roll bought himself quarters for house and shop and married the mayor's daughter.

But Roll was incorrigible. A few months later the clockmakers were complaining that Roll was again selling clocks and watches. Times were bad, they added, and Roll was taking bread from their mouths. To this Roll responded that since he was selling the work of Augsburg makers, work that they did not seem able to sell for themselves, he was in fact only helping them. Many of these clocks, he pointed out, went to customers outside of Augsburg. Back and forth the arguments went, for a period of years, without resolution. So long as the matter was under litigation, Roll probably went his merry way.

I do not propose to pursue the rest of Roll's biography in the same detail. He was a man of many parts, full of initiatives and surprises; and in spite of all efforts to confine him, he was continually slipping out in new directions. Among other things, he obtained the emperor's permission to run a *Glückshafen*—that is, a lucky-bag lottery—with clocks and watches among the prizes. This took some of the pressure off his *Kramerei,* for it was otherwise not possible to dispose of timepieces without displaying them under the noses of the clockmakers of Augsburg. In addition, he began in conjunction with a first-class mechanician to produce elaborate and precious clock-globes, which he then sold to the imperial court, among others.

This was a repeat of his earlier courtship of the Bavarian crown, with mixed results. At one point he made the mistake of selling two of these globes to the emperor Rudolf II and his brother. He made the brother pay more—fifteen hundred thaler against twelve hundred—perhaps with the thought of doing the emperor a favor. Little did he understand the psychology of the nobility. The emperor learned of the difference and was persuaded that Roll had sold him the lesser of the clocks and that his brother's was much better. Else why the difference in price? Roll tried to explain: the clocks were identical; he was simply doing the emperor a favor. But that was a dangerous argument: if the clocks were the same, then Roll must have been overcharging the archduke.

Emperors are not ordinary customers. Not only don't they think like other people, they don't have to swallow their grievances. Rudolf told the city of Augsburg to put Roll in jail for this *lèse-majesté*, and there he languished for some weeks protesting his innocence and asking the emperor to take pity on his employees if not on him. This Rudolf deigned to do, with the observation that this ought to teach Roll to do better by his ruler in the future.

Indeed. In the years following, Roll sold the insatiable Rudolf some immensely valuable pieces, at reassuringly extravagant prices, and, to lesser representatives of the south German nobility, slightly lesser pieces at reassuringly lower prices. His troubles with the Augsburg clockmakers continued on and off, and his connections with the mighty stood him in good stead. He was at the peak of his business activity and wealth when he died in 1592, only forty-six years of age, a victim probably of the plague. He left behind a widow and five children. Four years later the widow married another clockmaker, Matthäus Runggel, some nineteen years Roll's junior. Runggel survived his new wife, who died about 1620. Five years later Runggel remarried, presumably to a much younger woman, in the best tradition of *das alte Handwerk*.[16]

For all the efforts of the clockmakers of Augsburg to persecute and prosecute Roll, some of them had clearly connived at his horological enterprises: first Johann Reinhold, who did the mechanisms for his clock-globes; then Hans Marquart, who joined them in finishing an especially complicated clock-planetarium known as the Great Clock; and then an unknown string of others who helped Roll get the Augsburg timepieces that he sold in Italy. Some pieces, of course, he made in his own shop, which with about twenty-five employees was something of a small manufactory. But the economics of the trade were such that in Augsburg, as elsewhere, there were always masters who needed help disposing of their wares and who turned to the more active merchants, in and out of the clock trade, to sell them.

More often than not, it was the goldsmiths and jewelers who engaged in this gray market, just as some clockmakers turned an extra livre or thaler now and then by selling silver or jewelry. These were complementary trades, often taking on one another's work, and a little trespassing was only natural. The guilds remonstrated against these violations, but largely in vain, for proof and cooperation were hard to come by. Every once in a while, a search hit pay dirt and someone was "caught with the goods"; but like successful drug raids, these searches did little to change the habits of buyers or sellers.

We have the story of one of those raids, at Blois in 1636. An officer of the crown, at the request and in the company of two members of the clockmakers' guild, visits the premises of Isaac Gribelin, painter in enamel and goldsmith. By order of the king, he says, show us all clock- and watchwork in your shop. So Gribelin shows them four enamel watch cases; a silver watch that he says is his; another silver watch that he says belongs to a certain doctor; and another that he says belongs to the Comte de Brion's valet. They make him open every cabinet and chest and box in the place, but they find nothing. Undaunted, and perhaps informed by someone, the clockmakers ask the marshal to go up and search the attic, which he does; and there, behind a door, he finds three pine boxes filled with watches, in all twenty-three of gold and four of silver. A small fortune. What's more, these watches all bear the names of established makers of Blois or Paris, some of them deceased. Forgeries? If you ask the signatories, they will deny them. That doesn't mean they are telling the truth. Even those signed by dead men may well be authentic. There have always, in all times and places, been delays between the making of a movement and its casing, sometimes delays of years. And Blois has always seen a lively trade in movements made by one maker for another.

That, in essence, was Gribelin's defense: he was simply following the custom of the trade. As the report of the search put it: "The said Gribelin objected to us that goldsmiths have always dealt in clockwork, and that with the consent of the clockmakers, who have tolerated said traffic; indeed that there was once a time, during the stay of the Queen-Mother in Blois, when one watchmaker and one goldsmith did between them the greater part of the trade in clocks and watches, and this led the said Gribelin to believe that he might continue to engage in this traffic until the litigation between goldsmith and clock- and watchmakers was settled."

This argument seems to have carried some weight with the court, for when the matter came to trial, the watches were returned to Gribelin with the injunction that he sell them all within a month, but outside of Blois, and that he desist from such activity in the future on pain of a fine of three hundred livres and confiscation of the contraband timepieces. A mere slap on the wrist. It was surely not unconnected with Gribelin's reputation as an artist and the quality of his clients, among them the brother of the king.[17]

It is a time-honored principle of historiography that reiteration of law is proof of ineffectiveness. Laws that work do not have to be confirmed and reconfirmed; penalties that deter do not have to be reinforced. The record of the watch trade in Blois makes it clear that Gribelin's discomfiture, such as it was, changed nothing. In particular, the fact that the fashionable people wanted cases with watches in them, rather than cased watches, led them to order their watches from the enamelers, who had no trouble finding watchmakers to supply them with movements. In 1640, therefore, Nicolas Lemaindre, watchmaker to Marie de Medicis and master of the corporation, and his colleague Salomon Chesnon secured from the lieutenant-general of the district an ordinance giving them the widest powers of search and seizure, not only in the shops of suspected violators, but on their person. Let them or their employees but go to a house or cabaret or other place of mercantile assignation, and they would be compelled to turn out their pockets, by force if need be. There were some prosecutions under this decree, but in the long run, it was as impotent as those that preceded it.[18]

All of these exceptions, exemptions, and violations, though, were as nothing compared to the limitations imposed by political fragmentation. In most of Europe sovereignty simply did not run far enough to enforce these guild monopolies against cheaper competition. The Geneva watchmakers' *corporation*, for example, could impose its will on the workers in the rural parts of the republic, but it could do nothing to prevent watchmakers from setting up on the northern shore of the lake, at Nyon, Vevey, and Lausanne; and these in turn could do nothing to stop watchmaking from taking root in the guild-free mountains to the northwest. In England guilds were essentially powerless by the time the clockmakers of London succeeded in splitting off from the smiths, so that almost from the start the functions of the Clockmakers' Company were more ceremonial than professional. It could speak for the trade and was able on occasion to persuade the government to soften its fiscal demands or protect the English manufacturer against foreign competition. But we shall see that such measures were never very effective, and in the meantime the company had little or no power over the conditions of production and competition within its own jurisdiction. In effect, anyone could practice in London who was willing to pay a small fee, as well as some who were not. So empty was the content of the company that it was not always possible to get makers to accept the honor (burden) of office—a situation that would have been unthinkable in Augsburg. Outside London the trade was free and open, as it was more or less in Holland; and this freedom was both symptom and condition of the active clock and watch trade in both countries.[19]

14 The Ups and Downs of International Competition

IT WAS IN THE "UNINCORPORATED AREAS," those free of guild pretensions and controls, that the clock and watch industry was to enjoy its most rapid development. In the late seventeenth century, the British took the lead. French manufacture was crippled by the exodus of some of its best practitioners fleeing a wave of anti-Protestant bigotry and persecution. By way of contrast, the British trade was not only open to talent, but a happy conjuncture brought forth a succession of highly gifted and inventive makers who provided technical and commercial leadership to the rest of the profession. Five names stand out.

First was Thomas Tompion (active 1671–1713), trained as a smith but apparently a "natural" watchmaker. His talent and insistence on quality brought him the attention and favor of the rich and powerful, from the king on down. Friend and collaborator of Robert Hooke—when the hot-tempered Hooke was not execrating him as a "slug," a "Rascall," and "a clownish churlish Dog"—Tompion was the first to realize the possibilities of the balance spring and raised the quality of the traditional verge watch to unprecedented levels by a great many small but cumulatively important improvements. His work was characterized by the finest detail, by polished, high-numbered pinions and beautifully cut wheels of uniform profile, which made for evenness of movement, hence greater reliability and precision. His watches, wrote a French admirer half a century later, "were so little changed after sixty years of going that the pivot holes were still the same and had not had to be rebushed." Some of this reflected the precocious British recourse to special-purpose machinery such as wheel cutters, for in these matters Tompion was only *primus inter pares.* Take any first-class London watch of this period—by Joseph Windmills, say, or Daniel Delander—and compare the train with that of a comparable watch made anywhere else: the difference is a revelation. One reason why they all look so good is that the movements that went into the watches of different "makers" were themselves often produced by the same man to the same standards.[1]

Daniel Quare (active 1671–1723), the second great name and Tompion's rival, was the inventor of a device that seems trivial to us today, but was of great importance in a country where in winter people "get up at night" and in an age when candles were the standard source of domestic illumination and there were no matches as yet to light them with. This was the *repeater watch* (invented between 1680 and 1686)—namely, a watch that would, when the pendant was depressed, sound the time on a bell in the back of the case. The earliest repeaters generally sounded the time to the last quarter-hour, but within about twenty years half-quarter repeaters appeared. The English makers of repeater watches tried at first to protect them from imitation by locking the repeater train from view, but it was not long before continental makers learned to open the secret catches, and once they saw the device they were able to make their own versions. Still, the British held pride of place, Quare above all, and his name appeared with one or another spelling on a large number of continental forgeries.

Another improvement was the introduction around 1700 of the so-called *dumb repeater* (*répétition à toc*), which sounded the time in such a way that it could not be heard by others. The earliest form inserted a pulse piece (*sourdine*) between hammer and bell; this was a small, sliding rod that transmitted the impact to the outside of the case where it could be felt by the finger. (See Figure 23.) In later models the bell was simply eliminated and the hammer made to strike against the case itself. Such watches had the advantage of politeness: one could learn the time without being obvious about it, or could learn the time in one's bedchamber without disturbing the sleep of others. (With the abandonment of the dumb repeater in the early nineteenth century this problem revived, though much mitigated by the omnipresence of visible timekeepers—on the wall or mantel or, eventually, on other people's wrists. That is the secret, of course: don't look at your own watch—look at the other person's. But that ploy has lately been largely vitiated by the adoption of digital watches, which are much harder to read unobtrusively.) Until very recently the dumb repeater was reputed to be an invention of the great French maker Julien Le Roy and assigned a date of around 1740. It is now clear, however, that the British were making such watches from the early years of the century.[2]

The third important figure was George Graham (active 1695–1751), Tompion's employee, then his partner, and later his successor. We have already met him as the inventor of the cylinder or "horizontal" escapement, so called because the scape wheel lay parallel to the two plates rather than perpendicular, as with the verge; and as inventor (1715) of the dead-beat escapement in pendulum clocks, which eliminated all recoil and which, in combination with Graham's mercury-compensated pendulum (1725), made possible astronomical regulator clocks that kept time within a fraction of a second a day.

It was George Graham who, more than anyone else, was responsible for Britain's horological preeminence in the first half of the eighteenth century. The leading French makers, the Le Roys in particular, contested this, pointing to the special frictional weakness of the cylinder escapement, and argued that French watches were superior to British because larger. The British, who wound their watches from the back and had to open them to do so, solved the dirt problem by placing their movements in a double case and covering them with an interior metal "dust cap," all of which meant that much less room for the wheel train. Conversely, the French wound through the dial and hence could close off the movement. They needed only one case and usually dispensed with a dust cap. The Le Roys asserted that French movements were to British as 64:27, a ratio presumably derived by cubing four and three—that is, by assuming a ratio of four to three in each dimension. This is much exaggerated, however, and besides, a good cylinder watch did not need to be as big as a verge, precisely because of its horizontal layout. In any event, the Le Roys' pride on this score was undermined soon enough by the energetic efforts of French makers to compete in smallness and thinness: from the 1740s on, as we shall see, fashion decreed slimness *à tout prix,* and the devil take precision and solidity.[3] (See Figure 24.)

The fourth notable was John Ellicott (active c. 1728–1772), the son of a first-class maker. He married scientific interests to horology, was elected to the Royal Society, and produced some of the finest astronomical regulators of his day, using a compensated pendulum bob of his own invention. In an effort to reduce the wear on watch cylinders, he made some of them with ruby stone lips, anticipating here the stone cylinders of Breguet and other French makers of the early nineteenth century; and his center-seconds (sweep seconds) watches with arrangements for stopping the movement at will, although not the first watches to show seconds, gave expression to a new level of precision.

Finally there was Graham's student Thomas Mudge (active 1738–1794), completing a lineage of three generations of great makers.[4] Mudge, as much as or more than any maker who ever lived, had "golden fingers." His work, which was characterized by extraordinary detail and finish, was *sui generis;* like that of his contemporary Harrison or, later on, Louis Berthoud and Breguet, it could and did serve as a model, but it was not easily copied. Mudge first came to the attention of horological connoisseurs when he made a complicated watch for John Ellicott that was intended for the king of Spain. The watch was damaged, and Ellicott had to confess that he could not repair it, whereupon the king insisted on learning the identity of its maker. From then on, Mudge had *carte blanche:* the king stood ready to buy anything interesting he had to offer. And Mudge did make some interesting pieces. He made what may have been the first watch with equation of time; the first with "perpetual calendar" (taking account of the varying length of the months, including the long February of leap year);[5] the first with remontoir in the gear train; and what may have been the first with minute repeating. His center-seconds cylinders, along with those of Ellicott, were the peak of pocket precision before the invention of the chronometer. Mudge is best-known, of course, for his invention of the lever escapement (1769–70), which came to combine precision and reliability as no other and swept all escape mechanisms before it by the end of the nineteenth century. But his own version was too difficult to come into general use; he himself made only two.

In 1771 Mudge gave up making civilian watches for ordinary customers and chose to devote the rest of his life to the production of a marine chronometer. He turned out three, like Harrison using a special, idiosyncratic version of the verge escapement. It was the wrong tack to take. Mudge's chronometers performed well, because everything he made performed well. But they were costly to make, hard to copy, and over time did not match the precision *cum* reliability of Earnshaw's spring detents. Who is to say why Mudge persisted in what proved to be a dead end? My own feeling is that like other great artists—Ferdinand Berthoud and John Arnold, for example—he was a victim of his own prowess. He felt that he could do whatever he wanted, that no task was too difficult for his skill and ingenuity, and he did not find it in him simply to change his approach and work along an apparently successful line invented by someone else.

Thanks to the work of such men as these, British horology enjoyed a century of recognized superiority and commercial dominance. The only maker outside London whose work could compare in quality and prestige was France's Julien Le Roy, and it is no coincidence that his name was forged as much as Quare's had been. (To be sure, it is always very difficult to know whether a watch is a forgery or an outside piece, purchased by the artist in question from a contract supplier for resale under his own name. There is good reason to believe that many pieces signed Julien Le Roy that have always been deemed false, many of them in elegant enamel cases, were in fact made for him in Geneva and, though not original, are genuine.)[6] It was Julien who revived the French watch trade after a half-century of stagnation—the "onion" years—and restored Paris' reputation for quality and artistry. As Voltaire said to one of Julien's sons: "The Maréchal de Saxe and your father have beaten the English." For a Frenchman, there could be no higher praise.[7]

Much as Julien did for the reputation of French horology, however, output never came close to matching that of Britain. French watches surviving from the second quarter of the eighteenth century are *rarissimes,* and I have the impression that demand for French timepieces never really picked up again. Alfred Franklin, that indefatigable collector of the small details of social history, cites a French publication of about 1750 to the effect that even "a coach driver would not wear a watch that wasn't English."[8] If the story is true, it is the more impressive because the import of timepieces from Britain was prohibited. The king had so decreed, according to Savary's *Dictionnaire,* for two reasons: first, out of consideration and care for the French clock and watch industry; and, second, because British watches did not appeal to French taste anyway.[9] So much for the logic of commercial dictionaries.

In the early eighteenth century it was the British who posed the great threat, and French tariff arrangements reflected this: over and against the absolute prohibition of British watches stood a duty of only two livres on each piece sent from Geneva to Lyons.[10] Twenty years later, in 1767, a Paris watchmaker boasted of a national resurgence: "The English, whose industry was once so superior to ours and who supplied us with watches that we paid for with large transfers of funds, not only no longer sell them to us, but import a lot from France." In fact, however, these imports were not of French but of Genevan manufacture, as the author implicitly conceded: "The watch industry [of Geneva and Switzerland] has swallowed ours to such an extent that one can hardly find a small part that has escaped the near-general shipwreck."[11]

In short, what triumphed in France was not so much the French watch as the French style of watch, which others could make about as well and cheaper. For the collector the typical Paris timepiece of the post-1750 period, almost always made in Geneva though signed as from Paris, is cased in gold, often with laid-on decoration in golds of different colors, yellow, red, green, and white, or centered with miniature polychrome enamel portraits (made in Geneva by the thousands). Such watches were more adornments than instruments. They were not the sort of thing intended for a wide market, but there was some compensation in the profusion with which they were given as gifts in elegant society. The marriage *corbeille* of a noble lady would be filled with them, each more lavish than the next, more than she could possibly use; so the custom arose of keeping some and distributing the others among friends.[12] Fewer gilt brass and silver watches have survived, but that only reflects their cheapness: people threw them away more easily. We must assume that at the time they sold in much greater numbers than gold-cased timepieces.

Of these imports, an unknown but substantial proportion came in illegally at prices no French maker could or wanted to compete with. Where watches were concerned, smuggling was not a crime but a business. It entailed some risk, but the volume was so large and the probabilities of success so well established that shippers could insure themselves at modest rates against loss by seizure. One device of the Geneva houses was to conceal the watches in messy, malodorous loads—fish was a favorite—that customs officials would be reluctant to search. So profitable were these operations that fishmongers in Geneva tried to cut in and force a role as intermediaries—to no avail.

The collapse of the French manufacture in the second half of the eighteenth century suggests that whatever credit one gives to such artist-craftsmen as Tompion, Graham, and Mudge, British supremacy from about 1675 to 1775 can be explained only in part by the accident of talent. The French may have been short of innovators in the fifty years following the introduction of the balance spring and the Revocation of the Edict of Nantes, but once Julien Le Roy came on the scene, France did not lack for horological genius. The list of great names includes Pierre Le Roy (son of Julien), Jean-André Lepaute, Ferdinand and Louis Berthoud, Jacques-Antoine Lépine, Antide Janvier (clocks only, but what clocks!), Robert Robin, above all the incomparable Abraham-Louis Breguet. From the 1780s on, with his series of complicated *perpétuelles* (self-winding watches), Breguet led the way in almost every branch of horology. Once he established his shop and trained his team, he himself did little of the work. But he designed his pieces, and so fertile was his imagination, so chaste his taste, so resourceful and inventive was he, so insistent on the highest standards, that to this day his watches are the pride and joy of those collectors who can afford them, especially those who can afford to wear them. The crowned heads and wealthiest businessmen of Europe beat a path to his door, waiting sometimes months or years for the watch of their desire and specification.

After Breguet, watches were at best copies of his design and conception, and while he lived and worked, Paris was the watch capital of the world. Only in the manufacture of pocket chronometers did Britain enjoy greater prestige. But Breguet could not and did not revive the French watch industry. He himself worked for the rich. Even his *souscription* watches, the cheapest he sold, so called because they were batch-made and paid for in advance, cost a minimum of six hundred francs—a year's pay for a Paris laborer (say, about eight thousand of our dollars). Over a period of about fifty years, from the early 1780s to 1823, Breguet made approximately five thousand watches, or, more exactly, had them made by his own workmen to his specifications and standards. Alongside him, a few other Paris houses made some hundreds or thousands more, usually in imitation of his features and design. Many of these pieces were signed with the name of the maker, followed by the proud reference *Elève de Breguet*, for it was a title of honor to have worked for the master. All of this, however, was a drop in the large and growing stream of watches coming in from Geneva, Neuchâtel, and Switzerland.

Even Breguet relied on Geneva makers to supply movements and fill out his assortment, for there was no way his own carefully chosen team of "artists" could meet the demand that swelled with his reputation. Besides, there were watches that he did not want or deign to make in his own shop because they were incompatible with the high seriousness of his undertaking: musical watches, automaton watches, superthin pieces; but customers wanted them. These, too, he bought from Genevan or Jurassien makers, who felt themselves honored to have anything they made accepted by Breguet.[13]

We shall never know how many watches Breguet bought from outside and sold under his name, for they were not given numbers in his own series (an old device for distinguishing between inside and outside work, going back probably to the first numbered watches, those of Thomas Tompion). Many of them, in my opinion, went out under the signature that most collectors consider the hallmark of a false Breguet: *Breguet à Paris*. I have no doubt that most of these were indeed frauds: Breguet was the most imitated and plagiarized watchmaker in history, to the point of having to scribe a secret signature on his dials in order to protect his name and his customers. But some of these *Breguet à Paris* watches are too good to be fakes; they could sell on their own merits, and I think Breguet sold them alongside the house product.

All of this is by way of showing that it takes more than a great name to make an industry. It takes efficient techniques, low costs, and access to large markets. So with eighteenth-century Britain: nowhere was demand for timepieces so large and growing so rapidly. Partly this reflected higher income per head. (Clocks and watches, after all, were still expensive luxuries in the seventeenth century, accessible only to those who had much to spare from the purchase of necessities.) Wages were half again as high in England as in the richer parts of the continent—northern France, say, or the Rhineland—and the incomes of the middle class proportionately higher; only Holland could show a comparable standard of living. England, moreover, like Holland, was a relatively urbanized society: about 25 percent of the population lived in cities and towns around 1700, as against 10 percent or less on the continent. Nowhere was the middle class so numerous and strong: Napoleon's later gibe about a "nation of shopkeepers" was not unmerited. (It never occured to him, though, that shopkeepers were not intrinsically unworthy opponents.) Since timekeeping was a characteristically urban concern, no nation was so time-bound in its activity and consciousness.

Meanwhile, time awareness was diffusing through the rest of British society—through the countryside, which was being integrated into a national market and transport network, and down the economic ladder into the working classes. Cottage industry was the great agent of change. As much as half of England's woolen cloth came from rural manufacture as early as the fifteenth century. To be sure, the great majority of cottage workers were not time-bound; they fixed their own hours, much to the despair of the merchants who gave them their raw materials and waited impatiently for the finished product. But their role generated a large number of complementary activities of a quasi-urban character—distribution and collection within the manufacture, retail trade, transportation—and these did require attention to time.

Take transportation. The eighteenth century built an extensive road network, much of it in the form of toll ways (turnpikes) financed and operated by private enterprise. This risktaking aspect reflected a rational response to demand: the growing economy and developing links among areas of specialized production ensured that these highways would be well used. (In this respect, the contrast with less-developed France is striking. When Arthur Young visited the continent in 1787–1789, he was impressed by the solidity and breadth of the best French roads, some of them better than anything in England and all of them built by public authority; but also by their emptiness—there was nothing like the heavy stream of vehicles that he knew from England, and little of the complementary eating and sleeping accommodations that had sprung up so numerous at home.) Along with roads went a dense grid of coach services, competing fiercely not so much in price as in speed. It took two days to go from London to Bristol in 1754; thirty years later the fastest coaches did it in sixteen hours.[14] Drivers were given locked clocks to hold them to their pace, while the passengers, with their own appointments to keep, checked their watches. The advertisements of the day reflect the heightened time consciousness: they boast of ever-faster schedules, announce precise departures and connections. What they do not talk about is horses driven to exhaustion and death by the effort to save minutes in a prerailway age. Thus, the locomotive, when it came in the 1820s, found a society already prepared for it psychologically and horologically by half a century of record and leg breaking. The builders of the Liverpool–Manchester line thought that most of their business would take the form of freight. They could not have been more profitably wrong.

The contradiction in the eighteenth century between time-free (domestic workers) and time-bound (employers and their agents) gave rise to growing tension as demand increased. There is perhaps nothing that hurts more in business than profit forgone, and the unwillingness of cottage workers to devote themselves unremittingly to their tasks was a growing source of frustration to merchant-manufacturers who could not fill their orders. Nor did it help to pay higher wages: the typical worker, who had a strong preference for leisure, could simply earn what he felt he needed in fewer hours.[15] Matters were complicated, moreover, by the worker's tendency to embezzle some of the raw materials confided to him by the merchant capitalist. He saw no dishonesty in this, but rather a "perk" of the trade. The merchant saw it as theft pure and simple.

These internal contradictions of the *putting-out system,* as dispersed cottage manufacture has been called, could be resolved only by bringing the workers together in a place where the employer could directly oversee their performance. This was the factory, a new mode of industrial organization that had discipline as its *raison d'être* and central power generation as the condition of its success. England had known factories as early as the sixteenth century, but based as they were on the existing hand-powered technology, they could not compete with the cottage mode, which shifted plant and equipment charges to the work force. In the eighteenth century, however, the combination of technical innovations that we know as the Industrial Revolution made the factory both feasible and dominant. From the 1770s on, then, an increasing number of workers found themselves employed at jobs that required them to appear by a set time every morning and work a day whose duration and wage were a function of the clock.

Nothing was harder. These were people who were accustomed to work at their own pace, to take their rest and distraction, or for that matter relieve themselves, as and when they pleased. They could work very hard when they had to—at end of week, in harvest time—but Sunday was holy, Monday was holy, and Tuesday was often needed to recover from so much holiness. Coming as they did from cottages and fields, they felt the factory to be a kind of jail, with the clock as the lock. Their unfamiliarity with or resistance to the new discipline found its most frequent and powerful expression in their inability or reluctance to show up on time. To combat this, employers reserved the harshest fines for latecomers and absentees—that was the stick; while a favorite prize for good workers was a clock—that was the carrot. For the vast majority of workers who had no timepiece, employers sent around wakers to tap on windows in the dark morning hours. Many of these workers put in fourteen-hour days and longer and dragged themselves to the mill still half asleep. Small wonder, then, that one employer found the sound of the one o'clock bell insufficient to bring his workers back to the machines after the noon break. The workers said they missed the single peal. He made the clock sound thirteen times.[16]

The frictions that necessarily attended this breaking-in process were exacerbated by the same asymmetry that agitated the textile towns of medieval Flanders: the employer knew the time, whereas the employee had to take his word for it. How could the worker know that he was working only the hours he was paid for and that the employer was not in some way slowing or setting back the clock so as to steal additional labor? We have occasional reports of such practices, but I remain skeptical: it is not an easy matter to fix a tower or wall clock so that it keeps good time from day to day but runs slower during working hours, and keep the fixing secret. One could, of course, adjust the length of the pendulum rod morning and evening. But that is best done with a nonpublic clock—one placed, for example, in the manager's office—and such limited access could only arouse suspicion, as it did in the Middle Ages. Besides, such "fiddling" will not work very well if there is more than one shift, if one group of workers is waiting to take over from another: the "fiddling" will out.

What matters, however, is that many workers *believed* they were being cheated in this manner, and their beliefs were no doubt confirmed in many instances by conflicts between the time of their own mill clock and that of other clocks in the area. (There is nothing that will so quickly undermine confidence in authority as evidence of error.) The workers' concern to defend themselves led numbers of them to buy their own watches. These were usually cheap turnips that sold for a couple of pounds—that is, about a fortnight's factory wages for a skilled adult male. A thrifty man could afford that; if not, he could pool his savings with those of other workers or buy the watch on time. Employers, we are told, were not happy to see their men thus armed against them. Perhaps; although given the inaccuracy of such pieces, they must have seemed more a pretext to dispute than a serious basis for disagreement.[17]

I T WAS THIS PRECOCIOUS, rapidly growing, and socially diversified demand that made the eighteenth century Britain's period of mastery in the history of watch manufacture. Hundreds of master craftsmen, employing thousands of out-workers, produced tens of thousands of watches a year—from 150,000 to 200,000 by the last quarter of the century.[18] Much of this was for export: one witness to the House of Commons estimated that almost three-fifths of the output went abroad. That share seems high to me and may have been intended to persuade the House that the watch industry earned a great deal of foreign exchange. Exports were partially (largely?) balanced in any event by a swelling stream of imports, mostly from Switzerland and much of it smuggled.

This large output represented about half of the European total (more or less equivalent to world production).[19] It was made possible by what was probably the most highly developed division of labor of any branch of British industry. (Adam Smith would have found a much more impressive illustration of this principle in watchmaking than in pin manufacture; but of course he wanted to show that even so simple an object as a pin could be made more cheaply by breaking up the work into simple, repetitive steps. He did note, however, that thanks to division of labor and ingenious machines, a watch movement that had cost about twenty pounds in the middle of the seventeenth century could now be had, and much better made, for twenty shillings.)[20] From Prescot to Liverpool, eight miles as the crow flies, the countryside was dotted with the cottages of springmakers, wheel cutters, chainmakers, casemakers, dialmakers—every specialty that went into the making of a watch. Labor there was considerably cheaper than in London, which did most of the finishing.[21] Even so, some of the work had been mechanized. Brass sheet was not hammered, but rolled. Pinions were not cut from bar or wire and filed to shape; instead, pinion wire was drawn and drawn and drawn again, up to fifty times for the finest work. The wire, far cheaper and more accurate than anything that could be made by hand, began as soft steel. It was subsequently cut and filed to size, then tempered and finished. Along with these short cuts went the use of drills, presses, lathes, wheel cutters, fusee cutters, and other special-purpose machines that could turn out uniform, if not identical, components by the batch. Perhaps the first to move in this direction was Tompion, many of whose parts were interchangeable *in the rough*.[22] This was at the end of the seventeenth century. Half a century later British watchmakers were ordering *ébauches* (rough movements) from specialist firms in Lancashire, which worked from paper patterns using their own Lancashire gauge. All of this was very impressive, especially to would-be competitors. There was, however, an Achilles' heel: each shop had its own caliber and none of the shops was very large. The effect was to differentiate product and tie customers to a given supplier, the more easily as workers like nothing so well as the familiar. Fragmentation of the market in turn blunted the advantages that specialization might otherwise have yielded. This failure was not yet a serious problem in the mid-eighteenth century, when Britain's chief rival was Geneva, which worked on similar lines. But we shall see that it became a serious liability in competition with the nascent watch industry of the Jura.

Prescot also made superb tiny files, the kind that watchmakers use—"the best in the world, at a superior price, indeed, but well worth the money, from the goodness of the steel and the exactness of the cutting." The steel was crucible steel, invented by a clockmaker named Benjamin Huntsman around 1740 and still a British monopoly in 1795 when John Aikin wrote those words.[23] It was the best, most homogeneous steel in Europe, and watchmakers everywhere bought these files if they could afford them, just as they did British springs. At about that time Geneva offered a prize to anyone who found a way to make springs of comparable quality: dependence was especially painful in time of war and blockade. The answer was not found until after the turn of the century, when thanks to British émigrés and some industrial espionage, steelmakers in France, Switzerland, and Germany learned how to make crucible steel.[24]

In all this, the role of the master watchmaker, the man who signed the timepiece, was to plan the watch, order the appropriate parts, and supervise the work of assembly, adjustment, and finish. He set the standards, and his name was the warrant of the watch's quality. But most pieces that went out never felt his touch. As R. Campbell's *London Tradesman* (1741) remarked: "The Watch-Maker puts his Name upon the Plate, and is esteemed the Maker, though he has not made in his Shop the smallest Wheel belonging to it." (In fact, if we are to judge from the number of misspellings of signatures on watch plates, it is by no means certain that even that part of the work was done in the master's shop; and if done there, not the master but a specialist engraver did it.)

All of this, however, was not enough. The growth of demand for quality timepieces put a heavy strain on the better-known shops, which could not keep up, while the new interest in time on the part of those of modest means generated an ever larger market for cheap work—the kind that was simply not acceptable to a master trained to older standards. To meet this demand, an increasing number of makers began to sell as their own the finished work of others. This was a violation of the craft ideal and was forbidden by guild regulations, at least in London; but there is reason to believe that even the greatest makers were not above such augmentations. Sometimes these outside watches were signed with a pseudonym, by way of protecting the seller's own reputation; thus imaginary characters named Tarts and Wilter were "responsible for" thousands of watches of Dutch style or manufacture. Sometimes the seller put his own name on, but without the usual serial number or with a serial number that was clearly out of order. Often the seller just signed, for some of these outside pieces were no whit inferior to his own.

It has long been assumed that most, if not all, of these associated watches are forgeries: historians of watches love to tell how Tompion, confronted by one of these imitations brought in for repair, threw it on the ground, stamped on it, and offered the man one of his own: "Here, Sir, this is a real Tompion!" But the more we get into the business records of watch and clock houses, the more we learn of such purchases of outside stock. The great Breguet depended on them; why not the great Tompion? Otherwise how can one account for the occasional fine piece signed with his name but manifestly not made in his shop—number wrong, no secret signature under the cock, un-English details?[25]

This practice must have seemed innocent enough—at least in the beginning. It made it possible to reconcile demand with the constraints of apprenticeship and workshop practice. It also spread the work and, as in Blois and Augsburg, gave business to those masters who, for whatever reason, could not prosper on their own. Above all, it enhanced the profits of the leaders of the profession.

Yet there was another side to this coin. The sale of watches and clocks as commodities rather than as personal creations opened the door to the pure trader, who would as soon sell a foreign timepiece as an English one, and with an English signature at that; as soon a bad one as a good one. The premonitory symptoms were already apparent. In the eighteenth century, so great was British prestige and so rich the market that the Dutch and Swiss developed veritable industries for the manufacture of pseudo-London timepieces.[26] Some of these were costly and highly decorated objects. Fashion as always was an important consideration, and some of the styles produced on the continent were simply not available in Britain. The art of painting on enamel, for example, whether on cases or on dials, remained a French and, more and more, a Swiss specialty. But most of these imported watches were mediocre and cheap, produced by labor that in mountain Switzerland and Savoy (the hinterland of Geneva) cost perhaps a half or a third of the British.

How much foreign work came in, we shall never know, because much of it was smuggled. This, we shall see, is a recurring motif: smuggling is by nature a risky operation, but watches lend themselves to it better than almost any other commodity. They are small, easily concealed, valuable, and command a far wider market than other precious objects. Genevan and Neuchâtelois *contrebandiers* throve in the loose border controls of the eighteenth century. We have already seen them at work in the trade to France; they were almost as busy in England. To all this, the authorities back home turned a benevolent blind eye. What else could they do? These ministates had to export just about everything they made. They had to export to keep their people fed, and their markets were doing their best to keep them out.

Smuggling has left few traces in the historical record; that is the way of illicit operations. From time to time, though, one gets a glimpse of the trials and tribulations of clandestine commerce. Here is a candid letter of October 1789 from a Neuchâtelois in Paris to his father and brothers in the Jura counseling against a foray into the London market. Times are bad, Paris is in the throes of revolution, and no one is paying his bills. Still, he prefers to stay away from England, where it would seem the import trade was necessarily and customarily one of contraband and concealment:

As for the London project, I don't think it can pay us, especially not to establish ourselves there, because the English place too many obstacles in the way of foreign trade, especially the jewelry trade. They have recently ruined several houses by confiscations [of contraband goods]. If you establish yourself, you necessarily run the same risks, because you can't remain unknown if you have to do business and to use methods such as those used lately by those known to have foreign merchandise. You have to keep the goods at friends' houses; you don't dare carry them about in town or show them to people you're not sure of. I am convinced that such a trade cannot be profitable, especially for someone who first has to make acquaintances.[27]

As the above plaint makes clear, the British were exercised about this low-cost competition, which was beginning to hurt the branch of manufacture that added more value to raw materials than any other. John Grant, a first-class London maker, testified to Commons that "sometimes a Watch sold for 150 Guineas is intrinsically, with respect to the Materials, not worth 20s—the Materials are increased in value, in many Instances, by Labour, a Thousand Times."[28] Grant was trying to persuade Parliament to repeal a tax on clocks and watches, so he may have exaggerated somewhat, but only somewhat. A pound of balance springs was worth far more than its weight in gold.

The clock- and watchmakers might have added, as further evidence of their contribution to national wealth, their role in the development of industrial technique. Their machines were the models for the bigger metalworking tools used to build the equipment of the rapidly changing textile, iron, and engineering industries, and their craftsmen constituted the biggest pool of skilled mechanics in the kingdom. The want ads that filled the newspapers of Lancashire (cotton) and Yorkshire (wool) asked for experienced clockmakers; and the wheelwork of the new spinning machinery was known as clockwork. For once, the clock and watch trade, normally an all-conquering seducer of labor, was having trouble holding on to its people against offers of higher pay in other branches.

As early as 1764 the Clockmakers' Company of London was being pressed to do something about the influx of contraband watches from Geneva. These complaints resumed in 1780, and a number of prominent makers, hitherto independent, joined the Company by way of solidarity and reinforcement. The difficulty was aggravated in 1784, when the government imposed an excise duty on gold and silver. British makers, who were obliged to use 22-carat gold, had trouble enough competing with the French (20-carat) and Swiss (18-carat), without paying a tax in addition. In 1787 the Clockmakers' Company proposed that the government put a heavy import duty on watches and clocks, arguing among other things that the French (meaning primarily the Genevans) used cheaper labor such as women and children at gilding, and used gold of lesser fineness. The effort to compete, they said, would debase the quality of British work. These arguments carried and the government imposed a duty of 27.5 percent.[29]

Even that would not be enough. But it served for the moment, especially since revolution and war (from 1792 on) on the continent soon made watch shipment far more risky and costly.

15 Multum in Parvo

G ENEVA IS A SMALL PLACE. It measures, inclusive of attached country districts, about a hundred square miles—less than London proper, little more than a seventh of Greater London. A fifth of that is under water. In the middle of the sixteenth century, its population stood at 12,500; two hundred years later it had doubled, to about 25,000. By way of comparison, London at the latter date numbered nearly 700,000 inhabitants, and Paris over half a million.

Person for person, though, the Geneva of those days, a tiny republic in a world of monarchies and principalities, was probably the most productive and creative city in Europe.[1] This it owed to its role as a place of refuge for Protestant victims of Catholic persecution, which brought it some of the most independent, best educated, and most highly skilled subjects of the countries around. Some of these families—the Facios (Fatios), Fazys, Turrettinis, and others—were living testimony to the power of heredity, for they produced distinguished artists, businessmen, savants, scientists, statesmen, and writers generation after generation, for hundreds of years.[2]

Among the refugees who came to Geneva were watchmakers from France, where Protestants constituted the elite of the horological profession. They came to a city that had once boasted a strong jewelry manufacture, now blighted by religious passion: Jean Calvin had no use for ornaments and vanities. But the same puritanical regime that condemned jewelry was willing to make an exception for watches: if Calvinists were not interested in time and its measurement, who was?

The Huguenot refugees from France brought to the infant craft the skills and secrets of Europe's most advanced watch manufacture. There was a time when a certain Charles Cusin, of Autun in Burgundy, was credited as founder of the Geneva *fabrique* (in the collective sense);[3] but he was surely one of several exemplars, for he came on the scene only in 1574, at a time when a small watch industry was already flourishing. The historian of the *fabrique*, Antony Babel, was able to find in the records evidence of sixteen other watchmakers active earlier, and these surely constitute only a sample.[4] By 1585 the watchmakers of the city were already numerous and self-conscious enough—also sensitive enough to outside competition—to think about incorporating as a guild.

Whatever the facts, it is clear that the Genevans took to the new craft with singular aptitude and by the seventeenth century were making some of the most elegant and complicated watches in Europe. They seem to have made a specialty of so-called *form watches*—watches cased to look like animals or flowers or skulls (as a reminder of mortality) or anything except a watch; also *crucifix watches* (*montres d'abbesse*) for the Catholic clergy and pious laymen (manufacture prohibited by a regulation of 1566, but business is business); also *lunar-movement watches* (what we now call astronomical watches), particularly popular among the Muslim infidels with their lunar calendar (again, business is business).

Geneva watches were apparently cheaper than those of other centers, or their supply was more elastic, for we find watchmakers of Blois contracting for quantities of Geneva timepieces for export to Turkey. We have a compact of 1632, for example, between Pierre Cuper, "merchant watchmaker of Blois," and Anthoine Arlaud, "merchant watchmaker and bourgeois of Geneva," calling for a mixed lot of "good and faithful merchandise," to be delivered at Marseilles in one year's time, thence to be shipped to Galata near Constantinople, the shipment to include striking clocks, lunar-movement watches, and simple timepieces. The Arlauds made something of a specialty of selling watches to the Turkish market: forty years later Anthoine's son Abraham, who had spent some time working in Constantinople, agreed to furnish fifty lunar-movement watches to a certain Sébastien Chappuis, who also ordered forty from another Genevan named Jean-Anthoine Choudens; and a few years later the same Abraham promised to deliver "as many lunar-movement watches as he can make in the space of one year."[5] We may assume that all of these watches were of the same size and caliber, so that component parts could be made by specialists using repetitive techniques. This did not make these parts interchangeable, but it did make possible a greater than ordinary division of labor and prices that defied competition.

A word in passing about this Genevan tie to the Turkish market: the Genevans, Neuchâtelois, and Swiss in general early gave evidence of extraordinary mobility, a willingness to live and move abroad that made them the migratory workers and mercenary soldiers par excellence of early modern Europe.[6] Most of this reflected the pressures of poverty and the lack of employment at home, especially in the mountain districts. But such factors will not explain the ready travel and expatriation of young craftsmen and traders who spent some of the best years of their lives, sometimes decades, far from home. Listen to Jaques Savary des Bruslons, whose *Dictionnaire universel du commerce* is the nearest thing we have to an encyclopedia of the practice and composition of trade in the mid-eighteenth century: "Its [Geneva's] connections with the merchants of Marseilles and Amsterdam open to it [the commerce] of the two seas, and there is scarcely a corner of the earth, however distant, where the nations of Europe send their ships, where one does not see Genevans, whether as commission merchants or private traders; there are even some who freight entire ships on their own account, or who buy them outright for voyages to the Levant or what are commonly called the long routes."[7] This from a city two hundred mountain miles from the nearest salt water! Some of this extroversion may have been the natural consequence of specialization: Geneva was a small entity, a crossroads without much territory. Its fortune, if fortune there was to be, lay in its openness to the great world outside. To a city of merchants, the ocean was simply an extension of the land; to the young, ambitious Genevan, opportunity lay where the wind blew.

There was also the element of adventure and escape. The Geneva of the seventeenth century was no longer the strait-laced republic of Jean Calvin, but it was still sober, hard-working, humorless. Constantinople must have seemed, at least to some, a place of liberation, a kind of cruise ship. It certainly seemed that way to a visiting minister by the unlikely name of Léger who reported to the Compagnie des Pasteurs in 1652 that the Genevan watchmakers in Constantinople were living "in complete license, without religion or any restraint, so that they are a subject of great scandal." Perhaps the Compagnie should assign a pastor there "to bridle this wild youth, so given to great excess."[8] And it was not only the youth that went: married men went also, often at some cost to domestic harmony and happiness. Take Jaques Sermand *neveu,* a watchmaker working in the 1650s and 1660s. By the time he was thirty, he had made several trips to Constantinople. In his thirty-first year he went again, this time taking with him all his clothes, linen, and tools—presumably to stay a while. He died on the way. Why he went and how he died, we do not know; but he left behind a widow and children, and it is the widow's testamentary deposition to this effect that is our only record of Sermand's last project. A more famous expatriate (partly a case of reflected glory) was Isaac Rousseau, father of Jean-Jacques. Rousseau already had one child when he left, and he returned a few years later on the pressing insistence of his wife. Jean-Jacques writes, in a rush of modesty, that he was "the sad fruit" of that return. But he proudly dubs his father *horloger du sérail,* which for a good Genevan may have been as close as he could get to such French honors as *horloger du Roi.* Perhaps that is what Isaac told the family, by way of justifying his departure.[9]

As elsewhere in Europe, watch production in Geneva was initially organized in do-it-all workshops where master, journeymen, and apprentices lived and worked side by side. Beginning in 1601, all of this took place under the watchful and defensive eyes of a guild, which aimed as usual at protecting its monopoly against outsiders and at limiting competition among masters. Most of the regulations were standard: limits on the size of the shop, on the number of journeymen who might be employed, on the number of apprentices who might be trained; restrictions on the place of work, which was required to be confined to the house of the master; prohibitions on the use of such labor-saving devices as dies for the stamping of designs on watch cases or of watch dials.

The rules of Geneva, though, had a local flavor. The city was a place of refuge and drew a constant if fluctuating stream of emigrants from lands of persecution. Indeed, as we have seen, the watch industry itself owed much to the injection of skill and experience it received from this source. But that was in the sixteenth century. In the seventeenth the refugees were not only not needed, but not wanted. The guild excluded all newcomers (*habitants*) and their children, even those born in Geneva (*natifs*), reserving the trade exclusively to citizens (*citoyens*) and burghers (*bourgeois*). Also excluded were women, even the wives and daughters of masters.

These regulations were normative, an expression of the intentions and interests of those who had climbed the ladder of social and economic promotion and wanted to pull it up after them. They were ill-suited, however, to an industry in full expansion where demand for labor was outstripping supply, and the more enterprising masters quickly found ways to bypass them. One evasion, already tried in other European centers, was to divide production into specialties and thereby secure higher productivity while making the most of less costly labor. This process was already well under way by the mid-seventeenth century; witness a contract of 1654 between a *compagnon horologeur,* that is, an artisan watchmaker, and a *marchand horologeur,* or merchant watchmaker, for delivery of twenty-one rough movements (*montres blanches*).[10] At a time when production averaged one watch per man-month, this represented a substantial order. Once again, we may reasonably assume that these "blanks" were alike, made to some standard pattern, and included components purchased from outside specialists. And we know that, as at Blois and Augsburg, the day of a journeyman came cheaper than that of a master.

Thus was born the system that the watch trade has known by the name of *établissage* (after the worker's bench, or *établi*). The *établisseur,* normally a skilled watchmaker whose work commanded respect in the marketplace, was the organizer of production. It was he who decided on the watches he wanted, gave out raw materials to subcontractors (especially such costly materials as gold and silver to casemakers), advanced money for wages, collected and redistributed the various components as the work proceeded, supervised the finishing processes, and took the finished watches to market, either selling directly to consumers or to merchants for sale in distant places. Sometimes the merchant took the finished watches on consignment, so that the *établisseur* bore the risk until payment; and sometimes he bought outright and took his chances on finding customers. The price reflected the risk thus assumed.

The effect was to expand capacity without diluting the highly profitable privilege of mastership and hence access to the Geneva market—what the economist calls monopoly rents. On the one side were all the preparatory and ancillary processes: roughing out of plates, pillars, and wheels (manufacture of *ébauches*), chain- and dial-making, gilding, polishing, and the like. On the other was the fine work of assembly, *repassage,* adjustment. The latter continued to be reserved to citizens and burghers; the former was gradually opened to the lower-status groups within Geneva (mainly immigrants and women) and to the peasants in the country districts around.

This progressive devolution of even the less agreeable and dangerous tasks (the poor women who were "permitted" to gild with mercury paid for their jobs with their health and life) did not take place without resistance from the more conservative masters. Indeed, our major source for these developments is the repeated efforts of the Geneva guilds to retard this trend or so to channel it as to reserve to members all the really lucrative stages of manufacture. Their reiterated regulations and interdictions, however, are evidence in themselves of the limits of control in an area of highly fragmented political sovereignty. (The little Arve River, today some five tramway stops from downtown Geneva, was then the boundary of the Republic. Carouge, now one of Geneva's busiest industrial quarters, was in foreign territory.) There was simply no way the authorities in Geneva could keep maverick masters and *compagnons* from going outside to train and employ low-wage workers; the less so, as well-to-do peasants were ready to pay room, board, and cash to get artisans from the city to come out and teach their children. Here the hierarchical division of labor within Geneva itself played a role: those journeymen who found themselves underpaid within the *fabrique* had good reason to seek early promotion to the equivalent of mastership outside. The Geneva watch guild tried to keep a list of these renegades for eventual sanction should they ever try to return; but such reprisals were paralyzed by the evident interest of the more successful masters in the bootleg products of these country shops. They needed both rough and finished movements to fill their orders, and they were not inclined to be fussy about the source.

In all of this, Geneva was not radically different from older centers. Where Geneva innovated was in the range of its expansion, both in space and social scale. Starting in those districts that were part of the Republic, spreading thence to contiguous areas such as Carouge and Gex, the watch industry reached out by the early eighteenth century into the mountains to the south and southeast, then part of the duchy of Savoy. There it found an extremely poor population squeezing a living from an infertile soil and compelled to rely more on livestock than on cultivation. Such a pattern offered two advantages to industry: much idle time, especially during winter, when the snows lay heavy and long and communications were cut; and the relatively soft hands of the herder as against the calloused fingers of the plower and digger. This latter may seem a small matter, but Joan Thirsk has already told us of the importance of this consideration for the location of rural industry in England.[11]

These rural outworkers provided the human infrastructure of the extraordinary expansion of the Geneva watch manufacture in the eighteenth century. They came to outnumber those employed in the *fabrique* proper, but they earned far less. A French document of the late eighteenth century sketches the relationship in terms that anticipate modern concepts of dependency and unequal terms of trade: "Geneva had relegated the production of the rough watch movements to the mountains of France and Savoy, so that it paid six days of an *ébaucheur* with one day of a finisher; and at the same time it had given up all the low-wage operations, and it paid for them with the high earnings of the jeweler and watchmaker. That's why Geneva is so prodigiously rich; that's why 25,000 inhabitants of Geneva have more money than the 450,000 citizens of the [duchy? one word illegible] of Savoy and the 800,000 citizens of the neighboring departments."[12]

This class and geographical division of labor and profits obviously worked to the advantage of the master watchmakers of Geneva. We do not know their incomes, but we do know that they were men of substance. Even the journeymen (*compagnons*) were men of substance, for they earned more than skilled workers in other trades, and they made it a point to hold themselves aloof from lesser artists and artisans. They were the cream of a labor aristocracy: they took their long weekends (some of them had country houses), their afternoons of diversion, their outings in the hills. They had their clubs; they crowded the theaters of Carouge (spectacles were still banned in Calvinist Geneva); they dressed in silk stockings and powdered wigs. (One master lured a journeyman watchmaker he was trying to hire with the promise of a new hat with gold border and a new peruke.) Under the circumstances the apprenticeship fees were higher in watchmaking than in other crafts. At the beginning of the seventeenth century, one had to pay more to become a goldsmith; by its end, the order had been reversed and would-be watchmakers were paying twice as much and more for the privilege of instruction.[13]

THERE IS AN AXIOM in economics that is related to the physical law of conservation of mass and energy. It states that there is no such thing as a free lunch. Everything costs. There is also a law in economics that derives from this axiom, namely that every situation bears the seed of its own reversal. This is the law of nemesis: nothing good lasts indefinitely, because others will want to share it.

The privileged prosperity of the watchmakers of Geneva was no exception to this rule. All those monopoly profits were bound to stir envy and draw people into the industry. New centers of manufacture began developing outside the Republic, sometimes as a result of strenuous efforts by foreign governments to put an end to the drain of exchange and specie that went to pay for Geneva watches. Most of these artificial attempts to transplant a shoot of this flourishing tree to other climes quickly failed—so, for example, the quixotic campaign to establish a "New Geneva" in Waterford, Ireland. Constance, at the western end of the Bodensee, was a more realistic site: some hundreds of artists were persuaded by the Austrian government to move there in 1785–1787 and stayed on so long as Geneva continued to discriminate against *natifs* and the local authorities gave them special encouragement. But this was a hothouse flower that faded as soon as Geneva turned democratic and Habsburg enthusiasm slackened. Joseph II died in 1790, and the new king did not know his vision. In 1792 the exiles began returning home, and by the end of the century the Constantian *fabrique* was only a memory.[14]

The most vigorous effort along these lines was made by the French, who had good reason to bemoan the loss of their own watch manufacture to Genevan competition. Paris tried repeatedly to attract a critical mass of experienced watchmakers who might provide training for indigenous apprentices and serve as the nucleus of a reborn national industry. Every time there was political trouble in Geneva, which was increasingly caught up in quarrels over the civil status and political rights of immigrants and their children, the French sought to profit from disaffection and proscription to recruit mechanics and artisans. None of these attempts succeeded until the 1790s, when the long-standing dispute between conservatives and democrats became exacerbated by the ideological issues raised by the French Revolution. In 1793–94 agents of the revolutionary regime in Paris succeeded in persuading some hundreds of Genevans and Jurassiens to leave hearth and shop and settle in Bensançon, on the other side of the mountains. The immigrants were offered all kinds of aid: travel expenses, living and working accommodations, loans for raw materials and equipment as needed, exemption from military obligations. Needless to say, performance fell short of promises, while the reception by the local population, Catholic rather than Protestant and already aggrieved by the hardships of political upheaval and war, was anything but friendly. Even so, the immigrants stayed on, if only because it would have been psychologically difficult and sometimes politically hazardous to return home. In these early years the new center stumbled along, relying heavily on the import of components from the very *fabrique* it had split off from. A few years later, in 1798, Geneva was annexed by France and the whole effort became moot, until 1815, when Geneva regained its independence. Besançon, then, was more a dream than a real competitor. Not until the mid-nineteenth century was it to develop behind high tariff walls into something like a French watch industry.[15] By that time, as we shall see, Geneva had long since been passed by the mountain manufacture of Neuchâtel and the Swiss Jura, which started as a few cottage workshops at the beginning of the eighteenth century and grew into the world's most successful rural *fabrique* by the end.

At home, the effect of wider markets and increasing scale was to give ever greater power to mercantile intermediaries, who came to control the industry and reduced the guild masters to a position of dependence. They took over, in effect, the very putting-out system that the watchmakers of Geneva used with their rural subcontractors, and incorporated the watchmakers in it. Seen in retrospect, this development was implicit in the distribution of market roles: it was the middleman who knew and served the buyers and was in a position to dictate the content, character, and rhythm of work to the producer. As demand grew, as watches "caught on" in new places and with new sections of the population, this rhythm changed. Where the old shops had been content to turn out two or three timepieces a month, month in and month out, now the trade became seasonal—frenetic in anticipation of the holiday season and the great European fairs of Leipzig (serving central Europe) and Beaucaire (serving Iberia and the Mediterranean), painfully slow in between. "We work by leaps and fairs," was the way one observer put it.[16] As a result, the nature of the commercial tie became crucial to the independence and prosperity of the clock- or watchmaker. A hard merchant could exploit the slow periods by squeezing his suppliers one against the other to extract the maximum of concessions. A good merchant could help his *établisseurs* with advances in high season, purchases for stock in the quiet periods.

A good illustration of the latter is the career of Jules-Samuel Jequier, maker to the house of Bovet (kings of the China watch trade), and a man of great religious faith. (I cite Jequier here because we are fortunate enough to have his own testimony on this subject. He himself worked in Fleurier, in the Jura, and hence was not part of the Geneva *fabrique;* and he lived about half a century after the period we are now examining. But the principles of organization of production and sale were the same in the two areas and periods, and Jequier's career could just as well illustrate the experience of some of the Geneva houses of the late eighteenth century.) Recollecting his rise to independence and success as a watch manufacturer (*fabricant*), Jequier wrote that as he advanced in years and his family grew (he had fourteen children), God provided generously for his needs by opening new sources of income, among others, "the manufacture in my spare time [*à temps perdu*] of a certain number of Chinese watches authorized by Bovet, this in a time of crisis [civil war and foreign intervention in China], in order to keep my father busy, who had no work because of the crisis." As for Jequier's regular work, which he undertook at his own risk (*à mes risques et perils*), Bovet was always ready to buy it, and Jequier took this as a sign of divine providence: "This work was so much blessed that as soon as any of it was ready, an urgent order arrived [from China], and Bovet was only too happy to find these things ready and to be able to ship them forthwith."[17]

Thanks to such ties, numerous *établisseurs* moved on into trade and became merchant-manufacturers, sometimes by entering into partnership with friends or relatives who could act as agents in major markets. Abraham Vacheron, founder of the firm that later became Vacheron & Constantin and now prides itself on being the oldest such house in Switzerland, is a good example. Others began as merchants and integrated back into manufacture, as did Czapek & Patek, later Patek Philippe. Where technical skill was lacking, one could hire experts (appropriately called *visiteurs*) to check the work. In the course of these changes, the old hierarchical workshop dissolved, and more and more craftsmen (including unsuccessful *établisseurs*) came to work in isolation as *cabinotiers* (from the word *cabinet*, a small workroom), linked to the larger production process only by the agents of the merchant-manufacturer bringing fresh work in and taking finished pieces out.

At the time, the response to this structural transformation was mixed. Some rejoiced in the evident prosperity of the *fabrique* as a whole; others lamented the loss of innocence and standards. Writing at a bad time (1795), Isaac Cornuaud, the Genevan Cato, denounced the effect of irregular demand and mercantile greed on work and workers. In the haste to prepare assortments for the great fairs, the merchants took up such poor work as hurt the reputation of the entire *fabrique*, while the craftsmen straining to meet these demands learned bad habits of both work and leisure. A highly variable pattern of employment, argued Cornuaud, has a way of encouraging the grasshopper mentality: profligacy in good times and penury in bad; and bad habits, in turn, render the watchmaker that much more liable to domination and abuse.[18]

The indignation of Cornuaud has since given way to regret, modified by a somewhat lachrymose evocation of the good old *cabinotier,* that marvelous artist working quietly at his bench and turning out watches such as no machine can make. We should not be deceived by such sentimentality. Some of it reflects an anti-technological bias, and some of it is nostalgia. Nostalgia changes with time, but it is always with us, reflecting as it does a universal tendency to selective memory: "Those were the days, my friend; I thought they'd never end." In this instance, it also gives expression to a widespread prejudice against middlemen as essentially parasitic and unproductive. The fear and resentment of the trader go back to the earliest agricultural communities: the peasant who earns his bread by the sweat of his brow can hardly be expected to understand and respect what seem to be the easy and exorbitant profits of the merchant. This attitude has carried through to the present, has spread from country to town and city, and has been explicitly or implicitly cherished by such widely different groups as the landed aristocracy on the one hand, craft and industrial workers on the other. It is perhaps the one point on which just about every system of socialism agrees; to this day, the national accounts of certain socialist countries—the Soviet Union, for example—do not treat trade as productive of value, and hence do not include it in the national product.

Yet we shall see that these intermediaries of the Geneva watch trade earned their place and profits. It was the craftsman-artist who made the reputation of the Geneva watch, but it was the merchant who brought him the orders. No one else knew so well what would sell and what would not. It was surely the merchants who asked the makers of the early eighteenth century to sign many of their watches as from London, by way of exploiting the reputation of the dominant center. Nor were they above a little honest (and dishonest) imitation—the sort of thing the Taiwanese have excelled at in our own time. Thus, two Geneva merchants, Georges Achard father and son, copied in 1806 the watches of Higgs & Evans, London makers for Spain and Spanish America (they often signed their watches Higgs y Evans), then sold them at 115–130 francs each where Higgs & Evans were asking 200. The same house bought one of George Prior's watches for the Turkish market and turned out six hundred copies at 116 francs delivered in Smyrna, against 168 francs for the London product.[19]

The more prosperous the industry, moreover, and the wider its fame and markets, the more indispensable the services of the middleman. The vast bulk of watch output went abroad, much of it to areas not rich enough to support fixed agencies. The watches were carried from place to place—the kind of selection that would appeal to a range of tastes and purses. Someone had to take them, traveling for extended periods at considerable expense and some personal hazard. In the seventeenth century the maker would often undertake that task himself. In the eighteenth, no serious producer could spend that much time on the road, and no single shop could produce the assortments required. On the contrary, the search for efficiency had long given rise to a division of labor among shops and even localities by type of watch. Hence the need for intermediaries who could draw on the work of many shops.

The cost of such an assortment could be enormous. We have, for example, the inventory of Philippe Dufalga in 1767 showing watches to the value of 120,000 florins—equal to forty years' salary of a well-paid master, or more than a million of our dollars.[20] Dufalga, to be sure, was selling some very expensive pieces—complicated watches adorned with precious stones and polychrome enamel worth a year's wages and more. But his selection was not untypical of the range of Geneva work. Under the circumstances, one of the critical advantages of these merchant-manufacturers over the traditional master was access to bank credit. By "bank" is meant here not the joint-stock, corporate giant one thinks of today, but merchant bankers who combined credit with trade and specialized in the discounting and creation of commercial paper. They were prepared to open lines of credit to the entrepreneurs of the watch trade against which these could draw for the purchase of materials or payment of wages; these drafts were covered by remittance of bills of exchange and acceptances from the buyers of the watches. The bills ran anywhere from thirty to ninety days, but the bankers were prepared to discount them in advance when endorsed by the signature of their client. In this way the merchant-*établisseur* could draw on resources far exceeding his own, could afford to contract for rich assortments of watches for such crucial opportunities as the major trade fairs, and could sell watches for credit that he could not have disposed of for cash.

By the late eighteenth century this bank credit to the merchants was a major determinant of the prosperity of the industry. Easy money meant lots of orders to the *cabinotiers* and suppliers of components; tight money meant that almost everyone was idle. When the lenders of Geneva sneezed, the city fathers of Cluses (Savoy) took cold and officials in Turin took notice. Bank credit was also the biggest factor in promoting differentiation within the *fabrique*. Lenders and investors, noted a brochure of 1781, "placed their funds with the richest of the merchants; those whose only assets were their honesty, their activity, and their industry were more or less left to their own devices; above all, these families would not lend to the Artists [the master craftsmen]."[21] So the rich got richer. At the top of the trade were a handful of great merchant houses, "veritable pillars of the manufacture. They alone," wrote Cornuaud, "did a wide enough, sustained enough, safe enough business to provide a solid basis for this major branch of our export trade."[22]

Lesser houses had to scramble. Their merchandise was of mixed quality, and so was their clientele. They could not always close a deal for cash or even credit. Instead, they improvised. The traveler who found himself stuck with an array of unsold watches toward the end of a trade fair or a long journey—often "clinkers" much the worse for handling—was in no mood to carry these back to Geneva. So he traded them if necessary for cloth, garments, toys, or such other merchandise as the other party had to dispose of, usually for similar reasons. Then, when he got back home, he had to dispose of this stuff, and who better to take it, willy-nilly, than the watchmakers themselves? They were waiting impatiently for wages long since earned, and something was better than nothing, even if estimated above its market value. The practice gave rise to cruel anomalies. Workers too poor to buy necessities sometimes found themselves stuck with the most inappropriate luxuries. "It was not at all unusual," wrote Cornuaud, himself an old casemaker, "to see poverty covered with silk and velvet."[23] Which recalls an old nursery rhyme:

Hark, hark, the dogs do bark!
The beggars are coming to town;
Some in rags, and some in tags,
And some in velvet gowns.

The mercers and drapers of Geneva raged at this unwonted and unwanted competition. And the watchmakers wept for the dear, dead days beyond recall; also for the honor of their profession: "There is no hiding it: the corps of merchants and master watchmakers is not what it once was. The door has been opened to people who would have done better in the lowest occupations, the trade has been infiltrated, I fear, by unworthy types, without principles, dishonoring the commerce that they are ruining." The workers were losing, according to this source, 15–50 percent of their nominal wages. "Any man who contracts for work and then barters his wares and unloads on the worker does not deserve the name of merchant. Must the worker pay the bill?"[24]

It was a rhetorical question. In 1768, under pressure from France and some of the Swiss cantons, the bourgeois of Geneva reluctantly conceded the right of *natifs*—that is, Geneva-born children of immigrants—to sell the goods they made as guild members of divers professions. One year later, however, a further edict reminded them that they were not allowed to sell other merchandise or use it to make payment in kind—another way of saying that the truck system was legitimate but reserved to the privileged elite.

C LEARLY, the life and work of these merchant-manufacturers were not a round of unalloyed gratification and profit. The very recourse to barter and truck payments is evidence that not every watch could be sold and that success depended on quickness, suppleness, and nerve. Such distant markets as China, at their best extremely profitable, posed special problems; the high rates of return were in part a function of the risk. Contrary to widely held beliefs concerning the omnipotence of European imperialism, the writ of the Western powers did not extend to Canton or Peking. Charles de Constant, a Genevan resident in China in the late eighteenth century, tells of an English supercargo who brought a lot of handsome clocks that aroused the Chinese appetite; but he was asking more than the mandarins wanted to pay. So the Chinese officials ordered him to leave and take his clocks with him, selling them to no one else; whereupon he sold them for the price offered. This, in fact, may still have yielded him a profit, but others were less fortunate. The same Constant, writing in 1792 of the Macartney mission to Peking, noted that the English had good reasons to try for happier trade arrangements; for how could they go on putting up with things as they were, "without any fixed rule, without set tariffs, vexed and robbed in every way, liable for everything, even unforeseeable accidents, slaves of a frightful monopoly that always makes the European the victim"?[25] A merchant had to be diplomat as well as trader to survive in that kind of environment.

He also had to be tough and personally courageous. These merchants spent long, uncomfortable hours rolling on the seas or bouncing along the miserable roads of eighteenth-century Europe worrying about a chest strapped in iron, studded with impressive nails, and fitted with the kind of lock that is easier to break than to pick—just the kind that would discourage a petty thief but light up the eyes of a highwayman. These were filled with watches and jewelry to the value of tens, even hundreds of thousands of today's dollars. In Italy, which because of location became almost a hunting preserve for the Geneva watch industry, the plague of banditry was such that authorities sought to repress it by terror. J. B. Vacheron, traveling there in 1818 (things were no better in the eighteenth century), wrote: "We have encountered from time to time arms and legs nailed to posts, as a sign to travelers that brigands had been executed there because they had committed murder. I confess that this disgusting spectacle would have decided me to turn back if I had been alone; but being with Degrange,[26] and both of us very well armed and resolved to defend ourselves to the death rather than give up our assortments, we did not hesitate to go on ahead."[27] But then in the tranquillity of his hotel room, he was appalled by his recollected temerity:

> How much do all these reflections and these repeated chances make me want to go home, to lead a peaceful life free of all these tribula-

tions! To this end, I beg of you, dear friend, not to order any more watches. Be content to finish what is already in the works, and send it to me quickly so that I can, if that's possible, busy myself with placing it. I shudder every time I think again the sad thoughts that have afflicted me on this latest perilous journey! If I had been stopped and killed—the same thing—on this road, what would my family have become and my old father who has no one left but me? This thought alone is more than enough to make me swear off travel completely, and I assure you that that is my intention, even if I have to take up the file again.[28]

Back to the bench if necessary! Shortly thereafter, in April 1819, François Constantin joined the firm and took on the job of traveling in Italy. Vacheron stayed home.

The maritime equivalent of these cutthroat highwaymen were the Barbary pirates, who operated along the sea route from Marseilles to the Levant. They were better, in a way: they held their prisoners for ransom, rather than killed them. Swiss apprentices and watchmakers contracting to work in such cities as Constantinople and Smyrna often required their employer to commit himself by contract to pay the going price for their release if captured. But watches were not to be ransomed.

To all of these ambulant threats, moreover, were added the ordinary hazards of transportation—what the French call *accidents de parcours*. It was a happy trip that concluded without the vehicle's overturning—not the sort of thing to help even a well-packed watch. What if one overturned at a ford, just the kind of place where that sort of thing would happen? We know of one such accident in 1820: one hundred seventy watches soaked. The timepieces had to go back to Geneva to be salvaged: "Imagine how much they must have rusted by the time they got back!" And then a word of advice: "If anything like that happened to us, don't hesitate to throw them into a bath of oil. At least then, all we would have to do is clean them; otherwise they would be done for, as happened to these others."[29]

THIS INJECTION OF ENTERPRISE, this search for and cultivation of markets anywhere and everywhere, this readiness to sell and buy whatever would sell—in short, this commercial opportunism—was what converted the Geneva watch industry from a promising local specialty into the second largest in the world. In 1685—that is, a century after the Geneva watch industry took off—there were about a hundred masters who employed some three hundred workers to produce about five thousand watches. (That works out to about one man-month per watch, which is about right for *das alte Handwerk*.) A half-century later, in the 1740s, the number of masters had risen to 550—a huge number by comparison with other, tighter guild towns such as Augsburg. It was about this time that the Geneva *fabrique* grudgingly opened the gates to *natifs* and even *habitants*, initially by exception and subject to constraints. There was no other way to get the work done. By 1760, with the putting-out system in full swing and the two-stage organization of production well established, Geneva counted eight hundred masters, employing some four thousand workers in the city and an even larger, but unknown, number outside. A generation later, on the eve of the troubled Revolutionary era, the numbers seem to have been about the same.[30]

Those were the happy years, from 1750 to 1785, when Geneva could hardly make enough to satisfy demand. One French source of the period speaks of annual output of two hundred fifty thousand pieces worth forty-four to forty-five million francs, numbers apparently derived by multiplying estimated labor force by estimated productivity.[31] But these figures are probably too high, conceivably three times too high: our source had reason to exaggerate. A more conservative "guesstimate" for peak output is that of the economist Léonard Simonde de Sismondi, who gives eighty-five thousand watches a year for the period 1781–1786.[32] Then, in the mid-1780s, perhaps owing to domestic political strife between citizens and *natifs*, production began dropping, even before the revolutions and wars of the 1790s interrupted trade and exacerbated long-standing grievances. By 1801–2, according to Sismondi, output was down to some sixty thousand pieces, worth less than half the value of the earlier output. Work force fell by an even greater proportion, from almost five thousand to less than two in 1803. The city began emptying as craftsmen and their families sought better opportunities elsewhere. It was not to recover its earlier prosperity for another generation.[33]

This protracted decline would seem to indicate that the trouble was structural even more than conjunctural. The primary source of difficulty was the rise of a competitive watch *fabrique* in the Jura, able to undersell Geneva as Geneva had come to undersell everyone else. The Jurassiens, we shall see, paid lower wages, worked longer hours, used women and children more. They also made a cheaper watch, driving the Genevans into the upper reaches of the market. Profits were higher there, but demand was more elastic: sales rose faster in good times, fell faster in bad. This was the sector that was most hurt by the depression of the 1790s, the more so as luxury watches called for large advances to cover the cost of gold and jewels and frightened lenders were pulling out. The Republic tried to help by opening the public purse, lending almost half a million florins by 1797—to no avail. The market was simply not there, and the state soon found itself owner of a stock of elegant but unsalable timepieces. Private efforts met with similar results: a Comptoir National, or Etablissement Patriotique d'Horlogerie, funded by public subscription, distributed thousands of pieces to *cabinotiers* and spent hundreds of thousands of florins in an attempt to provide employment and sustain the *fabrique*. Within a matter of years it found itself holding a huge inventory and owing almost seven hundred thousand florins to the government.[34] (See Figure 25.)

Private enterprise, meanwhile, sought its own solutions. In a desperate effort to remain competitive, many *établisseurs* tried to save on quality and wages, pressing their workers to do more for less. It was a no-win situation: the Genevan *cabinotier*, who had a well-developed and highly privileged sense of a fair day's wage for a fair day's work, did not take well to that kind of squeeze. He balked; or he gave less. It was not a time for politeness or open hands. New merchants came to the fore, from the trade rather than the production side, market-tough, with no loyalty to older standards. The effect was to sacrifice reputation and long-run prosperity to momentary advantage.

This was surely the wrong tack. A better strategy was suggested by the sarcastic author of the *Avis du Compère Perret* (1794), but this was the kind of advice more easily given than taken:

> I wish we could learn to be a little thicker and a little more stupid. I ask why the Neuchâtel *fabrique* is going like all hell, while our

tongues are hanging out a foot long. I'm told it's because they know how to make lousy silver watches [*de mauvais péclots d'argent*]. Well in that case, why don't we do the same? Oh yeah, we wouldn't be making four *décimes* a day! But better that than nothing . . . Of course, those mountain people don't take a ten-o'clock break [*la sèche*], don't pause for coffee after dinner [the midday meal], don't frequent the exchange, don't go to see plays at Châtelaine . . . and they work twelve hours a day. They stick around with their wives and kids, and they're all of them the better for it.[35]

In short, a horological morality play: vice and virtue, profane and sacred love. As with men, though, so with watches: there was room for both. In the long run the Geneva *fabrique,* with its high-fashion, high-priced fancy goods, came back.

But Neuchâtel took a lot of room.

16 Notwithstanding the Barrenness of the Soil

I T SNOWS A LOT IN THE JURA. The major centers are three thousand feet above sea level, the mountain villages even higher. When the streets are dry and gray in Geneva, they are dirty white in Neuchâtel, bright white at La Chaux-de-Fonds. Even in summer one can drive along the shore of Lake Neuchâtel in sunshine and see the *massif* to the west shrouded in cloud. This is not the kind of climate that would ordinarily draw settlers, and for centuries the only inhabitants were clerics seeking the spiritual blessings of isolation and tranquillity. The Jura, it is said, is singularly propitious to theological contemplation. The first peasants moved in in the late Middle Ages, drawn by invitation and by the prospect of empty spaces for pasture. As late as the seventeenth century, this was a frontier wilderness.

The ground yielded little. The major farm crop was cattle, bought lean in the winter markets and sold fattened in the spring. Jura stockmen made it a point to attend all the fairs in the area looking for good buys, some of them riding, some of them hiking for days over sketchy roads and upland paths. These were people of uncommon vigor and stamina; otherwise they would have long since abandoned the mountains for the gentle climate of lake and plain. Their peregrinations brought them into contact with horse traders, peddlers, and merchants; they sold their livestock, for example, to butchers from Basel, Strasbourg, and even Paris. These contacts were a school for commerce and laid the basis for the later development of export industry.[1]

Before the coming of the railroad to the Haut-Jura, winter spelled closure. The first heavy snow of fall, as early as October, was an exhilarating experience: the countryside breathless-white, little children building snowmen and throwing snowballs, parents and older children rushing preparations for the months ahead. That first snow, though, was only a warning. It came and went. When the next real snow came, it stayed: today's cover became tomorrow's base. Then the roads and paths would be closed for months (the snow still lies thick on the passes in May), the markers buried, the pack animals stabled. This was a time for reflection, togetherness, some boredom, some religion, and much work—in particular, for those repairs and improvements that made the mountain peasant an extraordinary jack-of-all-trades: smith, carpenter, cabinetmaker, basketmaker, cooper, leather worker, mason, tinker, tinkerer, and inventor. There were few professional artisans on this barren soil—some smiths for heavy iron work (though every farm had its own little forge) and a variety of casual and itinerant day workers: cobblers, spinners (for home-grown flax), and seamstresses (for those, mostly townspeople, whose incomes permitted them to hire labor). One source of demand for these specialists was the growing commitment of women and children to home industry. Lacemaking was the preferred activity, not because the Jura needed or could afford lace—on the contrary—but because they could sell it outside the area.[2]

When watchmaking was introduced in the early eighteenth century, it swept all else aside. It paid better than needlework, or than agriculture for that matter. It drew in all the members of the family except the very young and very old; and it called upon just those skills that centuries of isolation and self-reliance had fostered. Many simply gave up farming altogether, rented their land to outsiders from Bern and elsewhere, and devoted themselves full-time to the making of wheels and pinions, plates and arbors, dials, chains, and cases.[3] It was a marriage of convenience, but also of love.

The role of a few innovators seems crucial here: these people might presumably have learned some other trade or simply sunk into the poverty that had afflicted much of upland Switzerland for centuries.[4] Popular legend assigns the credit to Daniel JeanRichard (1665–1741) of La Sagne, who as a teen-age lad offered to repair the English watch of a horse trader. He got it going again, and this success moved him to learn more about watches— among other things, to go down to Geneva and bring back tools and accessories. He began by making simple pieces, which he sold one by one to customers in Burgundy, the Franche-Comté, and the bishopric of Basel. Apparently he found some of his best customers in the Catholic church, in cathedral chapters, abbeys, and convents. With success came increasing scale. JeanRichard was soon employing his sons and a corps of apprentices, most of whom later went out to establish their own shops in other villages.[5] Both he and his pupils, moreover, drew on outside suppliers, at first Genevan specialists but increasingly cottage workers in the mountains. The subdivision of tasks made it possible for workers to acquire skills rapidly, while the familial character of this cottage industry meant that women and children were doing work that in Britain and France was reserved to highly trained journeymen. The very size of the pool, moreover, made it possible to spot and promote a substantial number of especially gifted workers to those tasks (layout, finishing, springing, adjusting) that required special knowledge and higher skills. The result was an industry that had it both ways: an abundance of skilled labor at low cost (compare Japan today) and a cadre of ingenious and highly qualified artisans who were not only capable of the best work but were technically creative. The system of *établissage*, then, was almost as old in the Jura as the watch industry itself.

Yet however much this story of an indigenous origin of the watch industry may gratify the Neuchâtelois and confirm their sense of achievement, it seems clear that JeanRichard was only one of the pioneers of Jura watchmaking. Protestant refugees from France furnished an important contribution, bringing with them not only their experience and skills but a network of commercial connections indispensable to an export trade. In addition, there already existed in the mountains a small, scattered production of domestic clocks, which was abandoned in favor of watchmaking as the new branch gave evidence of its profitability. And

then there were the foreign models. Geneva was close by—a
school for apprentices and a source of émigré craftsmen, tools and
supplies, devices and designs to copy. Geneva, as we have seen,
was not happy to be used in this manner, and Geneva craftsmen
were wont to look down at their mountain imitators—so much
so, that when the French sought to win away Swiss watchmakers
for their own manufacture, they thought to put Genevan dogs
and Neuchâtelois cats in different places.[6] Paris was far away
and much more seductive. Some of the best talent left for France
and stayed there: the most prominent examples are Ferdinand
Berthoud and Abraham-Louis Breguet.[7] (Like the Hebrew patri-
arch, Abram-Louis changed his name when he left his native
land: Abram became Abraham, and it is thus that Breguet is generally known.[8] But even those who stayed maintained personal
and commercial ties with their mountain home, and many more
went back to practice the arts they had learned from an older center.

By the last quarter of the eighteenth century, the watchmakers
of the Jura were producing their own horological innovations; an
example was Abram-Louis Perrelet, inventor of the self-winding
watch and teacher of Breguet. The first Jura watches, of which
very few had survived, were crude imitations of the French *oig-*
nons. Within a generation, though, they were imitations of
English, Dutch, and French work—watches for everyman, and no
longer crude. Not elegant either; that was left to Geneva. But they
were good enough to sell anywhere, and some of them were ingeniously complicated. JeanRichard, for example, went on from his
first, simple timepieces to more complex mechanisms—repeater
watches, for example, some of them in rock crystal cases to allow
the owner to enjoy the play of the mechanism. Workers of the
Vallée de Joux in particular gave proof early on of an aptitude for
the manufacture of repeater and calendar movements, which they
sold not only to local makers but to the Geneva *fabrique.* These
were skills learned when young, at the family bench, and they became rooted over time in the *terroir.* Two centuries later the best
watch firms in Switzerland are still going to the Vallée for their
complicated timepieces, just as Benson and Smiths in London
and Lange in Glashütte (near Dresden) did in the nineteenth
century. Even so hard-to-please and self-sufficient a house as

Patek Philippe goes to Piguet in Le Sentier for its perpetual-calendar mechanisms.[9]

The Jura watches were also cheap—cheaper than those of Geneva, which liked to look down upon them as shoddy work. Some of them *were* shoddy, for the Jura made just about the full range of qualities, down to ephemeral tickers for soldiers and country bumpkins. Yet low cost was more a function of low wages and efficient organization than of lower quality. The dispersion of manufacture in cottage shops made for heavy reliance on repetitive piecework, which lent itself in turn to special-purpose tools and machinery. Here, too, the Jura could draw on French and Genevan examples. In both the latter, watchmaking had built its early success on hand tools: hand and bench vises, pliers, bow-driven lathes and drills, and above all, the file. Then in the eighteenth century, copying British technique, French and Genevans adopted such machines as wheel and fusee cutters and centering and depthing devices. In Geneva the result was a gain in productivity from two rough movements a month to two, four, and then six or more per week.[10] (This was for crude work. In finishing and adjusting, the file remained as indispensable as before.) But the Jura makers went their predecessors one better. They bought their tools and improved them; invented their own; and went on from individual toolmaking (each watchmaker his own) to production by specialist toolmakers for general sale. In so doing, they created for the first time on the continent an equipment branch to match that of Lancashire and generate new devices and techniques.

Along with this, the Jura moved to mechanize the rough work and, so doing, took a very different path from Geneva, with its low-wage hand labor in Savoy. The first move in this direction actually took place across the border in France, where Frédéric Japy of Beaucourt (pays de Montbéliard), after three years of apprenticeship in La Chaux-de-Fonds, opened a small shop in 1771 and began making rough movements for watchmakers in Neuchâtel. Japy's business seems to have flourished, and in the best Jurassien fashion, he was soon taking in apprentices. We do not know what kind of tools he used, and we have no reason to believe that they were any different from those generally employed in the industry. But Japy had bigger ideas, and in 1776 he took a step that was to change his life, that of the region, and the character of watchmaking in the Jura. He purchased for the enormous sum of six hundred *louis d'or* from Jean-Jacques Jeanneret-Gris of Le Locle (or contracted with Jeanneret-Gris to build for him, on his [Japy's] designs) an assortment of machines for the mass production of a standard *ébauche*. (Swiss and French historians have different versions of this transaction.) By 1780, we are told, Japy was employing and housing some fifty "apprentices," plus numbers of journeymen, and turning out 43,200 pieces. This figure seems too high to one biographer, who suggests that we halve that to get closer to the truth. But even 20,000 represented an enormous output in the technological and economic context of the day. Where Japy, like other rough-movement makers, had been selling his work by the piece, like so many eggs, in the markets of Neuchâtel, he was now able to supply at unbeatable prices a steady stream of standardized material to the *établisseurs* of the district. As sales grew, so did the size of the plant, which became in the 1780s a power-driven factory. By 1795 Japy was making over a hundred verge-escapement *ébauches* a day—40,000 a year, or about ten times as much as Voltaire's *fabrique* at Ferney (1770) had turned out with twice as many workers. Ten years later annual output was up to 100,000, at the remarkably low price of two and a half francs—a day's wage for an unskilled worker, a tenth or less of the total value of the finished watch. Meanwhile, the Neuchâtelois were getting uneasy about their dependence on a French source of supply and were founding their own movement factories, beginning with Fontainemelon in 1793. Fifty years later, this enterprise, together with Japy, supplied the trade with half a million movements.[11]

The availability of uniform, if not identical, *ébauches* meant savings at all subsequent stages of manufacture, from the finishing of the movement to its casing. The implications for costs and competitiveness were the subject of a prize-winning essay at Geneva in 1818:

> In the mountains of Neuchâtel the workers work longer, and women and young people play a bigger role than with us. To be sure, the cost of living is higher there, because the land itself is poor and everything has to be brought in from outside. Their real ad-

vantages consist in the meantime in the orderly coordination of the
various work processes. For every kind of watch they use the ever
fully identical rough movements of Japy. The moving parts are always neglected in the work that comes from Savoy, so that [in Ge-
neva] the *repasseur* has to make the necessary corrections. For that
reason, [in the Jura] the various parts and accessories [*fournitures*]
are easy to match and, what's more, at the cheapest price imagin-
able; for since everything is planned out in advance, every piece is
made the same and costs proportionately less than with our movements, which we must constantly vary. They make there an endless
series of cases to fit one and the same plate size, knowing full well
that any and all watches of the appropriate caliber will fit. The
same holds for dials, hands, springs, and so on, with everything
costing less than variable parts would. Thus, a worker knows, when
he takes a rough movement in hand, what he has to do with it. He
loses no time checking and fitting; he works more surely and that
much better. The same holds true for the workers who take over
from him, so that without any more division of labor [than we], but
by means of a different distribution, the Neuchâtelois turn out better watches for less money . . .

Since things are so arranged there that there is no dead time,
the worker accustoms himself to a steady pace. This time is well
paid, but this too makes possible lower prices, since the worker
is being paid only for actual work time and not also for waiting
time.[12]

By the early nineteenth century, then, the Jura industry was
turning out silver watches for about twenty francs on the average
(gold watches cost six times as much). It is hard to compare this
with the Geneva price, which we must infer from estimates of
output and overall value. These show a mean price of one hundred fifty francs around 1815, but this figure includes a prepon-
derance of gold watches. The fact, however, that Geneva concentrated almost entirely on the higher-priced pieces (one source says
that they made up nine-tenths to eleven-twelfths of total output)
and left the lower ranges to the Jura more or less tells the story.[13]

Both Geneva and the Jura could undersell the British, in spite
of the fact that they had to import such raw materials as brass
and steel (the latter from England), and that wages, at least for
skilled adult males, had now more or less caught up with those in
Britain. It was not only that they pushed division of labor at least

as far, or that they made far more extensive use of women and children. They were also beginning to make a different, cheaper watch.

It was cheaper to make because it had no fusee wheel and chain. These were indispensable in the traditional, verge watch; but the cylinder escapement, unlike the verge, did not require compensation for the declining force of the unwinding mainspring; the cylinder ran slower when spring tension was greater and vice versa. It was a commonplace, moreover, that if one made the mainspring long enough and used only a portion of it, one obtained substantially even force over the length employed. The French were the first to exploit these possibilities: they began making cylinder watches with *going barrel*—that is, with a toothed barrel geared directly into the wheel train. And when the Swiss began making cylinder watches, they followed suit. (The British, we shall see, persisted in their old ways.)

Such a watch was also thinner, and hence took less metal to case. The verge watch, with vertical escapement, full-plate layout, and fusee wheel, was relatively thick—say, two centimeters at the least, depending on complications and the character of the case. Slimmed down, it lost much accuracy, for slimness called for a small crown (escape) wheel, and the fewer the escapement teeth, the greater the effect of any irregularity in cut or train. Big verge watches were no problem so long as they were hung around the neck or suspended from the belt, as in the seventeenth century. But when it became customary to wear them in vest or trouser pocket, thickness became an inconvenience. The new pocket fashion seems to have come in in the last quarter of the seventeenth century and to have been connected with the introduction of the balance spring: now that watches could be more accurate, it paid to find ways to minimize shaking and shock—at least, this was the counsel of Huygens.[14] It may also have been a useful precaution against watch theft, which seems to have assumed epidemic proportions at that time.

Fashion abhors bulges. Over the course of the eighteenth century the trend was to ever thinner watches, especially by those makers who catered to a more modish clientele. Nowhere was this trend more marked than in France, perhaps because of the value placed on sartorial elegance (more than in England?)—even more, probably, because of greater willingness to sacrifice time-keeping to fashion; so that at mid-century, French verges were characteristically thinner even than British cylinder watches, with their horizontal escapement.[15] And when the French finally did get around to using the cylinder or the comparable virgule escapement, they learned to make an entirely different kind of watch—flat, wide, designed to slip easily in and out of the pocket. In short, a modern watch. (See Figures 23 and 24.)

The innovator here was Jean-Antoine Lépine (L'Epine), who in 1770 invented the layout that has been known by his name ever since: separate bridges rather than full plate, balance and cock set to the level of the bridges rather than on top of the plate, going barrel instead of fusee.[16] (See Figure 26, left.) Lépine's model was soon copied with modifications by Breguet, and between them, they set the standard for all the others. Beginning in the highest circles of the aristocracy and working down through the bourgeoisie, the word was soon out that anything else was out of style. Even the republican leaders of the United States got the message: we have a letter from George Washington in Mount Vernon to Gouverneur Morris in Paris that spells it all out:

> Mount Vernon, Nov. 28th
> 1788
>
> Dear Sir,
>
> I had the pleasure to receive your letter of the 12th by the last mail.—I am much obliged by your offer of executing commissions for me in Europe, and shall take the liberty of charging you with one only.—I wish to have a gold watch procured for my own use (not a small trifling nor finically ornamented one) but a watch well executed in point of workmanship and about the size and kind of that which was procured by Mr. Jefferson for Mr. Madison (which was large and flat).—I imagine Mr. Jefferson can give you the best advice on the subject, as I am told this species of watches, which I have described, can be found cheaper and better fabricated in Paris than at London . . .

To this end Washington sent him a draft on a London bank for the substantial sum of twenty-five guineas, saying he would send more if necessary, "for I would have a good watch." Some months later Morris replied:

Paris 23 Feb. 1789

His Excellency Genl. Washington.

Dear General,

Upon my arrival at this Place I spoke to Mr. Jefferson on the Subject of your Watch. He told me that the man who had made Maddison's was a Rogue and recommended me to another, Romilly. But as it might happen that this also was a Rogue I enquired at a very honest Man's Shop, not a Watch Maker, and he recommended Gregson. A Gentleman with me assured me that Gregson was a Rogue and both of them agreed that Romilly is of the old School and he and his Watches out of Fashion. And to say that of a Man in Paris, is like saying he is an ordinary Man among the Friends of Philadelphia. I found at last that Mr. L'Epine is at the Head of his Profession here, and in Consequence asks more for his Work than any Body else. I therefore waited on Mr. L'Epine and agreed with him for two Watches exactly alike, one of which be for you and the other for me.[17]

It is not given to many of us to be immortalized by having our name enter the language. In the watch industry a *lépine* is any open-face watch. Yet as with other precursors (Harrison, Le Roy), it was not Lépine's model that became the standard for his successors. Rather it was the version devised by Breguet that set the style, within and without—movement, case, dial, and accessories. (See Figure 26, right.) It set the standard for half a century and even today invites imitation. A Breguet watch was to the older style of timepiece as an aerodynamic car of the 1980s to the boxy vehicles of our grandfathers' generation: there is merit in both, but one is unmistakably old hat and the other modern.

For reasons that we shall consider later, the British were slow to respond to the challenge. At their best, their eye was fixed as before on precision and reliability. Their finest pieces were their pocket chronometers, and for those who could not afford these but still wanted an accurate watch, British makers preferred the relatively new (1782) duplex escapement.[18] (See Figure A.3.) In both instances they used full-plate layouts with fusee, and produced a substantial timepiece whose weight and size—up to half a pound and a good inch thick—bore witness to its seriousness. More modest and less demanding customers could still have the verge watch of the old style or, increasingly, the new rack lever of Peter Litherland, both of these with fusee and full plate.[19] (See Figure A.4.) The cylinder, the pride of British watchmaking in the eighteenth century, was largely abandoned.

On the continent, however, the cylinder with going barrel—that is, without fusee and chain—was the answer to the demand for a cheap, thin watch. Whatever the French could do, the Swiss could do for less. Prices fell drastically around the turn of the century, once Japy at Beaucourt and the other mass-production *blantiers* began turning out *ébauches* for what had become a standardized caliber. Not long after that there appeared the first special-purpose machines for making steel cylinders and cylinder scape wheels.[20]

The Swiss also learned to make their watches thinner and thinner, until some works (movements and plates) measured little more than one millimeter and the whole watch only two or three. This was fashion pushed to an absurd extreme, and such wafers were never produced en masse; rather they became in the 1840s a kind of *chef-d'œuvre*, test pieces to train the best workmen to the highest skills of hand and eye.[21] Even so, the run-of-the-shop work was thinner than anyone else could or would make, and the very best Paris houses—even Breguet—found it necessary to import more and more of these movements from Switzerland.

Note that the verge watch did not disappear. It had the virtues of familiarity, and certain conservative clienteles liked its size and substance; so the hamlets of the Vallon de Saint-Imier and the Franches Montagnes continued to make verges right into the second half of the nineteenth century.

That was one of the great strengths of the Swiss industry: it was really a congeries of subbranches, of local *fabriques* specializing in watches of one or another variety or in one or another stage of manufacture. Whatever you wanted, someone somewhere could make. No run was too small, no order too special. As a result, the industry was able to cater to all markets, to experiment with novelties, to copy and exploit the inventions of others. We have already seen that the Vallée de Joux (especially Le Brassus and Le Sentier) was the center for complicated pieces. Le Locle and its satellite Val de Travers specialized in chronometers, among other things; La Brévine made chains for verges and other fusee timepieces; the Val de Travers, especially Couvet, made the best special-purpose tools.[22]

Fleurier and its environs made watches for the Chinese market. These were, to begin with (pre-1815), no different from the complicated, ornate pieces produced for a certain wealthy European clientele: watches that played music, automaton watches with little figures that moved, watches in heavily decorated cases of polychrome enamel with and without jeweled ornament. They were, in other words, portable versions of the ornate clocks that had always appealed to the Chinese mandarins. They were a specialty of Geneva, where even those British makers who sold to the Chinese had to come for their cases. By the second quarter of the nineteenth century, however, a specifically Chinese watch emerged, in part owing to the invasion of the market by the Jura industry. (See Figure 27.) It was typically fitted with center jump seconds—the Chinese apparently liked to see the sweep second hand skip around the dial—and had distinctively engraved and shaped bridges. The watches normally came in identical pairs (again a Chinese preference), which was not easy, since that meant duplicating the hand-painted enamel cases with their profusion of flowers or landscapes or portraits of pink-cheeked, saccharine blondes. As in other specialties, the entrance of the mountain Swiss was based on the production of cheaper models designed to sell to a wider market.[23]

Geneva, unable to compete with Jura prices, retreated into the upper end of the market and focused on the watch as jewel and ornament. It was Geneva that made the most beautiful and the most grotesque timepieces, fitted in cases of extravagant shape (musical instruments, fruits, animals, pistols that pretended to fire, pseudo-keys and pocket knives, and on and on); or painted with copies of great paintings or portrait miniatures; or set in small utilitarian objects such as snuff boxes, *bonbonnières, nécessaires,* spy glasses, and what have you; or lavishly set with jewels; or combinations of these. Some of these Geneva cases are of a quality that recalls the great enamels of the seventeenth and early eighteenth centuries. Others were crude copies of standard models. Some of the most skillfully painted were monuments of bad taste. All of them cost a small fortune and command high prices on today's antiques market. *De gustibus . . .*[24]

Musical and automaton watches deserve more than a passing mention, for they constituted a particularly remunerative branch of the Swiss industry. The musical watches played what passed for melodies (rarely recognizable, at least today) on the hour or on demand. The automaton watches showed little moving figures (the knife grinder, the tightrope walker, Moses striking the rock to bring forth water) that kept time with the balance or seemed to strike the hours and quarters like the *jaquemarts* (jacks) of a cathedral clock. One bizarre version showed a barking dog frightening a goose or swan: when the repeater plunger was pressed, the watch sounded the hours in simulated barks rather than chimes. Sometimes the two types were combined, making a musical toy that also happened to keep time. It goes without saying that these toys for adults were expensive; but they never lacked for buyers, nor do they now, for there always will be men who will be boys.

Costliest of all were the erotic watches, the latest in a line that went back to the Rubensian miniatures of rubicund goddesses on the backs of some seventeenth-century enamel watches. In those days, the prurient watch owner had to get his satisfaction from a little flesh and a lot of imagination, but the liberalism and license of the eighteenth century opened the way to far more explicit representations of sexual activity, often showing priests and nuns, by way of anticlericalism. These were still paintings, usually crude, but sometimes very well done; the Geneva erotica were the best. In the early nineteenth century, however, the new vogue for automata led to a logical transformation: from stills to animation. The rhythmic oscillation of the balance wheel was peculiarly suited to these simulacra, whose resemblance one to another would indicate that the moving figurines were cast by a few specialist suppliers. The tiny crudeness, even grotesqueness, of these representations would seem to limit their erogenic value: they are more amusing than arousing. But they must have made marvelous openers for conversation between the sexes, and I suspect that many a Casanova used his secret pocket peep show—these scenes were usually concealed behind special panels or covers—to test his partner's interest and open-mindedness.[25] (See Figures 28 and 29.)

Few of these form watches and automata had anything to recommend them horologically, which no doubt persuaded some that the Swiss did not take time measurement seriously. Yet nothing, as we have seen, could be farther from the truth. It was just that the Swiss were able and willing to make the kinds of timepieces that would sell, and they were perfectly aware that not everyone wanted a watch for the same reason. The watches they made for the U.S. market, often imitations of English models, reflect this desire to please. The fashion of the second quarter of the nineteenth century was *taille-douce* engraving on dials and cases. The British watches exported to the United States show unidentifiable buildings and landscapes, well done, attractive, but devoid of associations. The Swiss, on the other hand, would show the kind of scene designed to warm an American heart and loosen an American purse: an eagle screaming above a spray of stars; the military academy at West Point; a steamboat on the Hudson.

The same was true of watch movements. The Swiss learned that American buyers judged watches by, among other things, the number of jewels. The more jewels, the better the watch. This was by no means necessarily true, and indeed there was some reason to believe that there was such a thing as too much jeweling; at least, many of the best horologists of the time were convinced that not only did the center and third wheels turn too slowly to need stone bearings, but the lateral force of the train might easily crack them. (For this reason, some of the best watchmakers of the century, including Charles Frodsham in London, Henri Robert in Paris, and Edward Howard in Boston, refused to jewel to the center. Hence also the phenomenon of the sixteen-jewel watch: one center jewel for show on the back plate, but none for the other pivot in the dial plate.)[26] The Swiss, though, never hesitated. From the 1820s on, when the manufacture of watch jewels had become well established on the continent, the Swiss used them more than anyone else and made it a point to announce their number by engraving it on the *cuvettes* (inner back covers) of their timepieces. Indeed, for a number of makers the *cuvette* became a kind of billboard to specify the virtues of the watch: type of escapement, number of jewels, antishock devices, and so on. The marketing concept may not as yet have been defined, but the Swiss were already aware of the advantage to be gained by instilling pride of ownership.[27]

The Swiss were also the first to supply quantities of watches in smaller sizes, for women. Part of their advantage here was technical: just as it is easier to make going-barrel movements thin than those with chain-and-fusee drive, so it is easier to miniaturize them. (It is no accident that the smallest pre-balance-spring watches had stackfreeds rather than fusees.) But even more important, I think, was the quick Swiss perception that this was an untapped market—a perception that owed much no doubt to the Swiss experience with ornamental watches. (My own sense is that neither the Swiss nor their competitors thought as yet that women could really be interested in the time, except in the most approximate sense. Women's watches usually were fitted with inferior movements—inferior not only to those of men's watches but to what they could be.) The British were very slow to make for this market, and when they finally did begin around the middle of the nineteenth century, the Swiss were always far ahead: their watches were smaller and lighter; their cases were prettily decorated with enamel and simulated (or real) gemstones; or were painted with flowers or scenes of snow-capped mountains or the castle of Chillon. Often the watches were matched to brooches, so that they might alternately be worn around the neck or pinned to a dress. All of this was obviously much more attractive and romantic than the plain, well made, solid British lady's watch of those years.

In short, the Swiss made watches to please the customers; the British made watches to please themselves.

Low prices and energetic marketing translated into rapid growth. In 1752 an occupational census showed 464 watchmakers in Neuchâtel, almost all of them in the area of La Chaux-de-Fonds and Le Locle; ten years later, the number was up to 686, concentrated for the most part as before, but now showing up in villages all around: La Sagne, La Brévine, Les Ponts, Les Brenets. The next generation saw a combination of expansion and attraction: in 1792, the principality numbered 3,458 watchmakers, and roughly one quarter of the entire population consisted of outsiders, the vast majority of them drawn by the new industry.

All this time the *fabrique* was continually ramifying. Like Geneva, the major mountain centers were always looking for cheap labor to perform the preliminary and ancillary tasks of manufacture. The geographer Suzanne Daveau speaks of "a veritable hierarchy among the watchmaking districts."[28] (See Figure 32.) Thus, the *établisseurs* of La Chaux-de-Fonds moved outside the principality in the eighteenth century, down the Suze River into the canton of Bern, seeding the villages of St-Imier, Courtelary, and Cortébert; that is, they moved downstream in terms of topography and upstream in terms of stage of manufacture. Then, when these Bernois learned the trade, they became *établisseurs* in their turn, planting the industry in the hamlets of the Franches Montagnes, and that is when they started making real money. Meanwhile, Le Locle moved in the opposite direction, into the Val de Travers, the principal center of the lace manufacture, into Couvet, Môtiers, and Fleurier; and this last, as we have seen, became in its turn the center of production for the Chinese market.

Lacemaking posed no threat to the imperialist watch industry. Watchmaking paid better; what's more, it provided employment for the entire family, whereas tatting was defined as a female occupation. "From the age of eight or nine," the French noted at the beginning of the nineteenth century, "children can earn much more at it than the cost of their keep."[29] With division of labor pushed almost to the limit, with one hundred fifty workers contributing in one way or another to the making of a watch, there were jobs for everybody.[30] The only people disqualified were those whose hands perspired and stained the metal.[31] Even those whose vision was deficient could be helped by the use of loupes and water lamps, which focused the light on the work. A report of 1809 on the canton (in the French sense) of Courtelary (Val St-Imier) speaks of "at least six thousand workers in the watch trade, or more exactly, of the entire population in the watch trade . . . men, women, and children of both sexes, all of them from the age of eight occupied at some kind of industry, particularly watchmaking."[32]

The effect of this expansion of the mountain industry was a degree of prosperity for the watchmaking districts that far surpassed that of similar overpopulated areas both in Switzerland and outside. Already in the mid-eighteenth century, the affluence and chic of the Haut-Jura were a wonder to visitors:

They lack absolutely nothing of the necessities and even the amenities and conveniences of life. The neighboring provinces—the Franche-Comté, the bishopric of Basel, and the lower part of their own country—supply them with what nature has not provided and furnish the wheat, the wine, the vegetables, the fruit, and generally everything that is to be found in happier, more fertile countries. The assurance of being able to sell at a good price all sorts of commodities of necessity and even of luxury in these parts, which industry and trade have so enriched, brings in throngs of purveyors. Some of the inhabitants do a profitable trade in these various commodities. The same things hold for those articles that serve for the interior decoration of houses and for clothing. Visitors to the area are surprised to see very well furnished apartments and people of both sexes dressed with as much elegance and lavishness as in the big cities.[33]

Three quarters of a century later the watchmaker Jacques-Frédéric Houriet, in the memorandum he prepared for John Bowring, traveling reporter to the British government, gave expression to his satisfaction on the same point; looking back on the years since 1750, he wrote: "Meanwhile the population had increased three-fold, independently of the great number of workmen who are established in almost all the towns of Europe, in the United States of America, and even in the East Indies and China. It is from this period also that dates the change which has taken place in the country of Neuchâtel, where notwithstanding the barrenness of the soil, and the severity of the climate, beautiful and well-built villages are everywhere to be seen, connected by easy communications, together with a very considerable and industrious population, in the enjoyment, if not of great fortunes, at least of a happy and easy independence."[34]

17   Nor Could He Compete with Us

THE CONTRAST WITH BRITAIN could not be more complete. There the clock and watch trades were already in trouble when His Majesty's chief minister, William Pitt (the younger), looking for ways to finance the war with France, decided in 1797 to place a use tax on clocks and watches. Use taxes, it should be noted, are a particularly obnoxious charge because they keep biting. The effect of Pitt's levy was to drive timepieces into hiding and shrivel demand. Even those who could afford the tax refrained from buying; at least, that is what witnesses told the House of Commons. John Gregson, supplier to the carriage trade from his shop in Bruton Street, Mayfair, complained that business had declined by more than half. John Grant—"the clocks and watches that I make are of the best quality, none of the common sort"—said he had suffered a similar loss. Such makers as these sold off stock and simply stopped ordering work from the finishers (the equivalent of the Geneva *cabinotiers*): Whitfield Greenwell, specializing "in the repeating line, and the very best kind of work" testified that he had not been given a single watch to finish since the tax went into effect. Who wanted to have to deal with excise officers?

Fortunately, Pitt relented the following year. The story has it that a delegation of watchmakers convinced him of the error of his ways by dramatically contrasting the value of the materials that went into a watch and the price of the finished product. More persuasive may have been the observation that the tax was producing far less revenue than had been expected. For one thing, fewer watches and clocks were being sold. For another, it was not easy to persuade people to report their timepieces. Pitt was later to complain about the "evasion, fraud and meanness which have struggled to defeat the operation of the assessed taxes"—surprising naiveté for so canny a politician.[1]

In the decade that followed, the industry revived: war kept the Swiss rival at bay; it also generated new incomes and new demand. But by 1810–11 the hectic prosperity was over. The figures on watch cases stamped at Goldsmiths Hall in London show output of gold watches peaking in 1809, leveling off in 1810, and falling by 27 percent the following year. Silver watches held up better, with assays off only 15 percent, but by 1813 they were down by almost a third from the peak year of 1808. These were not easy times for a deferrable commodity: harvests were poor; food prices, at their peak; and those smaller men who had begun to think that a watch was a necessity rather than a luxury had little money left over for even a cheap timepiece. British watchmakers and craftsmen began petitioning Parliament for relief.[2]

The end of a generation of continental wars aggravated the crisis. For years the British market had been more or less protected from Genevan and Swiss competition by blockade and counterblockade. These artificial impediments to trade had caused great harm to the Swiss mountain manufacture, which lived by export. For a while, unemployment was so rife in the Jura that the public authorities tried to convert watchmakers into instrument makers. That should have worked, but it did not: the local industry had been built on private initiative and did not take kindly to development from above. Instead, the watchmakers kept busy on the land and on the roads, holding themselves in readiness for a change in market weather. When it came, output quickly responded, but no sooner had it picked up than it turned downward. The years of 1816–17 were marked by harvest failures and famine on a scale unknown since the seventeenth century. All Europe was affected, and the high cost of food reduced the income available for such postponable purchases as clocks and watches. Swiss makers and merchants had to flog their timepieces that much harder, putting further pressure on foreign competitors.

In Britain, meanwhile, peace brought demobilization, unemployment, reduced demand, renewed competition from European manufactures. This last was no problem in those branches (cotton and wool textiles, chemicals, iron and steel manufacture) that had been technologically transformed: there Britain's lead was greater than ever. But in watches, Britain had no technological advantage over the Swiss. Both countries used essentially the same tools; both had achieved their greatest economies by pursuing the division of labor; and while Britain still made the best steel, the Jura *fabrique* found compensation and more in lower wages and better organization.

This Swiss competition was felt in both the British home market and in third countries. In the latter, the Swiss were ready to supply either their own increasingly popular styles or, for conservative buyers who wanted watches like those their fathers wore, perfect imitations of the British product. British makers had apparently contributed to this debacle by selling uncased movements and parts, which the Swiss soon learned to copy; or by making complete watches on a contract basis and selling them either without signature or with the kind of pseudonym that anyone could appropriate. One such maker kept as a bitter souvenir of what he called his own folly the foreign copy of the watches he had once made. He had produced some twenty thousand pieces, he said, and had sold them for little more than cost; now his erstwhile correspondent was making them and he was bankrupt. Another recalled the man who first made watches with the name Wilter—a name that collectors today associate with so-called Dutch-style watches of the eighteenth century. In those days, before trademarks could be patented, a fictitious name like Wilter was anyone's game, and the first watches by "Wilter" were followed by others of greatly inferior quality, with sham jewels and fraudulent date wheels. The last such "Wilter" I saw, the man said, was offered to me at thirty-four shillings but was really good for nothing; whereas the earliest ones were worth the eight guineas they cost. And so, he said, one more market was lost.[3]

In spite of legal prohibitions, moreover, the British exported those machines, tools, and files that had laid the material foundations of British large-scale watch manufacture in the first half of the eighteenth century. Stop the diffusion of these instruments of production? As well try to stop the tides. All a good Swiss maker needed was one example of a British machine and he could make his own with some improvements thrown in. As for files, they were thinner than pencils and could be as easily concealed as the contraband watches going in the other direction.

Such watches, the British makers affirmed, were substantially inferior to the genuine British product. Perhaps so; although British makers were not above turning out their own rubbish. In any event, the leading London makers declared, Swiss pieces were "so much cheaper than any that can be made in this country, as to preclude all competition." All that was left of a once flourishing export trade was bits and pieces in protected territory: English-speaking areas such as the United States and imperial possessions such as India. Yet language and habit could not long hold back the flood: the American market was going fast. One watch merchant explained the rout by what would one day become the standard refrain of the hard-pressed British manufacturer: the customer was wrong. The Americans, he noted, "have always been fond of cheap articles . . . any thing of a decent watch there, is very little called for."[4]

Third markets, then, were gone, and as the Clockmakers' Company recognized, no legislative enactment could restore them. But perhaps something could be done to regain the home market. This, too, was falling to the Swiss, in spite of heavier protection than that enjoyed by almost any other British industry. The watch manufacture was an anomaly: at the height of the Industrial Revolution, when British techniques were far in advance of those of any other country, when Britain had become workshop of the world and pressures were building for the elimination of import duties in all other branches, the watch trade was calling for higher and more effective barriers. The irony was that even if its wishes had been granted, it would still have been in desperate trouble. Watches were too easy to hide and watch duties simply too hard to enforce. Smuggling had become an organized, institutionalized trade. Maybe prohibition might have helped, with body searches of passers-by and house searches without warning, but these were not a feasible recourse in Britain. In the meantime, anyone who wanted to smuggle without risk merely had to pay 10 percent of the value of the contraband as insurance. This was slightly higher than the premiums charged for shipments across the Jura into France, but was well under the British duty of 50 percent. Under the circumstances, one had to be something of an idealist to import legally, and the trade had few of those. No wonder that witnesses to the Parliamentary commission of 1817 thought there was scarcely a Swiss watch in Britain that had passed through the hands of Customs—except those that had been seized and then been thrown back on the market at bargain prices.[5]

Swiss competition was the principal source of difficulty, but there were domestic problems as well. Low-price competition induced British makers to seek their own cost savings, which in so labor-intensive an industry meant cuts in wages and the sacrifice of traditional quality standards. That kind of strategy had to be self-defeating, though, for the British watch industry lived on the claim (the pretension) to higher quality. The perceived difference between the British and the foreign product—that difference that was supposed to justify a substantial price gap—narrowed accordingly. What made sense, then, for the individual enterprise had a negative impact on the branch as a whole—an example of that potential contradiction between micro- and macrointerests that economist Thomas Schelling has wittily analyzed for us.[6] Not everyone was prepared to take this path. Yet except for a few high-prestige London houses that could count on a faithful carriage trade, those who were too proud or conscientious to change fell by the wayside. These were times of what Barraud contemptuously called "two-year watches," of fraudulent signatures, of movements with sham second hands. To deceive the assay offices, cases were made with only the back cover of proper fineness, since that was the part tested; the rest was made of inferior gold or silver. (Collectors have run into these watches: they show the stamp for eighteen carats, but after almost two centuries they have the tint of copper and show the pits and oxidation that good gold is immune to.)[7]

In the search for a scapegoat, much was made of the alleged malpractice of Jewish makers and traders. The committee of 1817 made it a point to ask just about every witness about this aspect of the problem, often in leading language that would have been thrown out of a court of law: "Do you know whether a considerable number of inferior watches are manufactured by the Jews?" "Do you understand that a great number of inferior articles are manufactured by Jews and other persons of a low description?"[8]

Such questions, presumably accompanied by encouraging look and intonation, elicited an occasional anti-Semitic response. Thus James Storer of Islington Road, London, argued for authentic signatures on watches, for "if we were all obliged to put our own names upon the watches, we should do away with the Jews; there are very few gentlemen would buy watches with the names of Moses or Levi on them."[9] But most witnesses, while agreeing that Jews were active in the manufacture and sale of cheap watches, recognized that non-Jews were doing the same thing; and some noted that there were Jews who made good timepieces. This preoccupation with unfair Jewish competition has yielded us some anecdotes of the shadier practices and "scams" of the day: the mock auction, for example, which was just a device to induce suckers to spend more for less; or the spendthrift's lament ("I spent six guineas for this watch, which is really worth much more; and now I'm broke and need money in a hurry, and you can have it for half that"); or the beat-out-the-Jew ploy ("This Jew is pestering me for this watch and wants to give four pounds; but I can't stand him, so I'll let you have it for three"); and similar "stings" designed, as always, to appeal to the greed in us. The essence of every con game is the victim's eagerness to get something for nothing.

Most of these reports, though, were hearsay—the kind of cautionary tales that told of other people's misfortunes. Witnesses made the point again that these practices were not confined to Jews; indeed a warning proclamation by the lord mayor of London implies that non-Jews (or Jews) could exploit anti-Semitic stereotypes to "set up" their victims. The game, designed to bilk the seafaring population, called for two men, one dressed as a sailor, the other "who personates a Jew." The sailor pretends to have paid twelve or fourteen pounds for a watch (a very large sum), which the "Jew" wants to buy. But the sailor says the "Jew" has robbed him before and shan't have it at any price; while the "Jew" whispers to the "mark" that he'll give him one pound profit if he buys it for him. So the victim pays six, seven, or eight pounds for a turnip worth a third or a quarter of that; and when he looks for his "Jew," the "Jew" is nowhere to be seen.[10]

Hard times make hard men. Manufacturers of every faith and persuasion were compelled to offer less and demand more of their workmen, while the latter took what they got or starved. Craftsmen who were used to getting two pounds a week for their labor now received a third of that or less for their efforts. Watchmakers, once the aristocracy of highly skilled labor, became a byword for poverty, to the point where some of them could not move to less expensive lodgings: no landlord wanted them. Some of the worst stories come from Coventry. There the British industry had attained its logical extreme of cheap labor, division of tasks, reduction of skill to the small amount required for one of 102 specialties. There were no qualified journeymen in the Coventry watch trade, except for a small number of indispensable finishers. Instead there were only so-called apprentices, beginners who worked for seven years in their own homes and were then put out to pasture while their places were taken by newcomers. When demand contracted, these unfortunates turned to manual labor or emigration (the inquiry makes repeated allusion to the danger such emigrants represented to Britain's competitive position in the watch trade). No job was too rough: mending roads, cleaning a mill dam, emptying privies. John Nightingale, a "very good workman at the watch trade," was one who sought and got this last job. The overseer of the poor pleaded with him not to undertake a task for which he was so unfitted; but Nightingale was desperate and insisted. For all his determination, he was overcome and could not complete the work.[11]

Taken together, the witnesses of 1817 presented a uniformly gloomy picture—to no avail. There was little the government wanted to do on the side of protection: the current of opinion and policy was moving the other way. Besides, the watch trade itself was of no single mind on this matter. Much depended on whose ox was being gored. The journeymen and subcontractors who produced watches and parts for the "manufacturers" and dealers complained bitterly, as did those makers who sold second- and third-class watches to miners in Wales and farmers everywhere— the so-called country trade. For them, foreign watches were tak-

ing bread out of the mouths of Englishmen and killing a manufacture that was the nursery of the highest forms of technical and
craft skill. "Name" makers and dealers, though, were understandably less exercised. (Except for Barraud, none of them appeared
in person to testify in 1817.) They were used to full order books
and had always bought in watches from outside to meet excess
demand. If sales were slower for the moment, wages were lower;
and better times would (and did) come. The leaders of the trade
comforted themselves with the conviction that the British made a
better watch, that anyone who wanted quality would be ready to
pay more for the British article; and with the thought that they
could always sell Swiss goods at a healthy markup to those less
demanding.

All of these sentiments show in a pamphlet of 1842 by Edward
J. Dent, sometime partner of John Roger Arnold and later to
achieve horological immortality as builder of Big Ben. Dent wrote
that the English watch continued to enjoy an unmatched reputation for "stability, soundness, and consequent accurate performance"—in short, the kind of watch that gives good service from
one generation to the next. Unfortunately, "the caprice of fashion" would cut its life short: "Its size, thickness, inelegance, and
the weight of the case, whether in gold or silver, present an insu-
perable objection to the adoption of the old servant, by the young
man who studies appearance." (The British, we have seen, were
not strong on watches for women.) The heavy British watch was,
of course, worth more: the gold in a British case, he wrote, was
worth seven pounds or more; that in a Swiss watch, only two. This
was surely an exaggeration, for the British were not above shaving
gold content, as we have seen. It was true, though, that the Swiss
had learned to make some very thin gold cases, meretriciously
adorned and most attractive when new, yet so weak as to show the
mark of every knock.

Dent went on to argue that this international division of labor
and market would last indefinitely: "As for watches of a superior
make, in which performance of strict accuracy is required and the
highest degree of perfection is attained, these the Swiss artist has
never attempted to manufacture to any extent. The reason of this
may be the consumption of time required in completing the adjustments for variable temperature, &c.;—a sacrifice so consider-

able, that the additional labour of the artisan would necessarily demand an augmentation of price on the instrument, and thus render it less suited for a general article of trade—the main object proposed by the foreign makers." So the British must continue to dominate in the manufacture of pocket chronometers and what he called "the strong lever watch"—that is, a watch "well adapted to what may be termed 'hard wear.'" "Such a watch," he asserted, "cannot be supplied by the foreign manufacturer under any circumstances; nor, if he were to make the experiment, could he, the quality of workmanship and weight of the case being taken into consideration, compete with us."[12]

Some of the British scorn for Swiss watches, especially those from the Jura, was no doubt justified. We have seen that the mountain Swiss, like the Genevans before them, produced the full range of qualities. The cheapest of their timepieces were notoriously bad—even the Swiss apologized for them. The market for watches was growing fast, and the new middlemen who were moving into the trade, many of them owners of variety shops, sellers of notions, or general peddlers, had no care for the technology of time measurement and no scruple about making and selling as cheaply as possible. In the long run they were right. Cheaper watches familiarized ever wider circles with the advantage, privilege, and status of owning one's own timepiece. The fact that most of these watches gave inaccurate and unreliable service did not necessarily discourage the user from buying another; it only taught him to buy a better one.

But that was the worst of the Swiss product. At their best, the mountain Swiss could make superb timepieces, in every escapement, with every complication. Dent himself bought some of them for resale and signed them with his name. It is ironic that at the very time Dent was penning his patronizing lines, a certain Jules Jürgensen, son of the great Urban, was producing at Le Locle lever watches as good as any English maker could make— and this, in spite of the fact that they had no fusee to equalize the force of the mainspring as it wound down. To be sure, Jürgensen's watches were expensive, hardly the kind intended to undercut the British in their home market.[13] But Jürgensen was not alone. There were others—the Courvoisiers, Louis Audemars, Charles Jacot, and many more whose names have not come down to us—

who were producing in larger quantity and for lower prices, while incorporating such elements as detached escapement, compensation for temperature, jeweled bearings, and other features of a high-quality watch.

It was the combination of this complacency of Britain's master artists and the stubborn conservatism of alarmed workmen that killed the efforts of the Swiss Pierre Frédéric Ingold to introduce into England in the 1840s the manufacture of watches by machine on the principle of uniform and interchangeable parts. The idea was not entirely new: indeed, it had already found partial application in the mass production of rough movements by such makers as Japy. The emphasis, though, must be on the word "partial": Japy's *ébauches* were shipped in the rough and required finishing and adjustment before the train could be assembled and made to run. Japy could work, therefore, to wide tolerances. His parts were similar, not uniform.[14]

What Ingold proposed was far more difficult: the manufacture of plates and wheels to a fineness such that one piece of a given kind could be exchanged for another. Such nearer identity effectively excludes hand finishing, which of its nature alters idiosyncratically everything it touches. In principle, the elimination of manual skill should save money. In fact, this kind of work demands much larger capital outlays for the kind of heavy-duty, special-purpose equipment that can do the same thing time after time independently of human intervention. Even then, normal wear and tear will change the character of the work—say, the depth of cut—with every pass; so that it may be necessary to readjust machines and replace dies at frequent intervals to ensure uniform results. Under the circumstances, it may not always be possible—more accurately, it may not always pay—to insist on true interchangeability. Much depends on tolerances and quality standards. It may pay, for example, to make the easy parts interchangeable while finishing and fitting by hand those pieces that demand near-perfect articulation. (Much of the manufacture of handguns is characterized by such selective interchangeability.) Or it may pay to produce run-of-the-mill work on the interchangeable principle, while lavishing the attention of skilled craftsmen on high-priced, individualized models.[15]

Ingold's machines have not survived. Indeed, it is not entirely clear whether he ever made more than one or two samples. We are told, however, that they made use of a highly versatile chuck to hold different tools at all angles, and that the careful use of guides and stops made possible accurately repetitive work. The aim was to take each blank and shape it as required in a single machine. All of this presumably called for much changing and setting, hence for not inconsiderable skill on the part of the workmen. I doubt whether such machines would have been capable of turning out interchangeable watch parts, but a second generation would surely have done better.

Ingold never had the chance. The Swiss did not need him; neither did Japy, who surely learned something from him but was working in a world of lower standards. In Paris he was spurned by the defenders of watchmaking-as-an-art. He then went to England (1839) and tried for several years to interest businessmen in and out of the trade. The English did need him—badly—and his efforts were almost successful. In a prospectus for a "National Company for the Manufacture of Watches" (1842), he spoke of cutting costs by 30 percent and turning out two to three hundred movements a day—that is, sixty to ninety thousand a year—at a time when the largest merchant-manufacturers were finishing a few thousand. Experienced watchmakers observed his prototype machines in action and spoke well of their performance. These and others joined with him in forming the British Watch and Clockmakers' Company, nominal capital £250,000. The leading figures of the trade held themselves aloof, but who wanted them? Britain finally had an answer to Swiss competition.[16]

Not so fast. Unfortunately for the hopeful Ingold, a limited-liability company required at that time a charter from Parliament, and there the opposition of the watch trade, led by the Worshipful Company of Clockmakers, made itself felt. The charter was denied.[17] A disappointed Ingold left England for the United States, where any projector could found a limited-liability corporation for the price of a registration fee. But he had no better luck across the Atlantic than he had had in the Old World. Swiss historians and chroniclers like to think that Ingold was at the origin of the machine manufacture of watches in the United States, an industry that got under way in the 1850s after some twenty years of experiment and effort. But we shall see that there is no evidence of any link between Ingold and the technique and equipment of the first successful American watch factory—what was later to be known as the Waltham Watch Company.

So THE BRITISH WENT THEIR WAY, and this at a time when the stakes were steadily growing in importance. The big changes here were, first, the general increase of wealth, which brought watches within reach of an ever-growing number of people; and second, the construction of railways, with their emphasis on timetables and precision. The railway companies themselves and their employees were destined to become a major market for watches, but even more their riders, who not only wanted to know the hour and minute in order to catch trains but found their entire consciousness of time altered by the requirements and opportunities of a railway world. Trains not only left, they arrived at a destination; and it was just as important to passengers to know when they arrived as when they left. Train schedules opened new possibilities for appointments, for work done within time limits, for long-distance comings and goings, hence for ordering of movement and multiplication of activity.

In the early days of the railroad, each company tried to live and work, as the coach lines did, with a multitude of local times, varying as always with longitude. But trains moved too fast for this, continually exposing passengers and crew to discrepancies and confusions. The invention of the telegraph, however, first tried in 1837 on the London and North Western Railway, made it possible to transmit almost instantaneously an exact hour and minute from the central office to every point on the line. The effect was to establish a standard time for all those served by a given network. The next step was to unify railway practice: on September 22, 1847, the British Railway Clearing House recommended that each company adopt Greenwich time at all their stations "as soon as the Post Office permits them to do so." Before the end of the year the General Post Office had given instructions to that effect, so that railway Britain and all activities linked in one way or another to rapid transportation were operating on a single time. The effectiveness of the change depended, of course, on the creation of a national time service, communicating precise time signals at regular intervals to clocks and stations around the country. It was the newly developed technique of electrical time-keeping that made this possible—that, and the enterprise of George B. Airy, astronomer-royal, who correctly grasped from the start the possibilities of what he called galvanism for the synchronization of a multitude of timekeepers.[18]

What was possible in Britain was also feasible in the major continental countries: wherever there were railroads, there soon developed standard times and time services, and these in turn prepared the way for a single national hour as reference time for all. Only the largest countries such as the United States or Russia ran into difficulty, because they spanned so much territory that no single hour would do. The Americans solved this problem a generation later, in 1883, with the institution of *time zones,* which led in turn to international agreements for a similar partition of the entire earth. The only difficulty there was the decision regarding the base measurement—where to locate zero longitude. The British and just about every other nation in the world favored Greenwich, which navigators everywhere had long taken as their point of reference. The French held out for the Paris observatory, whose longitude differs from that of Greenwich by slightly more than two degrees. So in October 1884 the International Meridian Conference voted to mark the prime meridian at Greenwich, while France remained odd man out for another generation. Finally, in 1911, the French bowed to ecumenical usage and voted to establish a legal time defined as "Paris Mean Time, retarded by 9 minutes 21 seconds." But this was nothing other than Greenwich Mean Time, without the word "Greenwich," which the French have trouble pronouncing anyway.[19] Thus were national susceptibilities salved—but not solved. The French had great difficulty living by British clocks and chose to set their timepieces one hour ahead. To be sure, British and French do not shift to and from daylight saving on the same dates, so that there are some weeks in the year when their clocks coincide. But with the best of will, it is hard to disagree all the time.

None of this, it goes without saying, took place without considerable psychological and social resistance. In a timekeeping society, we are all linked by habit and mind-set to the standard of measurement (which is one reason why political authority has always sought to put its stamp thereon), and there are always some who resist chronometric changes as though they were violations of identity and self-esteem. Just as the French sought to defend Paris time, so in an earlier generation, conservatives in Scotland, Ireland, and the West of England fought to preserve the hour of their locality, or even of a given clock, denouncing "railway-time aggression" and complaining that man was once again "usurping the power of the Almighty." One town alderman waggishly warned the Plymouth council that workers would now be able to *quit* work sixteen minutes earlier, then quickly reassured it by noting that they would also *start* work sixteen minutes earlier. Some deplored all the unnecessary confusion, concocting tales of brides waiting for grooms, hosts waiting for guests, ministers waiting for congregants (or the reverse)—though it was not clear in logic whether such difficulties, even if real, were an argument for retaining the older system. Even Airy, for all he did to create a unified national standard, was of two minds on the subject: in spite of all temptations to belong to other nations, he remained an Englishman.[20] In the meantime (no pun intended), watchmakers tried to help punctuality by producing dual timekeepers—what collectors now call a captain's watch—with two dials for showing different hours. These disappeared with the institution of national standards, but they have come back into fashion in our own day to enable jet travelers to keep track of time at home and at site and prevent them from calling friends, family, and business associates at ungodly hours.

THE UNIFICATION OF TIME, of course, enormously enhanced the value of time measurement, for it eliminated all manner of confusion and of pretext for inexactitude and made possible a far more efficient ordering of activity. Watches became that much more necessary to everyday life, as the statistics of consumption make clear. Where world watch output at the end of the eighteenth century was around 350–400,000 pieces a year, it was up almost tenfold, to 2.5 million pieces, three quarters of a century later.[21]

In the course of this boom, British output had if anything slightly diminished, from about 200,000 to 169,000 pieces. The only bright spot was Lancashire, with its characteristic Liverpool watch: large, ornately dialed and cased, with an ostentatiously jeweled movement. It was a watch for the new-rich merchants and manufacturers of the North and Midlands. Coventry, with pieces mass-produced by low-paid "apprentices"—really machine operators who would never learn the trade in the traditional way—was a lesser focus of growth. But London was going nowhere. The great shops that had once supplied the Chinese and Turkish markets folded one by one. The great names relied increasingly on foreign suppliers, while Swiss expatriates set up shop in England to supply quality work that a shrinking population of English artists could no longer produce.

Even the French manufacture was doing better than the British—not in quality, but in quantity. To be sure, the once-great center of Paris was now reduced to a handful of high-priced artists. One observer of the 1830s said that the capital was not making ten watches a year, although such prestigious makers as Breguet continued to finish and adjust imported pieces, presenting them as their own.[22] It was Besançon that was the great French *fabrique:* after years of ups and downs, promises kept and broken, disappointments and defections, the Swiss colony there had finally taken root, trained a new generation of indigenous craftsmen, and—behind a heavy protective screen—more or less taken over the French home market. The second quarter of the century had been a difficult time, with output fluctuating between 40,000 and 60,000 pieces a year, almost all of them of modest quality and price. But the fifties saw a second takeoff. Production passed the 100,000-mark around 1854, rose to about 300,000 in 1864, to 395,-000 a decade later. This represented about 12 percent of world production.[23] (See Table 1.)

Lesser centers hardly counted. The Dutch manufacture was long gone; the German, reduced to a few scattered makers. One focus of growth was Dresden, where Adolph Lange had introduced a new, high-quality lever watch (the Glashütter, after the suburb where they were made) that would become the staple of the *fabrique* and achieve a world reputation.[24] But this was artisanal production, numbering at that time in the hundreds of units. The mass-production industry of the Pforzheim area (Junghans especially) would not get into watches until later.

Meanwhile the United States was just getting under way: factory watch manufacture began around 1850 but took a decade to work out the problems of machine production. In those first ten years, the United States managed to produce some few thousand watches—hardly a threat to Swiss supremacy.

The big gainer was Switzerland, and within Switzerland, the Jura. Geneva had stood still, making the same 100,000 watches in the 1850s that it had made in the prosperous 1780s. Neuchâtel kept growing: output, estimated at 280,000 pieces in 1844, almost tripled over the next decade. Delegates to the Paris Exhibition of 1855 boasted that the canton was then making some 800,000 to 900,000 pieces per year at an average price of forty francs, hence worth 32–36 million francs in total.[25] Even more impressive were developments outside the old core. The Vallon de Saint-Imier, for example, was already punchmarking some 100,000 gold and silver watches a year at the beginning of the century, and these presumably constituted only part of a substantially larger output.[26] By 1870, the canton of Bern showed almost as many people engaged in watchmaking as Neuchâtel, and the manufacture had spread into the cantons to the north and east, into Solothurn and Fribourg.[27] These newer centers, along with the Vallée de Joux (canton of Vaud), were by then turning out more than Neuchâtel, using more modern equipment, and growing faster. (See Table 2.) Altogether the Swiss were producing around 1870 something more than two-thirds of the world's watch output by value.[28]

It was the Swiss, then, who were the great beneficiaries of the revolution in time consciousness and the democratization of the watch. The British simply missed their chance—and this in the very midst of an industrial revolution that was establishing their supremacy in almost every sphere of manufacture. The more the watch trade lost ground, the more the British consoled themselves with the thought that they were right and their customers wrong. Or they took satisfaction in small things—like the British secretary of legation in Switzerland who noted with gratification in 1857 that the Swiss were sending their watches overseas in British vessels.[29] If, in this field at least, Britain could not be workshop of the world, it could carry the work of other shops.

The economic pundits should have been paying attention. The watch trade gave warning of things to come.

18 Ah, But He Could!

THE BRITISH WERE WRONG on every count, not only on the future of the watch industry but on the ability of the Swiss to satisfy demand at every level. They saw the earlier Swiss productions, the cheap cylinder watches in particular, and scorned them. It never occurred to them that so diligent and enterprising an industry, highly emulative and competitive within its own ranks, would generate a flow of innovations and improvements that would enable it to surpass its predecessors in quality as well as in quantity.

The Swiss went about their task with single-minded intensity. The earliest prize competitions for the most reliable and precise watch took place in 1790 and 1792 and rested on the most approximate and adventitious tests: the competing watches were simply worn by the judges and assigned by lot; and verification was made against a sundial reading of high noon. These competitions were repeated after 1815 at irregular intervals, utilizing ever more systematic and demanding procedures. By the middle of the century, time trials were conducted on a continuing basis both at Geneva and Neuchâtel by the new observatories (Geneva in 1829, Neuchâtel in 1858), built in large part to serve the needs of the watch industry. These trials were not strictly competitive at first:

watchmakers simply deposited their watches and specified the duration and type of tests they wanted to submit them to, with a view to obtaining a certificate of performance. But they were implicitly comparative, and they made and unmade reputations. In the 1870s the procedure was standardized; the time trials became annual competitions; and the results were published for all to see and for the winners to repeat in their advertising.[1]

To be sure, the British were conducting similar trials at the Greenwich Observatory, going back to the early 1830s. But for half a century these were reserved for marine chronometers and deck watches, with a view to purchase of the best pieces by the Royal Navy. They had nothing to do with ordinary pocket watches and hence had no impact on the industry. If anything, they were conservative in their influence, because they helped preserve the notion that the only true chronometer was a timepiece with what the British called a "chronometer escapement"—that is, a detent escapement. The Swiss chose to stick with the original meaning of "chronometer"—an instrument for the precise measurement of time—and were prepared to grant this appellation to any watch that met certain standards.[2] As a result, they learned an important lesson: that detent escapements were not the only ones to keep good time and that one could achieve chronometer accuracy in less expensive ways more conducive to mass production.[3]

In addition to these time trials, the Swiss conducted a number of competitions in invention, offering prizes for improvements in escapements, compensation for temperature, isochronism, production techniques, and work procedures. Here local *sociétés d'émulation* played a key role. The earliest of these was the Société des Arts of Geneva, founded in 1776 by a savant (H.-B. de Saussure) and a watchmaker (Louis Faizan), which began at once to offer courses in mechanics (Marc-Auguste Pictet, in 1781) and to promote import substitution by offering prizes to anyone who could equal or surpass British crucible steel and the files it made possible. It was the Société des Arts that founded the first enduring school of watchmaking in 1823, followed by others both in Geneva and the mountain centers.[4] Interestingly enough, one of the active promoters of this development was the young Louis Agassiz, later to become a world-famous zoologist and geologist at

Harvard University and remembered today by a number of points on the map of Cambridge. Agassiz proposed to the Patriotic Society of Emulation of Neuchâtel (founded 1791) a plan to "put new life into the industry": the establishment of horological libraries, museums, and societies, the institution of annual exhibitions, and above all, the creation of schools. Much of this program, especially the schools, was realized in subsequent years, indeed was already under way—with or without Agassiz.

Ingold, then, was not a maverick. He may have been ahead of his time, but he was part of an established tradition of innovation and invention, not only in the design and character of timekeepers, but also in the methods of production. Long before the Swiss were ready to accept the desirability or necessity of interchangeable parts, their recourse to highly specialized operations furnished a stimulus to and opportunities for mechanization. For obvious reasons, it was the early stages of manufacture—the production of *ébauches*—that were the first to be affected, beginning with Japy in the 1770s and 1780s. The efforts to found similar firms in Geneva failed: the Genevois, we are told, did not take to machine-made products, much less to machines.[5] In the Jura, on the other hand, Fontainemelon (1793) was only the first of many mechanized (but not water- or steam-powered) movement makers, operating on a scale appropriate to the lower quality, longer runs, and more homogeneous assortments of the mountain *établisseurs*.[6]

The fine work of fitting and finishing was obviously less susceptible of mechanization. On the other hand, the very availability of cheap rough movements put pressure on the subsequent stages of manufacture, which accounted for more than 90 percent of the final price. (A verge *ébauche* by Japy, which had cost six francs fifty at the end of the eighteenth century, cost only one and a half to two francs in 1815.[7] The finished watch cost twenty francs or more.) There the division of labor (about one hundred *parties brisées* in 1870 as against fifty-four in 1830) generated a growing family of special-purpose tools (templates, gauges, depthing devices) which notably reduced the skills required and increased productivity. In the 1840s a new specialty appeared—the machine finisher, who turned *ébauches* into *blancs roulants* (running movements), ready for the final touches of *repassage* and *ajustage*, gilding and signature. One of these was Louis Japy of Berne (not the capital of the Swiss Confederation, but the French village, near Seloncourt in the Pays de Montbéliard), who turned, polished, and assembled wheels, pinions, and plates, then added the escapement and sold his *finissages* for perhaps a third again as much as the rough movement.[8] This was far less than the normal markup based on hand labor: the time of *repasseurs* and *visiteurs* came high.

Mechanized finishing, though, was essentially clean-up work done faster and cheaper. It was not manufacture on the uniformity principle. The first move in that direction (aside from paper schemes that went back to the eighteenth century) came in 1827, when the Geneva firm of Vacheron & Constantin began experiments with a standardized layout, with a view to mechanical shaping and finishing of the component parts. Some twelve years later, still working toward this end, the firm contracted with Georges-Auguste Leschot for the development of an assortment of machines to turn out uniform components for an array of watches of the same caliber but different sizes. Leschot was in large part successful, to the point where his equipment was soon turning out enough similar (though not identical) parts to supply not only Vacheron but others as well; also to the point where the firm made him a partner.

Leschot's contribution is generally defined, especially in the Swiss literature, as the introduction of interchangeable parts into watch manufacture.[9] Not so. The pieces produced by his pantographic drills and cutters were alike to fairly small tolerances, but they had to be made in soft steel and still wanted tempering, which altered their shape and yielded 25 percent or more of rejects. Leschot's machines, moreover, were quite small and light and called for hand feeding and adjustment by screw of the holding and working parts. So although they did yield substantial economies over the older techniques, they required skill and time to operate, and anything that requires skill is not going to produce the same result on every pass.[10] It was not until the invention and introduction of heavy presses and single-purpose machine tools working to much finer tolerances with tempered steel that something close to true interchangeability was achieved. This came in the United States in the 1860s and 1870s. More on that later on.

Leschot was not the only one working toward uniformity. The horological jury of the Great Exhibition of 1851 thought it worth remarking that Antoine Lecoultre [sic: for Le Coultre], a Swiss maker of Le Sentier, "is stated to make all his watches with the corresponding wheels of the same size, so that when any wheel is damaged, it can be replaced by a new one without any other trouble than that of putting it in."[11] The following year Pierre Dubois, France's leading horological publicist and reporter, repeated this account of Le Coultre in his *Tribune chronométrique:* "All [his] movements are made on the same model, so that they can be disassembled and the parts exchanged without any inconvenience."[12] We have similar reports of production of substantially uniform parts by Louis Audemars of Le Brassus and P. E. Jacottet of Travers. Clearly, mechanization and normalization were proceeding at several points—were in the air, as it were.

On the other hand, as we saw in the case of Ingold, it would be wrong to anticipate and to confuse these moves toward uniformity with the achievement of machine-made, truly interchangeable parts. Elie Le Coultre (son of Antoine) noted in his memoirs that his father's wheels were die-stamped [*se découpaient dans de belles matrices*], but then had to be finished by hand. The same was true for bridges and plates: the barbs and rough spots had to be filed smooth before an *ébauche* became a *blanc roulant,* ready for gilding. Some of these finishing steps were omitted in the manufacture of cheaper watches; what is interchangeable on one standard is not on another. According to Elie Le Coultre, Jacottet was turning out such rough components by the thousand. No doubt; but only some components: according to one observer (c. 1859), most parts of Jacottet's *ébauches* were made by hand.[13] For the rest, the product reflected these short cuts. As Elie put it, "Of genius in this fabrication, nothing; but lots of order, supervision, and with these, very good profits."[14]

As a result of these new techniques, the equipment and structure of the industry began to change. The middle decades of the century saw the appearance of larger shops, well equipped with machine tools and sometimes powered by water, or even steam. By 1870 one-tenth of the work force was employed in these "manufactories" or "protofactories."[15] With the concomitant reduction or devaluation of hand skills, moreover, it became possible to recruit labor outside the traditional watchmaking district. This was an old pattern: Daveau notes that from the start the rough-movement factories were established in peripheral areas and made use of an unskilled labor force.[16] Now the industry began for the first time to move outside the French-speaking districts, in and beyond the canton of Bern toward Grenchen (Granges) and Solothurn (Soleure), where a large reservoir of low-wage labor, including women and children, was waiting to be tapped. In the 1850s this was still a spotty, exceptional phenomenon. Two decades later, with the introduction of factory production, the spillover became a rush. (See Table 2.)

Along with new techniques (process innovation) went new features and capabilities (product innovation). The middle decades of the century saw the introduction of the fly-back chronograph (Nicole & Capt, 1862, but patented in England by Adolphe Nicole as early as 1844); the reinvention of the perpetual-calendar watch (Louis-Elisée Piguet, 1853);[17] above all, the reinvention of stem winding.

Watches, with very few exceptions, had always been wound by means of a key, which worked well enough, except that key winding was an awkward business, especially for older watch owners (a disproportionately large share of the constituency) whose hands trembled as they tried to fit the key to the winding square. There are very few French watches of the eighteenth century, when they were wound from the dial side, whose enamel is not chipped around the winding hole; just as there are very few key-wind watches of the nineteenth whose cuvette is not well scratched for the same reason. Further, although all watch owners tried to keep their keys about them, attached to their watch chain for example, I should like to have a penny for every time an owner wanted to wind or set his watch and found he had left his key at home. Under the circumstances, some very gifted watchmakers bent their efforts to find a better way. They found a variety of solutions: self-winding (Abram-Louis Perrelet in the 1770s);[18] pump-wind (Charles Viner); pull-wind (J. A. Berrollas' British patent of 1827); butterfly nuts on the winding and setting squares. But the technique that was to sweep the field was the one we are all familiar with today: a stem projecting to the outside of the watch and turned by means of a knob (crown) on its end.

The man who did more than anyone to introduce the new device was a Frenchman named Adrien Philippe, who invented his version of stem winding in 1842. I say "his version" because others had already built stem-wind watches, notably Thomas Prest, working for John Roger Arnold in London, from 1820 and the house of Breguet in Paris, working now and then to special order. But none of these earlier experiments had had any impact on the industry. It was Philippe who, after trying in vain to interest his compatriots in the new device, joined forces with a Polish émigré watch manufacturer of Geneva, the Comte de Patek, and made the stem-wind watch the last word in elegance and quality. The early Patek stemwinders proudly announced on the cuvette: "Invention et Exécution par Patek, Philippe et Cie." This partnership was the beginning of an extraordinary commercial ascension that was to bring the firm to the top of the industry in prestige and quality of production by the end of the century.[19]

Philippe's winding mechanism opened the way to acceptance, but he was only one of several working on these lines, and it was in the form devised by Antoine Le Coultre that stem winding was to establish itself over the next thirty years. It swept the field everywhere except in Britain. The irony was that, as noted above, the inventor of the first practical stem-wind watch was an Englishman, foreman to the most prestigious maker in London. Yet Arnold produced only about two hundred of these keyless watches in the next few years and then more or less gave up on them, although occasional watches with Prest's winding are found as late as the 1860s.

Was this failure due to customer resistance? Or to the inadequacies of the device? (Prest's stemwinder could be set only by opening the glass and using a key on the center square—always a tiresome procedure and an invitation to damage. But surely it would not have been very hard to remedy that deficiency.) Or to the fact that Prest's winding would not work with a fusee, which British watch owners had been taught to look on as the hallmark of a well-made timepiece? (Again, this could no doubt have been remedied, though admittedly at a price.) Or was it that Arnold and his successors did not have their hearts in the new product?

The story is told that Prest devised this mechanism in order to satisfy the special requirements of a British army officer who had lost
an arm in combat, and hence could not wind a watch with a key.
It may be that John Roger Arnold thought of a keyless watch as
some kind of orthopedic aid for handicapped people—hardly a
large and growing market.[20]

There is some evidence that conservatism of both sellers and
buyers decided the issue. In 1844 Adolphe Nicole, a Swiss maker
established in London, patented a stem-wind and stem-set device
that he used with going barrel (no fusee) in a snap-back case (no
cuvette, no hinges) for thinness and style. Nicole, setting a pattern
that later Swiss immigrants to England would follow, made
watches not so much for himself as for other makers, among them
the best in the profession. His work was superb. He could turn out
any escapement desired. His watches performed as well as any
then being made in London—to the point where they have been
put forward as examples of the best British work.[21] And he
brought with him from the Vallée de Joux the ability to make
those special-purpose calendar, chronograph, and repeater
watches that the British had almost stopped making. Yet Nicole's
work did not initiate a shift to stem winding as Philippe's did in
Geneva. He sold his small output to dealers catering to the rich.
And he had few imitators, though beginning in the 1850s stem
winding was also available for fusee watches (Johnson's patent).

It was decades before British makers of low-cost watches finally
gave up the fusee for the going barrel; and even when they did,
they clung to key winding. In the 1870s and 1880s, many watches
with stem and crown were also fitted with squares for key winding
and setting. One can understand that: given the conservatism of
many buyers, why not make a watch that would appeal to both
constituencies? What is harder to explain is the widespread practice of making key-wind fuseeless movements and then inserting
an extra wheel whose sole purpose was to make the watch wind
counterclockwise as before. That cost money and hurt performance. Some have argued that makers did this in order to use up a
large stock of *ébauches* for fusee drive. Possibly; but I know of no
such large stock, and the practice continued for years. I think this
was a psychological ploy; the British watch buyer simply preferred the familiar. (Even on the continent, more receptive to

changes in fashion, watch owners were slow to change. For decades after the introduction of the going barrel, makers felt it necessary to engrave an arrow around their winding holes to warn the user to turn his key to the right; while those who could afford to used "tipsy keys" with ratchet insert, so-called because they could turn squares only to the right and hence were proof against the errors of inebriation.) Only in the last two decades of the century did key winding become the exception in Britain, and after the turn of the century a mass producer such as the Lancashire Watch Company was still offering key-wind models. Their catalogues, as well as those of such top-flight makers as Smith, tried to educate buyers to choose the keyless watch, not only because of its convenience but because it did not have to be opened all the time, hence kept cleaner and needed less maintenance.[22] But they got little support from the repairers, who more than anyone shaped the tastes and judgment of watch buyers. "There is no watchmaker who would not rather wind three watches of ordinary construction than one keyless watch!"[23] On this point British preferences were running a generation or two behind those of the rest of Europe and the United States.

The history of stem winding illustrates a very important point—namely, that insofar as the British industry lost ground in the course of the nineteenth century, it was not for want of talent. The history of watchmaking in these years is punctuated by a long series of British inventions and achievements, both by native makers and by immigrants. As late as the middle of the century, some of the most gifted watchmakers in the world still thought it worth their while to set up shop in London (Adolphe Nicole and Victor Kullberg, for example) or at least to do a stage there (Sylvain Mairet). But the commercial opportunism and responsiveness were wanting. Meanwhile the Swiss were producing great craftsmen and technicians in ever-growing numbers: Antoine Tavan, Jacques-Frédéric Houriet, Sylvain Mairet, Frédéric-Louis Favre-Bulle, Isaac-Daniel Piguet, Antoine Le Coultre, Ulysse Nardin, and others too numerous to mention. In the eighteenth century some of the best Swiss talent had gone to Paris and London (Ferdinand Berthoud, Jean Romilly, and Abraham-Louis Breguet to Paris; Justin Vulliamy, Josiah Emery, Henry-Louis Jaquet-Droz, and Louis Recordon to London). That was where the action and money were. By the nineteenth century good makers were going to Switzerland to work and stay (Jules Jürgensen, Henry-Robert Ekegren, Edward Koehn, Albert Potter). That was where the skills and experience were.

At their best, the English were still making some of the world's finest watches. At the end of the nineteenth century, the winners of the annual chronometry competitions at Kew (later at Teddington) were invariably the great London houses: Frodsham, Dent, Kullberg, Usher & Cole, Smith. No victory carried more prestige. But these makers represented the tiny tip of the British industry, the last survivors of the great tradition, and they in turn relied heavily on a small number of semianonymous Clerkenwell suppliers, including such Swiss expatriate firms as Nicole, Nielsen and Hector Golay. This was especially true of complicated watches, where the British had more or less abandoned the field.

After the turn of the century, the tables turned. The Swiss decided to make a bid for the equivalent of the world championship and began sending their best pieces to England for testing. This was no small matter. It was necessary to spring and rate these delicate instruments to a fare-thee-well, then confide them, carefully wrapped, to the tender mercies of the post. This was both risk and handicap, but it was not enough to save the British. In 1903 Paul Ditisheim of La Chaux-de-Fonds made the breakthrough, winning first prize and setting a record of 94.9 points (out of a possible 100). Paul Ditisheim was the greatest horologer of his day, a savant as well as an artist in the best eighteenth-century tradition. Statistically, he was an outlier, four or five standard deviations above the mean—a record breaker who usually had to break his own records—and the British might well have cried unfair. The point, though, is that he was only the best of a substantial pool of first-rate competitive springers and adjusters—single-minded, combat-hardened veterans of the Swiss wars—next to whom the British were gallant amateurs. Beginning in 1907 Swiss watches took first prize in the pocket-watch category every year. On the prize lists Ditisheim was followed by such well-known names as Patek Philippe, Vacheron & Constantin, Zenith, Movado, Longines, and Omega. These were the producers, rich enough and ready to pay for the finest talent and the latest and best in testing equipment, quick to turn each prize into an advertisement. Behind them, anonymous, stood the artists whom Charles Thomann calls the *dignitaires de l'horlogerie.*[24]

Thus was British watchmaking robbed of its final consolation, stripped of its last illusion.

H OW DOES ONE ACCOUNT for Swiss success and British failure? The latter is easier to explain: high costs, conservative styling, obsolescent technique, entrepreneurial complacency, resistance by labor to innovation. There is no way to assign shares to so many convergent factors; they interacted, reinforced one another. Some would contend that the British could not have done things any differently; that their comparative advantage lay elsewhere, and they had no reason to stay in watchmaking; indeed, that their only fault was in sticking so long.

I am not impressed by such arguments, which are very popular among neoclassical economic historians—practitioners of the so-called New Economic History—whose motto is "Whatever has been, had to be." The market is perfect; businessmen know their best interest and are rational in their choices; and there is no such thing as failure, only fate.

The trouble with this deep and overly principled pessimism is that it does not take into account the possibility of changing the givens—that is, of altering the parameters rather than adjusting the variables. It will be clear from an examination of the American experience, for example, that opportunities did exist in the nineteenth century for a radical transformation of watchmaking technology—the kind of leap that we dignify ex post by the name "enterprise." The British could not compete with the Swiss on the basis of dispersed domestic manufacture, but they might have imagined another way of making watches. Given the structure of the British industry, however, such an initiative probably had to come from outside—from outside the industry, or from outside the country (as from Ingold), or from the trade side.

But most of those on the trade side lacked the technical competence to conceive and evaluate a new mode of manufacture; while those among them who had experience of production had severed themselves from it—and so far as they were concerned, a good riddance too! The merchants who put their names on these timepieces had little interest in messing with the sweat and dirt of manufacture; it was easier to leave that to the makers (the British equivalent of the Swiss *établisseurs*) and their tool-proud craftsmen. Joseph Sewill was the leading "nautical optician" of Liverpool—that is, supplier of navigational instruments and chronometers to the merchant marine—and bought his stock from a number of suppliers. When Tom Mercer brought in his first chronometer, Sewill liked it for quality and price, ordered five more, and told Mercer, "Don't stop delivering till I tell you to." With that offhand kind of business, why would Sewill have wanted to invest precious capital in buildings and machinery? When someone reproached him for not having manufacturing facilities, he hired a shop for a day and took his accuser out to see it.[25]

As for the producers, they were, most of them, old bench hands promoted to *établissage*. They were prisoners of the past, seized by what Tom Mercer saw as "an uncontrollable death wish."[26] Confronted by Swiss work—cheaper, better-looking, just as reliable—they were mesmerized. It never occurred to them—I would say, it could not have occurred to them—to turn the industry upside down and destroy everything they lived for. In this they were joined and reinforced by workers whose unequal struggle with new ways only made them cling tighter to old. We have the story of a small attempt to cut production costs in one characteristic first-class London watchmaking firm, Usher & Cole. When young John Francis Cole entered the business, fresh from courses at the British Horological Institute and a most successful apprenticeship, he thought to save money by eliminating unnecessary work. Among other things, he tried to persuade the finishers to stop polishing the arbors of the wheel train, which in the British full-plate layout were not visible even when the watch was opened for winding. He got nowhere—the men simply went on doing things as before—for the simple reason that the finishers could not conceive of doing things any other way: a well-finished arbor was a polished arbor, and that was that.[27]

If so small an innovation could not be imposed, how was one to introduce a new technology? The Lancashire Watch Company—England's belated contestant in the mass production of watches—tried to eliminate potential opposition by hiring the artisans in the arca. It thought these old-timers would be loyal to the interests of their new employer and accept innovations they would have otherwise rejected. Not at all; it had merely succeeded in bringing the enemy within the gates. One old toolmaker recollected the unequal combat: "They [the old workmen] were very much against any alterations to these old things. Now the Swiss, when they introduced the cheap watch, they didn't take the men who had been used to the good quality work, they trained up another lot which hadn't got the tradition, you see, so there was no prejudice behind them using those sort of things. Of course you can understand it, when you get a lot of very fine craftsmen who have always been used to very good work, they don't like anything cheapening anything."[28]

In the end, unfortunately, it was their labor that was cheapened. File makers worked their heart out to cut teeth by hand that could be cut better, faster, and cheaper by machine. Toward the end of the nineteenth century, fusee chain makers in Hampshire got two and a quarter pence an hour (the wages of an agricultural laborer a century earlier) for miniature masterpieces. In 1891 the *Horological Journal* complained that "the wages paid and the prices obtained for the work of our trade, do not bear comparison with those of any other whatsoever. Here you have a really skilled man doing, in many cases, beautiful work for prices that a bicycle repairer would have a good laugh at." As Alun Davies has put it, these were craftsmen "trapped by their own skills."[29]

IT IS HARDER to account for Swiss success, which is almost without parallel in the annals of commerce. How could so small a country, indeed one small part of that country, have dominated a major world industry for so long?

The British economic historian Roy Church stresses two factors. The first, he says, was the necessity for the Swiss to find a substitute for the staple textile manufacture, which was declining under the pressure of British and other foreign competition. The second was the need for Switzerland to export: the home market was too small, and only a responsive, opportunistic industry could survive.[30]

The first of these explanations is simply not correct. The Swiss

watch industry did not grow in an area given over to textile manufacture. Nor did that part of Switzerland engaged in textile
manufacture ever take up watchmaking to any extent. The textile
industry was concentrated in German-speaking Switzerland
(mainly in the Zurich area), and one of the peculiarities of Swiss
watchmaking is that it was reserved at first, with a few peripheral
extensions and isolated exclaves, to French-speaking districts. To
be sure, there was a shift of both labor and capital from lacemaking to watchmaking as the former declined; but these transfers
were local in import and occurred after Switzerland's horological
vocation had been discovered and affirmed. What is more, the
shift occurred not because lacemaking was declining, but because
watchmaking needed all the labor it could get, and even a flourishing lace industry could not pay wages like those of watchmakers.[31]

The second explanation is more plausible: the need to export,
as we have seen, made the Swiss watch industry much more sensitive than it would otherwise have been to market opportunity and
competitive pressures. There was no large home market in which
to take refuge behind tariff walls. On the other hand, market
pressure is not enough; the challenge is not its own assurance of an
effective response, as the British experience reveals.

To understand the Swiss achievement, then, something more is
needed, something that will account for the energy, creativity,
and enterprise of the swarm of people and shops that made up the
industry. Individual talent will explain some of it, but the British
had this in abundance in an earlier period and showed the same
virtue in other branches of manufacture right through the nineteenth century.[32] So did the French, with similar results: examples
of French invention, adopted and exploited by watch manufacturers in other countries, abound. (And why not? As we saw with
the spread of the repeater watch, it is next to impossible to sustain
a monopoly of an idea in this kind of industry.) Besides, individuals can and did move about, as we have seen. They will go to
where the industry is. No, what is called for to explain the Swiss
watch phenomenon is a model of *collective* effort and performance—something to tell us why a multiplicity of Swiss makers and
workers behaved as they did.

Consider in this regard the map of the Swiss industry. We have already noted that watch manufacture was for a century and a half largely confined to French-speaking Switzerland, and this in itself is an indication that we are dealing with a phenomenon that was partly cultural. The skills of watchmaking were transmitted by personal contact, and this is obviously easier among people of common language and customs. Such contact, moreover, was enormously facilitated by blood and marital ties; so that the history of the watch industry is punctuated, like that of nations, by the annals of great dynasties. In Switzerland, the same names recur again and again: Favre, Piguet, Audemars, Le Coultre (Lecoultre), Jequier (Jéquier), Courvoisier, Meylan, Golay, and others.

But there is more: not all of French-speaking Switzerland made watches. There was little or no watchmaking in the Alpine Valais (except for the area proximate to Geneva) facing south toward Savoy and Italy. Only the districts facing France along the Jura frontier made watches, so mountain air, snowy winters, and *s'il vous plaît* were not enough. Perhaps commercial ties made the difference: the villages of the Valais were off the beaten track, whereas the Swiss Jura was on trade routes linking Switzerland and south Germany to Burgundy and the Rhône valley. But so was the French Jura on the other side of the border, and except for a genuinely original clock industry in the area around Morez, none of the villages there went in for horology, except as satellites to the Swiss industry. Besides, as trade routes go, those across the Jura leave much to be desired. There are easier ways to get to France than through the principality of Neuchâtel.

Religion seems to have played a role here: on the Swiss side, Protestants; on the French side, Catholics. The lines of influence remained to be explored, but there is no mistaking the geography. Not only was watchmaking stunted on the French side, but it was confined largely to those villages close enough to the frontier to serve as cover for the smuggling of Swiss work into France. The one major exception was the large-scale manufacture of *ébauches* by Japy at Beaucourt, mentioned above. But Beaucourt proves the religious thesis. Japy was Protestant and his workers were Protestant. Beaucourt is located in the Pays de Montbéliard, a pocket of dissent in an otherwise Catholic region. It lies seventy-five to eighty-five kilometers by road from the capital of the Swiss watch industry in La Chaux-de-Fonds, separated by mountains and wooded country that have no connection with the watch manufacture. This in itself is extraordinary: for the same reasons that make for the diffusion of watchmaking skills via networks of personal relationship, the industry tends to expand spatially to contiguous areas. Because of the extreme division of labor, there are too many ties of interdependence to leap away and cut oneself off from a multitude of interrelated sources of supply. Here, however, the industry made a leap almost as long as the long axis linking its extremities at Le Brassus, at the southwest end of the Lac de Joux, and Bienne (German Biel), on the frontier of French and Alamannic Switzerland.[33] (See Figure 32.) Surely this implantation in Protestant country was not a coincidence.

Yet even if one accepts the reality of a link between religion and enterprise, the question still remains of the nature of the link. Was it the classical "Weberian" values of diligence and commitment that made the difference? There is some evidence of Weberian behavior in the Swiss Jura—of a personal ethic of work and duty conducive to intense application; but such evidence is still sparse and awaits verification in as yet unexplored family papers.[34] And even then, would we not have to explain more than the devotion and ambition of the leading entrepreneurs? The watch industry, after all, was one that demanded high levels of skill and initiative at many stages of manufacture at dispersed points of production. The influence of Protestantism, then, insofar as it counted, must be sought in a more general sphere—in its implications, say, for the intellectual and cultural level of the watchmaking population.

The key here, I think, was the contribution of Protestantism to literacy and numeracy. We have the testimony of contemporaries to the education and interests of these mountain folk, and there is no mistaking their special character. They could all read and write; and they could reckon—the girls as well as the boys. Their reading was not confined to the Bible. We have inventories after death that show small libraries of travel books, works on natural science and philosophy, histories and biographies—in short, the libraries of cultivated autodidacts. Le Locle and the country around were able to support one of the best bookstores in western Europe—one that would have done credit to Paris or London.[35]

The region, moreover, seems to have been a veritable hothouse of mechanical invention and scientific experimentation, a seedbed of Ben Franklins *à la jurassienne*. The banneret Frédéric-Samuel Osterwald, visiting the area in the 1750s, conveys very well this sense of mutual stimulation and emulation, of rivalry in resourcefulness and invention:

> In these mountains, then, necessity has certainly been a stimulus to talent, along with the freedom the inhabitants enjoy, the fine air they breathe, the need to take care of themselves with the little trade that they carry on with the outside. Every person in the old days was obliged to be a mason, carpenter, wheelwright, smith, etc., and to make for himself all his furniture. Those who enjoyed music made their own instruments. Besides, the nature of the soil required little work and offered little incentive to the owners to cultivate it; so they were easily turned in the direction of industry to provide for their subsistence and to dispel the boredom inseparable from their long winters. Today the education that they receive, the commerce that they have between themselves and the outside, the reading that they have so decided a taste for, their habits of travel, the emulation that they all feel so strongly—all of that serves to stimulate and develop these very talents that are so unmistakable to anyone who knows them. They have by nature a singular genius, a remarkable aptitude for all mechanical arts, all the more extraordinary in that one finds absolutely nothing comparable among the populations about immediately to the north or south. Indeed, one frequently finds in these mountains people who exercise certain trades without ever having been apprenticed.[36]

Singular mixture of institutional, spiritual, and physical ("fine air") explanations! Osterwald adds other factors: the very poverty of the villages, which had little common land and funds to protect and hence interposed no obstacles to immigration and settlement;[37] and the absence of guild constraints on recruitment of labor or entry into trade.[38] Neither of these considerations, it should be noted, was peculiar to the mountain watchmaking district, though they did set it off from older centers of watch manufacture in the lowlands, particularly Geneva.

When all is said and done, then, this extraordinary, unexpected achievement by what would have seemed destined to remain a poor backwater needs a multiple explanation. As always, it is the complex of factors and their interplay that matter. We are dealing here with a collective undertaking and achievement, the work of a population delineated by space, climate, political and social institutions, religion, values. It can be understood only as a cultural and social as well as economic phenomenon.

One can think of other examples of this process in other branches and places—in the spectacular success of certain centers of textile manufacture, for example (Roubaix, Łodz, Barmen-Elberfeld); or in the establishment of entrepreneurial patriciates joined by blood, marriage, and a common sense of status and accomplishment (Hamburg, Bordeaux, Basel, Boston). It is a pattern closely linked to personal effort and family enterprise, to a world in which name is important—but one that can be institutionalized so that it is compatible with the anonymity and universalism of the joint-stock corporation. This is the function of the schools, the learned societies, the museums, the exhibitions, the research institutes—the formal apparatus that defines, communicates, and generalizes the group's knowledge and values and thereby socializes recruits.

19 Not One in Fifty Thousand

IRONICALLY, the Swiss greeted the first American attempts at making watches with the same scorn that the British had had for Swiss efforts. Yet they soon had reason to recognize that the infant American industry was not the same as other competitors. It was built on different lines and used a new technology, which consisted in the production of assembled objects from machine-made, standardized, interchangeable parts.

To understand the implications of this new technology, we must first see it in a larger historical context. The effort to produce like objects in large numbers goes back to ancient times and early gave rise to two techniques, both of which continue to be important features of mass production. The first was stamping, using dies to shape some softer material that would then hold the imprint. Coinage is the best example. The second was casting, the use of molds to shape a liquid material that would harden and hold shape as it cooled. Chinese bronzes of the Chou dynasty are evidence of the superb objects that could be produced in this manner. Note that in neither instance was the end product identical. Anyone who has seen old coins knows how much they vary, partly because the strike is not always well centered, partly because dies wear with use and the impression loses sharpness. The same is true of castings: they leave the mold with rough edges and need finishing; and finishing alters them. Yet such objects do not require identity. Coins can be and were weighed to verify content, and bronze kettle covers can be exchanged one for the other within a wide range of tolerance.

That is the key, of course: the technique employed is a function of the tolerance acceptable. So long as the objects to be interchanged could be approximately alike, so long as some play was acceptable, as in panes of glass for a standard-sized window frame, or balls for muskets or cannon, or limbers for gun carriages, traditional techniques were adequate. Much more difficult to achieve was uniformity in articulated, moving parts of composite devices. These could rarely be cast, not only because molds change with use but because castings tend to crack when subjected to the tension and shear incurred by rapid movement and interaction. (That is why cast iron is best used in radiators, fire backs, grilles, pipes, frying pans, and other objects that remain still when in use.) Moving parts had to be made of such metals as steel or wrought iron or brass, and these had to be formed by working—by planing, turning, punching, drilling, squeezing, hammering, filing (or otherwise abrading), and so on.

To convert in this manner a succession of blanks into substantially identical pieces requires the use of machines (hence the verb "machining"). No hand-guided tool is capable of the repetitive motions required. The machines, moreover, have to be "smart." They must know where and how far to travel, which is another way of saying that they must make the same pass time and time again. They must be neat, for cleaning entails deformation. And they must be tough, for any wear of the cutting tool will change the result. All these requirements, of course, vary in rigor with the margin of tolerance. The smaller the margin, the more precise and consistent the machines have to be.

In Europe, machining for economy if not for interchangeability went back to the early eighteenth century; witness the cog wheel cutter of that Swedish genius, Christopher Polhem—a kind of technological isolate, far ahead of his time.[1] Other, later examples are the machine-made *ébauches* of Japy; Blanc's gunlocks, which may have been the inspiration via Thomas Jefferson for Eli Whitney's innovations in the manufacture of muskets for the U.S.

Army beginning in 1798;[2] and Henry Maudslay's mechanized production of the highly complicated Bramah lock (1790–91).[3]

In Europe, however, manufacture on the uniformity principle remained a rarity. It was reserved for uses where tolerances were large, and in each instance it seems to have remained the exception, without impact on its own, let alone other branches of production. (Historians remember what contemporaries were not ready to notice and copy.) It was not yet, in the first half of the nineteenth century, a general principle of fabrication.

That is just what it became in the United States, where it found a widening array of applications, to the point where Europeans came to call this technique the American system of manufacture.[4] There was a time when Eli Whitney was credited with the invention of interchangeability, but recent research is inclined to view his achievement as something less than that: the production of similar rather than matching components.[5] It took some decades more before the efforts of John H. Hall and Simeon North produced rifles made in separate shops with truly interchangeable parts. This took not only precision machinery but a large array of go and no-go gauges: without testing, there could be no substantial identity of shape and size. Long before the achievement of this standard, however, the uniformity principle was introduced into the fabrication of clocks, locks, furniture, and hardware. As a result, interchangeable parts came to constitute for American technicians and entrepreneurs a familiar, comfortable, even proper response to the problems of organizing assembly-type industry.[6] In matters of this kind, I would argue, the cultural matrix is at least as important as cost calculations. The United States had developed an American way of doing these things, and it was no accident that the watch industry found the answer it did to foreign industrial domination.[7]

The answer had been found far more easily in clocks than it would be in watches, for one simple reason: clock parts were much bigger and tolerances proportionately greater. Existing machine tools could do the work, the more easily as the early American shelf clocks were made of wood—wheels, arbors, pinions, and plates. The pioneer here was Eli Terry (1772–1852), who had learned to make clocks the conventional way, using the foot lathe and hand-operated cutting engine and finishing with file and penknife as required. Around 1802 or 1803, Terry undertook to make not a few clocks but a thousand and more, and fitted out a water-powered shop for the purpose in Plymouth, Connecticut. One local wit thought this goal so ludicrous that he offered in advance to buy the last of the series, thinking he would never have to pay. He paid. A few years later (1807–8) Terry launched a new venture that called for making four thousand clocks: it took most of a year to set up the machinery, a second year to make the first thousand, a third to make the other three. It was this factory that showed for the first time that mass production of timepieces was feasible and profitable. It was also the training ground for such other leaders of the clock industry as Seth Thomas and Silas Hoadley. Terry played the role of teacher to the trade, both in product design and production techniques, just as Maudslay did for machine-tool manufacture in Britain.[8]

One of Terry's pupils was Chauncey Jerome (1793–1868), who began as carpenter and casemaker, then moved into buying movements, casing them, and selling the finished clocks. Jerome, like the other batch producers of the time, confined himself at first to wooden movements—the easiest and cheapest to make; and like the others, he had his good times and bad, for demand for a postponable acquisition was very sensitive to general economic conditions. Then in 1837, his mind concentrated no doubt by a severe business contraction, he had an inspiration: Why not make a cheap, mass-produced brass clock? The raw material would cost more, but the clock would be smaller, more accurate, more reliable: no more swelling and jamming in damp weather.

People laughed when Jerome predicted that his new brass clocks would drive the wooden clocks from the market. Within a year they were laughing out of the other side of their mouth: the new clocks sold so well that Jerome was assailed with offers of money from investors eager to take a share of the enterprise, and several new companies were formed to compete with him. Jerome gives the impression that he could hardly keep up with demand, although his factory in Bristol, Connecticut, was operating on an unprecedented scale: the first run was forty thousand movements at $1.40 each.

Still, he must have had some clocks left over, because he decided to ship a few to England. This appeared at first glance to be a folly—like sending coals to Newcastle. People laughed—in Jerome's words, ridiculing "the idea of sending clocks . . . where labor was so cheap." Besides, there was a 20 percent duty on the import of finished timepieces. There was also a psychological barrier: it was English craftsmen who had brought clockmaking to America, and the prestige of the English product was such as to intimidate all but the bravest. London dealers, for example, could not forget that it was they, after all, "who made clocks for the world" and could not conceive a challenge from rustic "colonials." On one occasion, Jerome's representatives were driven out of a shop and told not to return: their "Yankee clocks" were "good for nothing or they could not be offered so cheap."[9] One should never underestimate the power of such prejudice, but neither should one underrate the rationality of a careful shopper. British bracket clocks sold for £5 (equaled $25) and up—a month's wage for a skilled worker; Jerome's shelf clocks sold for considerably less. To be sure, the British clocks were better: their cases were more stylish; their movements were made of better materials more carefully finished and adjusted. But the American clocks were good enough.

Jerome's agents finally persuaded one shopkeeper to take a couple of clocks on consignment. They sold immediately. The dealer tried four; they went off just as fast. So the man ordered a dozen and then two hundred, and other dealers decided that they too could make money on American clocks. Now Jerome sent off a shipload, invoiced at $1.50 each (why pay more duty?), to sell in England at a lucratively competitive price of $20. There then ensued one of the most amusing "second takes" in industrial history. In Jerome's words:

> I had always told my young men over there to put a fair price on the clocks, which they did; but the [customs] officers thought they put them altogether too low, so they made up their minds that they would take a lot, and seized one ship-load, thinking that we would put the prices of the next cargo at higher rates. They paid the cash for this cargo, which made a good sale for us. A few days after, another invoice arrived which our folks entered at the same prices as before; but they were again taken by the officers paying us cash and ten per cent in addition, which was very satisfactory to us. On the arrival of their third lot, they began to think they had better let the

Yankees sell their own goods and passed them through unmolested, and came to the conclusion that we could make clocks much better and cheaper than their own people.[10]

As with Swiss watches, so with American clocks: British industry failed to respond effectively to this foreign competition, which came not only from the United States, but after the middle of the century from France and Germany as well. The Germans, be it noted, were quick to recognize the possibilities of the new technology. What was eventually to become the largest clock and watch firm in Germany and one of the biggest in the world, Gebrüder Junghans of Schramberg in Württemberg, began in 1864 with the announcement that it would be making clocks *nach amerikanischem Prinzip* and advertised itself as a *Fabrik amerikanischer Uhren.*[11] The British went their way, making superb clocks in the old manner—bracket clocks, carriage clocks, and regulators—all of which command today special prices on the antiques market because of their high quality and scarcity. But that is the point: they became ever scarcer.

In the United States, the application of machine methods to clock manufacture was bound to influence watchmaking as well. Here there was no domestic industry to speak of: America relied almost entirely on imports from Switzerland and Britain, reserving to itself, and then only in part, the role of casemaker. There was simply no pool of cheap skilled labor to sustain a cottage watch industry like that of Europe.

The solution lay, of course, in the substitution of machines for labor, as with clocks; but as noted above, the smallness of watches and the degree of precision required entailed the application of far more exacting production standards than in woodworking and relatively rough metalwork. What is more, the demands made on a watch, in use and hence in manufacture, are far greater than those on a clock: just the act of wearing a timepiece exposes it to movements and shocks that disturb or even injure the mechanism, whereas a clock sits in one place and needs little maintenance. An early effort to mechanize watchmaking in the late 1830s and 1840s failed for this reason: the job was harder than anyone anticipated.[12]

The breakthrough came in the 1850s. The vision was that of

Aaron Dennison, mechanic and watchmaker, who was early struck by the shoddiness of many of the timepieces he had to work on:

> Within a year I have examined watches made by a man whose reputation at this moment is far beyond that of any other watchmaker in London, and have found in them such workmanship as I should blush to have it supposed had passed from under my hand in our lower grade of work. Of course I do not mean to say that there is not work in these watches of the highest grade possible to carry the finisher's art, but errors do creep in and are allowed to pass the hands of competent examiners, and it needs but slight acquaintance with our art to discover that the lower grade of foreign watches are hardly as mechanically correct in their construction as a common wheelbarrow.[13]

It was one thing to perceive the opportunity to manufacture watches by machine, another to realize it. Dennison joined in 1850 with Edward Howard, a skilled watchmaker, and Samuel Curtis, an investor and Howard's father-in-law, in founding what was later to become the Waltham Watch Company but was to operate under such a succession of names in its early years that even the historian of the firm makes no attempt to adhere to strict nomenclature.[14] The first task was to assemble a team of technicians and mechanics with the necessary know-how. Some of them were found in the United States: veterans of earlier, abortive watch ventures; a machinist from the Springfield Armory; and then two of those mechanical wizards who can make a venture— Charles S. Mosely and Napoleon Bonaparte Sherwood, both born machine designers. (Mosely later became superintendent at Elgin, which he helped make Waltham's most dangerous competitor.) Others were recruited by Dennison on a trip to England, where he had every occasion to observe the limitations of dispersed manufacture—at least the British variety:

> I found ... that the party setting up as manufacturer of watches bought his Lancashire movements—a conglomeration of rough materials—and gave them out to A, B, and C, and D to have them finished; and how A, B, C, and D gave out the different jobs of pivoting certain wheels of the train to E, certain other parts to F, and the fusee cutting to G. Dial-making, jeweling, gilding, motioning,

etc., to others, down almost the entire length of the alphabet; and how that, taking these various pieces of work to outside work-people—who if sober enough to be at their places, were likely to be engaged on some one's work who had been ahead of them, and how, under such circumstances, he would take the occasion to drop into a "pub" to drink and gossip, and, perhaps, unfit himself for work the remainder of the day. Finding things in this condition as a matter of course, my theory of Americans not finding any difficulty in competing with the English, especially if the interchangeable system and manufacturing in large quantities was adopted, may be accepted as reasonable.[15]

The British watch craftsman was indeed an individualist, proud of his skills, master of his time, not amenable to discipline or contradiction. (Tony Mercer tells the tale of one gifted artist who was so "upset" by his girlfriend's decision to emigrate that "he tied a piece of string round one of his vices [caution! British spelling] swearing that he would never use it till she returned." It was still there at his death, though he had long since married another.)[16] Dennison might have said the same things about the *cabinotiers* of Geneva, as we have seen; but he would not have described the Jura cottage industry in those terms.

Dennison and associates started work in 1850, but what had seemed simple and obvious proved exasperatingly difficult: it took three years to place the first watches on the market. In the four years thereafter, the company managed to put out about a thousand pieces, no more than a large shop in London and hardly worth the trouble of making special-purpose machinery and of mass-producing uniform parts. Clearly, there was more to watches than met the eye. On some parts it was necessary, for example, to work to one five-thousandth of an inch. The makers of machine tools of that day were simply not prepared to meet such tolerances. In the end, Waltham had to invent its own machines, and so did the other major makers. In so doing, they moved far from Dennison's original conception. His early machines were, like Ingold's, essentially tools writ large. The aim, as in Europe, was to take each piece and use a versatile cutting or drilling device to shape it as desired, and this called for constant change of cutters or bits, adjustment of stops, loosening and tightening of butterfly nuts—in short, work, fuss, and skill.

The second generation of American machines marked a start in an entirely different direction. These were heavily framed, single-purpose presses, drills, and lathes designed to loose the Gordian knot in the manner of Alexander. They were strong enough to shape hard steel (crucial, because tempering altered shapes), were fitted with fist-size levers rather than adjustable nuts, and used templates and fixed stops so as to minimize the need for human judgment. In cutting wheels, for example, the blanks were punched out in quantity by comparatively small presses: around 1880 one man with a twenty-ton punch could blank out ten thousand wheels a day. These were then finished on wheel cutters that turned the stack of blanks automatically as each tooth was finished. Pinions, once drawn like wire by repeated passes through profiled holes, were now similarly turned and cut; in this way they could be made of much harder steel. In other work, the blanks were fixed while a succession of cutters was brought to bear, sometimes by hand, but more and more automatically. Other pieces, starting as small rounds or bars, moved from one machine to another, initially by manual transfer, later by mechanical means.

There were, of course, some areas where human skill was indispensable. The fitting of balance wheel to hairspring, for example, required a fine hand, and in Europe the one was painstakingly adjusted to the other to ensure isochronous swings. The Americans had neither the time nor the skills. Instead they made large numbers of balances and springs as close to standard as possible; then carefully sorted these by weight and force. It only remained then to pair them by choosing from the right boxes or jars. The same technique was applied to the selection of jewels and pivots: instead of drilling jewels and then making pivots to fit, the Americans turned out an array of both, measured diameters to a ten-thousandth or even twenty-thousandth of an inch, sorted the pieces accordingly, and matched them as needed. No need, then, for fine tuning: just choose a target and let statistical distribution take care of the rest.[17]

All of this sounds simple enough, even inspired, in retrospect. But it did not come easily at the time, and there were inevitably tasks that seemed to defy mechanization. One of these was dial-making. The porcelain dials started rough, finished smooth, and in between had to be fired and refired at least four times. In spite of the importation of skilled British dialmakers, a deplorably high proportion cracked in the making; Waltham heaped up a small mountain of rejects outside the plant before it finally learned how to make them. No one ever succeeded in mass producing those beautiful white dials, and it is no wonder that the watch industry gave up on them after World War I. In their place there appeared metal dials that could be stamped out and machine finished— very handsome, very fashionable, very durable, but less legible.

The history of the American Watch Company (alias Waltham) illustrates the vicissitudes of those early, experimental years. The company's first design called for a fuseeless movement with two mainsprings; the idea was to have the watch run for eight days instead of one. No go: the watch ran fast at first, then slowed. So back to the drawing board: stick to going barrel (no fusee); forget the eight days; use only the first part of a long spring, that there be little change in tension over twenty-four hours. Little by little the designers got the bugs out, until by the end of the decade they had a good product. The effort cost them a lot of money and several liquidations and reorganizations.

In 1859 the newly named American Watch Company began production in the new plant at Waltham, and from then on the trajectory was upward. The Civil War was an unexpected stimulus: large compartmentalized structures such as armies depend on a knowledge of time to organize and synchronize action. In 1858 Waltham produced its fourteen-thousandth watch; in 1864 it was up to 118,000. By 1865, the so-called soldier's watch, the *Ellery,* accounted for 44.7 percent of unit sales and 30.4 percent of receipts. The last full year of combat—the accounting year ending January 31, 1865—saw the company earn 58 percent on sales, 164 percent on capital.[18] By American standards, this was the birth of an industry. By Swiss standards, it was just a small cloud on the horizon, no bigger than a man's hand.

The success of the American Watch Company was bound to encourage imitators: Howard in 1857, Elgin in 1864, Illinois in 1869, Hampden in 1877, Waterbury in 1879, Hamilton in 1892. These and about sixty others—easy entry, easy exit—were to produce over a period of three quarters of a century about 120 million jeweled watches plus at least twice as many so-called dollar watches or clock-watches.[19] Elgin alone made about fifty million; Waltham, almost forty (see Tables 3 and 4); the American watch industry was in effect a duopoly. This outpouring of timepieces was accompanied, needless to say, by a comparable flood of publicity. Where earlier watchmakers had been content to place discreet notices of their existence, or perhaps of new additions to stock, the mass-production giants published and distributed illustrated materials designed not only to sing the praises of their products but to awaken in potential consumers a desire for the prestige of watch ownership. Here too, America innovated—in scale, in the use of indirect psychological appeals, in vividness of presentation, in the popularization of memorable slogans. Among the best was the Ingersoll motto: "The watch that made the dollar famous."[20]

In order to understand the magnitude of this achievement, one has to keep in mind what watch manufacture was like under the old system. One can distinguish four stages:

1. The serial numbers of the most successful craft shops in the eighteenth and early nineteenth centuries ran into the thousands over a period of decades.

2. The shift to *établissage* and the extensive recourse to machine-made *ébauches* made possible a tenfold increase in volume: leading Swiss and British makers of the early and mid-nineteenth century, using the old hand techniques but pulling together material from dispersed outworkers and buying up finished movements and watches from subcontractors, numbered their lifetime sales in the tens of thousands.[21]

3. Mechanization in combination with putting-out raised the ceiling by another order of magnitude: the largest makers of the 1850s such as Jules Huguenin of La Chaux-de-Fonds marketed tens of thousands of watches a year.[22] I have seen some Swiss watches of the mid-nineteenth century with serial numbers in six figures.

4. Waltham increased that number tenfold—the third such jump in a century. Where output per worker-year in Switzerland averaged forty watches in the mid-1870s, Waltham was doing 150 in 1880, over 250 in 1900. By 1877 Waltham's serial numbers passed the one million mark; twenty-four years later, with annual output of over 600,000 pieces, the number was ten million.[23] That

figure, moreover, was far surpassed by the output of the giants in the cheap clock-watch sector. Ingersoll, for example, sold a million watches in 1898, most of them made by the Waterbury Clock Company, and would attain at peak a volume of twenty thousand units a day, over six million watches a year. In its first quarter-century, it made fifty million watches.[24] Ingraham, which moved from clocks into watches in 1911, sold about eighty million over the next half-century.[25]

Numbers are only one aspect of the American achievement. Quality was another. The Europeans refused at first to believe that a machine-made watch could do well, and it was only the painful pressure of competition that moved the Swiss to inquire into the matter. Swiss exports of watches and parts to the United States rose steadily in the 1860s to a peak of 18.3 million francs in 1872. The next year saw a sharp setback to 13.1 million francs, but that was the year of the great crash, and the watch business was bad everywhere. By 1876, however, exports had plummeted to 4.8 million francs—far below anything that could be accounted for by the depression—and the Swiss understood that the market had changed. (See Table 5.)

That year the Swiss sent a representative, Edouard Favre-Perret, to the American centennial exhibition in Philadelphia to see what the difficulty was. His findings threw his compatriots into momentary consternation: the Americans, he told them, were making by machine watches every bit as dependable and precise as the best Swiss instruments.[26]

Favre-Perret's report, delivered also as an address to his fellow watchmakers in a school auditorium in La Chaux-de-Fonds, has been described as the most famous speech in horological history.[27] He blamed some of the Swiss sales debacle on the Swiss themselves: instead of sending the Americans work of good quality, they had taken advantage of the surge of demand during the Civil War (and the ignorance of the American soldier) to send "the worst trash." The new American watch manufacturers could profit, therefore, from the disaffection of buyers who, burned once, had no intention of letting themselves get burned a second time.

But, he went on, there was much more to American success. Their productivity was higher; their prices lower—nineteen francs for the latest Elgin lever movement with visible pallets, as against several times that for a comparable Swiss piece.[28] Their watches, thanks to interchangeability, were easily repaired; one had only to write away for replacement parts. Perhaps the most astonishing thing for a Swiss audience that was not without some familiarity with machine tools and mass-produced components was Perret's account of the statistical-distribution approach to adjustment described above: "They arrive at the regulation of the watch, so to say, without having seen it. When the watch is given to the adjuster, the foreman delivers to him the corresponding hairspring and the watch is regulated." Sensation! Springing, after all, was the highest, most difficult form of the watchmaker's art.

Then Fevre-Perret clinched his argument with the story of a personal experiment:

> I asked the director of the Waltham Company for a watch of the *fifth* grade. A large safe was opened before me; at random I took a watch out of it and fastened it to my chain. The director having asked me to let him have the watch for two or three days, so as to observe its motion, I answered: "On the contrary, I insist on wearing it as it is, to obtain an exact idea of your manufacture." At Paris I set my watch by a regulator on the Boulevard, and on the sixth day I observed that it had varied 30 seconds. And this watch is of the fifth American grade—it costs 75 francs (movement without case). On my arrival in Le Locle I showed the watch to one of our first adjusters, who asked permission to "take it down"—in other words to take it to pieces . . . After a lapse of a few days, he came to me and said, word for word: "I am completely overwhelmed; the result is incredible; one would not find one such watch among fifty thousand of our manufacture."[29]

20 Who Killed Cock Robin?

A FTER THE INITIAL SHOCK, the Swiss reacted far more
sensibly and vigorously than the British in similar
circumstances. For one thing, they took the challenge seriously
rather than dismissed it out of hand. They were, as we have seen,
already moving in the direction of mechanized production. They
continued on this path with all deliberate speed, and they introduced instruction on the character and use of machines in their
schools of horology. For another, they refused to panic. The report
from Philadelphia was subjected to thorough criticism, as were
the new American watches, and the Swiss were quick to note that
hand rectification and adjustment continued to offer advantages.
Indeed, it soon became clear that American producers were also
obliged to retouch and finish parts and movements by hand.[1] A
deputation of Swiss experts to the Exposition Universelle of 1878
in Paris reported with carefully contained glee that Waltham had
ducked out of a test demonstration. Five like watches were to be
taken down and reassembled at random. French and Swiss horologers gathered at the appointed hour, but Waltham's representative was in conference. So long did he stay, that the Swiss had
to leave. But the French waited him out and were treated to an
exchange of movements from one case to another. The implica-

tion was clear: all this talk of interchangeability was puff and bluff.[2]

The fact was that with watches, as with guns, true interchangeability came much later than is commonly thought and than was proclaimed at the time. This is not to say that the Americans had not made an enormous advance by making parts so alike that replacements could be simply ordered by number and with a few retouches made to fit. The Swiss could not have done that. The first timepieces to be composed entirely of truly interchangeable (in the sense of randomly selectable) components were probably the cheap dollar watches of the turn of the century. They were simply not worth the cost of hand finishing and adjustment. Every part had to work, or else. But this, it should be noted, was at the point of assembly. Later performance was another matter: if a dollar watch ran, fine; if not, the buyer could send it back for replacement or throw it away. At low price levels, it is simpler to shift the task of quality control to the customer.

Jeweled watches reached this point much later. D. W. Leverenz, formerly head watchmaker of the Elgin National Watch Company, places the change in the 1930s. Before that time, he writes, "watch manufacturers claimed they were producing watches with interchangeable parts. This was not true. At the time of assembly, many parts were altered and adjusted by hand methods." To achieve real interchangeability on this level, it was necessary to build machines that would work to a tolerance of one one-hundred-thousandth of an inch; and to shave pivot holes fine enough so that jewels could be press-fitted rather than set. By these and similar methods, "all watch parts were made to fit any movement at the time of assembly without hand adjusting, filing, bending, etc."[3]

In light of their observations, the Swiss decided that the multiplicity of their enterprises and the competition and emulation that characterized their industry were the best assurance that it would not overlook the opportunities offered by the new technique. At the same time, the data on the performances of those Swiss watches submitted to the observatories for testing reassured them that, except for the best American pieces, the machine-made watch was still not up to the finest hand work. Such results, to be sure, told nothing of run-of-the-mill production; but they redounded to the credit and prestige of the Swiss industry and did much to sustain its confidence. They also allowed the Swiss to launch their own publicity counteroffensive, stressing the virtues of Swiss precision as well as style and reliability. The Americans may have invented the techniques of modern advertising, but the Swiss were quick learners.

The Swiss strategy, then, was to move toward interchangeable parts selectively, while retaining an interest in special finishing, complicated pieces, and high fashion. Their factories were built on a smaller scale than those in the United States. They did some work in the new way; other work in the old way; hired domestic labor outside the plant; subcontracted for special parts; gradually drew additional steps into the new mode of production. As one of the founders of the Longines works put it after two years of frustrations and disappointments: "Ask machines to do only what they can do: rough work, turning and filing, polishing and rounding off; but don't ask of them what they cannot do, and above all, don't think that you can do without the watchmaker."[4]

It took the Swiss two generations to move from the older to the newer mode of production. In 1870, of 40,000 employed in the industry some 5,000 worked in "large" shops and factories (the statistical threshold was ten employees), 35,000 at home. By 1905 the proportions had not quite been reversed: 38,000 in the larger units (ten and above); 12,000 at home, of whom 7,000 were outworkers for factory enterprises. The average size per unit remained substantially below that in the United States: 38 employees per firm in Switzerland in 1901; 529 per enterprise in the United States in 1900. These figures, moreover, understate the disparity: the Swiss average excludes all the smaller units, of which there were thousands. The United States had only thirteen watch-manufacturing enterprises at that date. The largest of these—Waltham and Elgin—employed thousands each, all of them in integrated plants producing the complete watch. Compare the data on the largest Swiss firms in 1905, all of which continued to rely to some extent on outworkers. (See Table 6.)

In the long run, this gradual adoption of the new mode of production proved extraordinarily successful. The strength of the Swiss industry, as before, lay in its versatility and flexibility. It continued to work for the widest range of tastes and markets and to generate new and interesting combinations of features for those who wanted something different. In this way the Swiss retained their effective monopoly of the market for novelty watches and complicated mechanisms, while producing some of the finest precision instruments for those who wanted to know the time and nothing but. No one could compete with them for ornamental cases; no one was so quick to take up the latest fashion—art nouveau or art deco.

Much of this fancy work was still done by hand tools, but toward the end of the century the Swiss succeeded in mechanizing production of ever more complicated movements and cases. Their machines, unlike those in the United States, tended to be versatile and were susceptible of all manner of adjustment, hence required some skill to operate; but the Swiss had an unequaled pool of skilled labor to draw on. They also used simpler machines to open the work to rank beginners, often women and children, paid half-and-less what a man made. The old-timers fought these innovations with a passion, but the dispersed, cottage character of the industry made organized resistance next to impossible. For the *établisseur*, it was precisely the machine's ability to curb the pretensions of the workshop aristocrats—those very few key craftsmen who dictated the rhythm of work and could make or break the reputation of the shop—that made the new mode satisfying as well as potentially profitable. In the 1870s the Swiss had learned from America; a generation later they were taking the lead in mechanical innovation, to the point where, even in matters mechanical, the United States watch industry became a Swiss dependent.

With all this, the Swiss never won back their monopoly of the American market. Not for want of trying: they made, for example, hundreds of thousands of copies, often deceptive copies, of American watches, called by trade names chosen to convey the impression of American manufacture. The Wm. Ellery of Waltham became Wm. Elley or Ellerty; P. S. Bartlett was changed to P. F. Barlett, P. S. Barzlett, and P. S. Bartley; Hampden appeared as Hampton; Waltham, as Walham. What was one to infer from such plausible names as the New Haven Watch Co., the New York Central Watch, the Ohio Watch Co., the Pennsylvania watch? What will archaeologists of the future think?

The American manufacturers of that day were not amused and got Congress to ban the import of such fraudulent merchandise (Dingley tariff of 1898).[5]

Most of these imitations were "deadbeat" tickers. But the Philadelphia Watch Company imported superb Swiss movements of the finest quality, perhaps finishing them in this country, and collectors are still arguing whether they were ever really made in Philadelphia. And Tiffany in New York; Bailey, Banks & Biddle and Caldwell in Philadelphia; Tilden in Providence; Spalding in Chicago; and comparable carriage-trade emporia imported the best watches that Patek and Audemars could make, placing their own name on dial and movement and often stipulating that no sign of Swiss origin appear. Tiffany even set up its own manufacturing plant in Switzerland and turned out a splendid watch of quite distinctive layout and finish; but the enterprise did not pay, and Tiffany ended by selling the plant and equipment to Patek, Philippe. The Swiss, as always, made and sold the worst and the best; they sold what the market wanted.

Yet in 1900 the United States was only the tenth largest market for Swiss watches: the Swiss could not easily compete with the far-flung network of American jewelers and watchmakers committed to selling and servicing American brands. The Swiss found compensation elsewhere. The most important customer was Germany, which bought more than ten times as much as the United States. Great Britain was almost as big an importer, and then, far behind in third place, came Austria-Hungary.[6]

In all these countries, the large and growing demand for timepieces was an accompaniment of urbanization and industrialization—indeed, of the adoption of all the values that we commonly associate with modern living. In this sense, the consumption of timepieces may well be the best proxy measure of modernization, better even than energy consumption per capita, which varies significantly with the relative cost of fuel, climatic requirements, and product mix. The need to know the time is independent of all of these, and data on consumption of timepieces have the additional advantage of quantifying the imponderable. They sum up, in effect, a whole bundle of new work and life requirements and the inculcation of the values and attitudes that make the system go.

Under the circumstances, use of watches was growing fastest in those countries that were on the steepest part of the development curve. Britain, whose industrialization lay largely behind it and whose domestic product was leveling off, saw watch imports level off as well, even shrink substantially in the early years of the new century. Germany, more buoyant, but now an established *Industriestaat*, increased its imports of Swiss watches very slowly, partly because its own watch industry was meeting an increasing share of the demand. Austria-Hungary, running hard to make up lost time, with major industrial development under way in Bohemia-Moravia and parts of Hungary, took 80 percent more watches over the period 1900–1911. The biggest gains, though, came in the newly developing countries: Russia, Japan even more (but as in Germany, restrained from 1907 on by home production), Argentina most of all.

From 1890 to 1913, the last "normal" years before war threw Europe into turmoil, Swiss watch exports grew by 5.6 percent a year, from 4.8 million units to 16.8 million. (See Table 7.) Since more than 95 percent of output was sold abroad, this growth was also that of production as a whole. Over these years, in spite of the appearance of important domestic watch industries in a number of countries, the Swiss probably accounted for well over half the world's output. By way of comparison, in 1890, when Switzerland exported horological products to the value of 104 million francs, its largest competitor, the United States, produced watches and watch cases worth 73 million francs. Ten years later, United States output was still valued at 73 million, whereas the value of Swiss exports had risen to 120 million.[7]

This domination continued into the postwar period, though demand was no longer so buoyant. Some of Switzerland's best customers had suffered heavy loss of capital, and prewar patterns of trade had been severely disrupted by new political boundaries. The war, moreover, had seen the institution of all manner of exchange and trade controls, which like most temporary measures gave rise to new vested interests and outlasted the circumstances that gave them birth. And the large Russian market, Switzerland's fourth most important, was simply closed, first by the October Revolution and its sequel of disorder, then by want of foreign exchange and by the determination of the new Soviet regime to build its own clock and watch industry.

As a result, Swiss exports of watches ran down from 18 million units in 1916 to 8.4 million in 1921.[8] This brutal contraction was felt the more keenly because many horological enterprises had augmented their earnings during World War I by the manufacture of such related products as shell and bomb fuses, which often incorporated timers. Now the bottom had dropped out of that market.[9] From this low point, watch sales did pick up and reached 14.4 million units in 1923, surpassing the 13.8 million units of 1913. But the composition of output had changed: many fewer gold watches; many fewer pocket watches; many more movements to be cased abroad behind the protectionist walls that were going up everywhere. Even more disturbing was the increased export of components (*chablons*) to be incorporated in the watches of foreign competitors.

The response of the industry to these new conditions was very different from what it had been to the American challenge of the 1870s. More was at stake now. Much more capital was tied up in plant and equipment, business units were larger, and production was more concentrated. These are the conditions that promote cartelization, and this was exactly what happened. The first step was the creation of a number of trade associations; among them: (1) the Fédération Suisse des Associations de Fabricants d'Horlogerie (later Fédération Horlogère, FH), founded in 1924 to defend the interest of the makers and sellers of finished watches; (2) Ebauches S.A., a trust created in 1926 to embrace all producers of rough movements; and (3) the Union des Branches Annexes de l'Horlogerie (UBAH), established in 1927 for the makers of components (such as dials, jewels, hands) other than movements.

The next step was the acceptance, beginning in 1928, of collective agreements governing output, pricing, and export policies of all producers in the industry, with provision for enforcement and compulsory arbitration. Special constraints were aimed at the elimination of *chablonnage* (export of components).

The final step was government intervention. The strongest of cartels has trouble maintaining discipline when times are bad, and times were very bad for the watch industry from 1930 on. Even with the best of will, many firms were no longer able to live with restrictions imposed in better years and had to realize inventory at almost any price. The Swiss government, therefore, faced with the prospect of disaster in one whole corner of the country, created in 1931 a super "holding": ASUAG, the Allgemeine Schweizerische Uhrenindustrie AG, which bought up a majority of the shares in Ebauches S.A. and in several of the leading makers of component parts.[10] This was followed in 1934 by a federal statute giving the watch cartel's private agreements the force of law and imposing new restrictions on output and technique. Components manufacturers were to sell only to Swiss makers; and Swiss makers in turn would buy only Swiss components unless foreign parts were available at prices at least 20 percent lower. Price markups at each stage were to be set by industry-wide agreement, and all exports of watches and movements needed official permits. To hamper the growth of competitive industries, government authorization was required for the sale of watchmaking machinery, dies, tools, and designs. Finally, no firm could expand, move, sell out, make acquisitions, or change the composition or character of its operations without prior approval of a federal commission composed largely of other watchmakers; while entry of new firms was subject to similar authorization.[11]

Enforced in this way, the Swiss watchmaking cartel was one of the strongest in history. The prohibition of export of equipment and knowledge recalled hoary mercantilist antecedents: Venetian interdictions of the sixteenth century to protect the secrets of their naval arsenal and their glassworks on the isle of Murano; and British efforts to sustain a monopoly of mechanized manufacture against continental rivals during the Industrial Revolution. These constraints had not worked for either country: there was no way to keep craftsmen from slipping away, and business, like love, laughs at locksmiths. Whether the Swiss cartel would have proved more effective and enduring than its predecessors is hard to say, for it did not have that long to run.

THE ACHILLES' HEEL of cartels is loss of market control, whether by secession from within or competition from without. The more successful the cartel, the greater the reward to going out or staying out. Yet the Swiss are more disciplined and law-abiding than most, and their watch cartel initially worked like a charm, for want of an outside challenge. In particular, the once dangerous American rival had lost its teeth.

Waltham, pioneer and bellwether, was in desperate straits. A long history of mismanagement had left it deep in debt and disarray. Labor was working far under capacity, in effect behaving like those cottage weavers of the eighteenth century who did only enough to earn what they felt they needed. (Waltham found this out the hard way: every time it reduced piece rates, the workers did just enough more to take home the same pay.) Supervision was no better: in 1917 the vice-president felt obliged to scold the foremen for not spending "enough time actively supervising the work for which they are responsible . . . Too much time is spent in the corridor, at the door, in the garage, visiting in other departments, and most of all sitting in your *own office.*"[12] Pilferage was rampant, and personnel thought nothing of bringing in their own property for repair in the company shops. Each shop was a little fief, staffed by nepotism and connections, working without regard to the needs of the company as a whole. The result was serious imbalances in output, a growing stock of unfinished movements, a lengthening line of frustrated would-be customers. Every once in a while some of these broken pieces found their way to the market, completed catch-as-catch-can by some worker who wanted to eke out his wages with a little private enterprise—again, just as in the eighteenth-century wool manufacture. Waltham, in short, was an enterprise out of its time—a modern factory that had gone back to an earlier mode of production.

This kind of mess cannot be kept secret. Dealers were complaining about the quality of the goods delivered; suppliers were worried by delays in payment; stockholders threatened revolt. Least patient were the bankers, who as always wanted to be paid first. In such circumstances, the standard cure is a reshuffling at the top, hastened in this instance by the nervous collapse of the general manager in February of 1921. His successor was "Silk Hat Harry" Brown, the treasurer, who had learned his watchmaking in the hotel business. Brown knew he was beyond his depth and called in a team of consulting engineers, who reported in astonishment that four thousand people were employed where two thousand would do. Before the year was out, the First National Bank of Boston, in cooperation with other creditor banks, took control of the company. Their man was Gifford K. Simonds, who had learned watchmaking as treasurer of the Simonds Saw and Steel Company of nearby Fitchburg. Charles Moore, the historian of Waltham, describes him as a "stimulating personality" with a "capacity for strong and active leadership."[13]

Simonds had been brought in to clean house. He immediately launched an economy drive which disposed of such superfluities as the cafeteria and swimming pool; also the doctor and optician. Pensions were curtailed. A speedup eliminated rest periods: the workers who threaded the hairsprings, for example, had been doing fifteen minutes on, six minutes off. They claimed they needed the rest. Simonds did not think so. Some workers did their job with a loupe, that small eyeglass that is the prime tool and symbol of the watchmaker and jeweler. Simonds found it "too quiddly to do work that way." (The New England elite has a fondness for "inside" words, understandable to peers but cryptic to everyone else.) What happened then is not clear, but one presumes that other magnifying tools were made available.

Not every change was so trivial or vexatious, but these were the things that people seized upon to complain and undermine the authority of the new manager. Simonds saved some money, but the bankers had had enough. In 1923 they reorganized the company. This time the prime mover was Kidder, Peabody & Company of Boston, who brought in a kindred spirit to run the new Waltham. This was Frederic C. Dumaine, who had learned watchmaking as boss of Amoskeag mills of Manchester, New Hampshire, probably the largest cotton manufacturing plant in the world.[14]

Dumaine was a banker's man—what Charles Moore calls a financial industrialist. He was a self-made man who had begun with Amoskeag as an office boy at the age of fourteen and had worked his way up through the ranks of this brahmin bastion to the post of treasurer, which in New England mills meant chief executive officer. Here his French-Canadian name stood out from the roll of Coolidges, Amorys, Appletons, Cabots, Jacksons, Lymans, Lowells, Wigglesworths, Gardners, Searses, Dexters, and Cuttings who had run Amoskeag for a century. (One would almost think this a list of Harvard University buildings and endowment funds.) Dumaine was more than a match for these Yankees. He combined shrewdness and intelligence with the ambition of the parvenu, and he ran Amoskeag with hard head and tight fist. During the still-prosperous years before and during the Great

War (Amoskeag had always boasted an extraordinary record of earnings and dividends) he accumulated a large reserve of cash and liquid assets, which he carefully buried in the accounts where neither workers nor shareholders could find it, let alone extract it. At the same time he nursed his personal fortune, moving into railways, shipping, and banking and establishing himself as one of Boston's most capable capitalists—"a financier among financiers," to use Alan Sweezy's phrase.[15]

It is worth pausing a moment to look more closely at Dumaine's handling of Amoskeag, because it throws light on his direction of Waltham as well. Amoskeag, like most other textile mills, had enjoyed exceptional prosperity during World War I; but this was followed by lean years, and management moved to cut costs by speedups and layoffs. The result was mounting discontent that fairly exploded in 1922, when wages were abruptly cut by 20 percent and the work week increased from forty-eight to fifty-four hours. The Amoskeag workers, long cradled by paternalism and traditionally the most loyal in New England, "hit the bricks" for the first time in the history of the company. Manchester was a one-firm town, and closure of the mills meant pain and hardship for every resident. All levels of public authority sought to mediate and intervene, but Dumaine was adamant: "Grass will grow in the streets of Manchester unless the workers agree to terms."[16] In the end, after months of bitterness, the reopening of other struck textile mills elsewhere in New England compelled Amoskeag to compromise; strikebreakers were no longer available. The company rescinded the wage cut; the workers swallowed the longer hours. No one won. The unwritten social contract that had been the basis for Amoskeag pride and performance was shattered and would never be restored.

It was a seriously wounded firm, then, that had to meet the intensified competition of low-wage Southern manufacture. Now was the time to use those carefully hoarded reserves to turn the company around, specifically to buy new equipment and move into such new, more competitive lines as artificial fibers. This was what the production men were pleading for, and this was what Dumaine had always promised in justifying his thrifty management: the fat years would pay for the lean ones.[17] But Dumaine was better at saving than spending. Rather than risk fresh capital (which is presumably what capitalism is all about), he chose to segregate what he already had while pumping the company for such additional funds as it was still able to earn. Not only was it decided not to purchase new equipment, but the existing machinery and plant were allowed to run down into disrepair and abandonment. Whereas in 1921 about 750 men were employed in the machine shop and 250 in construction, by 1929 there were only about 250 left in both groups, and in 1931–1935 less than 100. After 1927, painting was neglected, structural defects were ignored, machines cannibalized, tons of spare parts sold off as scrap.[18] This policy of systematic niggardliness was at once the sign and assurance of Amoskeag's demise.

With deferred liquidation in view (death by the drip method), Dumaine moved smartly to retrieve and protect those huge cash reserves. His strategy deserves to be remembered. In 1925 he created a new company called Amoskeag Manufacturing to buy operating plant and equipment, inventory, and some $6 million in cash from the old firm. The latter kept $18 million and received payment in stocks and bonds for the assets transferred. This bonded debt then served to pump out the rest of the cash and the first fruits of each year's operations. The workers protested that it was their labor that had created those reserves. To no avail. The years that followed saw Amoskeag slide inexorably into bankruptcy while Dumaine continued to drain such revenues as he could into the holding company. The liquidation of the enterprise occasioned so much outrage and scandal that there was a U.S. Senate hearing in 1937. The special investigator stood appalled: "The activities of the holding and manufacturing companies savor of nothing short of financial sabotage."[19]

This was the Monsieur Purgon to whose enematic mercies the fortunes of Waltham were now entrusted. A later historian was to write that Dumaine "brought Waltham the advantages of financial capitalism without destroying the efficiencies of industrial capitalism."[20] The "advantages" consisted in what the Germans call a *Sanierung,* the sanitizing of the company's balance sheets. In the two decades that Dumaine ran Waltham, he turned red ink into black, earned substantial profits, paid off just about all debt, made a tidy sum for Kidder, Peabody and a fortune for himself. He did this by cutting wages and holding fresh investment to a minimum, by squeezing the last turn out of machines that had been running for decades and the last die out of toolmakers who had been working even longer. Moore, who wrote his history of Waltham under Dumaine's eye, is reluctant to criticize, but his figures tell the story. Under Dumaine's management (1923–1943), the company spent $1,288,330 for equipment (an average of $65,-000 a year), or 1.5 percent per year of the book value of machinery at the start of the period. At the end of his term, fixed assets had been written down from $4.3 to $1.2 million and more than half of the cumulative cash net of $17.2 million had been spent buying up shares and notes, substantially increasing Dumaine's share in the equity. With exquisite timing, Dumaine sold his stock in 1944, just before the end of wartime orders. The *American Watch Worker,* organ of the union labor he would never negotiate with (he dealt only through intermediaries), lauded him as personifying "New England Industry," and then went on: "He represents a school of thought that built America."[21] With that kind of union, who's afraid of class conflict?

The economics of pumping and extraction was bound to hurt quality and performance. Dumaine took the first opportunity offered to impose an across-the-board wage cut, which brought him the first and only strike in Waltham history. It was a bitter struggle that lasted months, and when it was over, the company had won—but had also lost. The conflict exacerbated the legacy of mistrust and resentment that Dumaine had inherited. Waltham was still paying for it twenty years later. Meanwhile, Dumaine was as parsimonious in hiring new talent as in buying new machines. Waltham almost never went outside for skills and experience. Instead, it hired from among those who applied for jobs and showed, among other things, "a satisfactory general demeanor" (read: appropriate docility). Workers learned at the bench: "experience is the only teacher." Those few who would become watchmakers spent nine months at beginners' wages in Waltham's own school, whose teachers had learned at Waltham. Just about all promotion was from within. The whole system could not have been better designed to perpetuate routine and reward faithfulness rather than initiative and imagination. Even the tactful Moore has to concede that "under such conditions the spirit of competition languished and there was little incentive for a workman to excel either in comparison with his fellow workers or in relation to an established standard of quality."[22]

Yet as Moore notes, this was an industry that more than most depended on cooperation between labor and management. This was especially true in the interwar years when the shift from pocket watches to wristwatches entailed a steady miniaturization of calibers, with concomitant reductions in tolerances. Nothing could have been better chosen to strain the capacity of obsolescent machines and the patience of aging master craftsmen. An internal report of Waltham in 1935 hints at the scope of the problem: "There *must* be no more quibbling about tolerances on measurements or crying for larger tolerances. What we need *more than anything else* are smaller tolerances, strict adherence to *them*, and to the proved methods which are producing parts. A man must *not* say, my machine will not repeat [at low tolerance], and rest content. He must roll up his sleeves and help determine why his machine will not repeat."[23] So much for the "efficiencies of industrial capitalism."

Waltham was a basket case. Elgin and Hamilton, the other survivors of one hundred years of factory watchmaking, were doing much better. In the interwar years (1918–1940), for example, Elgin sold some 18 million watches, against 8.3 million for Waltham. Elgin made money. And Elgin "always kept [its] machines and tools in tip top condition" and spent large sums on the equipment needed to produce small calibers with interchangeable parts.[24] It also understood better than Waltham the importance of formal schooling as well as training: the Elgin Watchmakers College, built in 1920, graduated over six thousand students by the time it closed in 1960, and some of the best of them went to work for Elgin.[25]

Yet even the healthy American makers were losing ground, not only to the Swiss watch firms but to those American assemblers (Bulova, Benrus, Gruen, Longines-Wittnauer) who bought their movements in Switzerland and merely cased them in the United States.[26] (See Table 8.) These last, by the very nature of their enterprises, were market-oriented rather than production-oriented. They knew the value of publicity, the power of a slogan thousands-of-times repeated ("Bulova Watch Time"), and the might of the new mass medium, radio. The advertising budget of any one of these exceeded that of all American makers combined.

As marketing gained in importance, technique lost. The watch industry of the twentieth century was operating at the upper bend of the S-shaped growth curve, and gains to further mechanization were leveling off after the big leap forward of the third quarter of the nineteenth century.[27] Market penetration now depended on product rather than process innovation. Here no one could match the Swiss, with their smaller production units, shorter runs, their large pool of draftsmen and designers, their excellent schools of horology. In particular, they were far quicker than the Americans to seize the opportunity presented by the shift of taste from pocket watches to wristwatches. Not only were they better able to design smaller calibers and place them in production, but they were prepared to invest enormous effort in attaining standards of accuracy comparable to those achieved earlier by the larger pocket watch. Such companies as Rolex and Omega played a major role here, winning prizes for their small pieces in the chronometric competitions and stressing once again the special virtue of SWISS MADE.

This lead made all the difference; the Americans were always playing catch-up ball. Like British watches of the nineteenth century, American timepieces of the 1930s continued to enjoy a reputation for solidity, reliability, and accuracy. But they were thicker than Swiss watches, and the Swiss somehow managed to convince the public that thinness—even a difference of a few millimeters—is chic. (It is also a source of fragility and reduced accuracy, but in these matters, profile is more important than content.) American dials and cases were less elegant than Swiss; even when they tried to copy, their mimetic appearance lacked "flair" (compare these with nineteenth-century copies of classical furniture). American watches were also boring—just timepieces, with none of the useful and amusing complications (self-winding, chronograph, calendar work, and so on) that appealed to newer clienteles.

The only thing that kept the Swiss at bay was the imposition of ever higher duties on imported watches. Even so, imports of Swiss watches and movements almost quadrupled from 1934 to 1940, and these were only those legally entered and duly recorded. High duties made contraband flourish like the psalmist's green bay tree. The American manufacturers were outraged by the ingenuity of those smugglers who were stopped, such as the baker on a German liner who used his ample girth to conceal 780 seventeen-jewel movements in a special belt, and they rightly assumed that for every watch caught there were dozens getting in. Dumaine of Waltham estimated the volume of smuggled imports in the early thirties at one to two million movements a year. He had reason to exaggerate the scourge, but his figures do not sound far-fetched.[28] Two thousand belt-loads would have done the trick. Watch movements, especially the new style of small and thin calibers for wrist wear, were just too easy to hide, so that the government rarely made a seizure unless tipped off by an informer. Informers in turn expected to be rewarded, and the usage was to pay them from the proceeds of the seized goods, which were sold at auction. These auctions, though, simply increased the competitive pressure on the American industry, which made every effort to halt them, thereby effectively paralyzing enforcement. It was a classic "catch-22" situation—damned if they did and damned if they didn't. As for the Swiss authorities, they reacted to American protests by shrugging their shoulders. Smuggling had been a way of life to the Swiss watch industry for centuries. Besides, as they pointed out to the Americans, the cure was obvious: just lower the tariff. It was precisely the advice the Genevans had given the French 150 years before.

The interruption caused by the war simply enhanced the Swiss domination of world markets. (See Table 9.) Everywhere else production of civilian timepieces more or less stopped while manufacturers devoted all their resources to producing war material. Meanwhile, demand scarcely diminished, for there is nothing so conducive to an interest in time as military service, unless it be factory employment. I myself spent most of my first army pay for a Swiss-made wristwatch with luminescent hands. (We didn't know then that radium-coated objects were potentially carcinogenic. The workers who made them, of course, were in far greater danger than the wearers.) Reveille in my camp came at 0445 hours, and we had carefully measured times to wash, dress, and do everything else we did. With such a timetable, the only way to free oneself for other things was to know the hours and minutes. Looking back, I can see that I was like one of those medieval monks for whom lateness could be a grievous sin. He risked loss of salvation; I risked kitchen police. He had his bells; I had my trusty watch.

When the war ended, military orders for watches dropped abruptly, but unlike the years after the First World War, there was no depression, and long pent-up civilian demand pressed hard on Swiss watchmaking capacity. In the late 1940s merchants were ready to pay a premium for Swiss watches, just as dealers and customers paid "under the table" for still-scarce automobiles. For a brief moment, the Swiss share of the world production is estimated to have surpassed 80 percent. It fell back rapidly as the crippled national industries of Japan, Germany, France, and the Soviet Union returned to production. Even so, the early fifties still found the Swiss with over half the world market. There they held, with some ups and downs, through the years that followed.

21 The Quartz Revolution

R EMEMBER THE LAW OF NEMESIS? Success invites emulation and challenge. The extraordinary Swiss domination of the watch manufacture had been a provocation to potential competitors from the eighteenth century on. These now mounted a new assault, more serious than any before.

The first intimations of trouble came from the United States, where the moribund Waterbury Clock Company, soon to be renamed the United States Time Corporation, came under new management and turned into a tiger. As the manufacturer of the old Ingersoll line of dollar watches, Waterbury had initiated tens of millions of Americans into the pride and trials of pin-lever timekeeping. But Ingersoll had managed the difficult feat of running a thriving business down into receivership; and although Waterbury had taken it over and had brought out in 1933 the immortal Mickey Mouse line, it had had to be content with a *succès d'estime*—the watches were priced too low to bring in much money. (Mickey made record sales in Berlin in 1945–1947, where the going price to Soviet troops was $500 a watch. But the Russians were paying with other people's money, printed on plates that the U.S. government had obligingly made available to them; and none of that went to Waterbury in any case.) Waterbury was nearly defunct when in 1942 a group of businessmen led by Norwegian refugee Joakim Lehmkuhl acquired a majority of shares with a view to converting the plant to the production of fuses. This it did very profitably for a while, but with the coming of peace, sales plummeted from $70 million to $300,000 and Lehmkuhl cast about for a new line.

The answer was the Timex, a superior *clock-watch*—that is, an unjeweled, clock-type watch with pin-lever escapement. Instead of ordinary steel, Lehmkuhl's engineers used armalloy, a new hard metal developed during the war that gave service as good as that of many stone pallets and bearings; and its designers packaged these movements in plain, clean cases that contrasted sharply with the chunky, manifestly economical lines of the old Ingersolls. The first, simple models sold at $6.95 to $7.95—a lot more than a dollar; but then, the dollar was not what it had been, and the Timex looked the money and more. With success, the company added complications: sweep seconds, shockproofing, waterproofing, calendar indications, self-winding—all the special features that made Swiss watches so attractive to a wide range of clienteles. Rebuffed by the traditional trade channels (jewelers, watchmakers), United States Time sold these watches in every available retail outlet: drugstores, supermarkets, variety stores, airport shops, hardware stores, tobacconists—a quarter of a million points of sale at the peak. They were selling not elegance or prestige but cheap time, and at those prices, people could afford not one watch but two or three or more, with faces and straps to match a variety of outfits.

The Timex was not the world's most accurate watch. It kept time within a minute or two a day. But this was accurate enough in a world of radio time announcements and telephone time. All one had to do was adjust the watch each morning, and it would give close enough readings for just about any activity. And Timexes were made for just about any activity. Timex salesmen convinced retailers of their solidity by slamming them against counters and dunking them in pails of water. Television advertising featured Timexes undergoing "torture tests": watches fastened to the hooves of galloping horses, to a spinning ship's propellor, to a high diver at Acapulco, to the pontoons of amphibian airplanes. There is even the story of the Timex that was still going after five months in a man's stomach. The man, a New Yorker, had swallowed it when confronted by a mugger. Greater love . . . When the watch was removed, of course, the time was no longer correct, but the surgeon generously allowed that the watch may have been wrong at the time it was swallowed.[1]

The makers of conventional jeweled watches were horrified. Was that any way to treat a watch? Neither their egos nor their products could stand up to that kind of abuse. Timex couldn't have cared less.

Here was value for money. The whole success story rested, in the last analysis, on the productivity of Timex's workers, who were equipped with the most efficient machines that money could buy and five hundred toolmakers could design. In this area, Timex represented the culmination of two centuries of striving: parts were standardized and made interchangeable not only within plants but among plants; and machines were automated as much as possible so as to reduce the human element to a minimum. As one of the managers put it: "When we put a watch together with relatively unskilled labor, it must be able to run accurately the minute the last wheel is put in place. We can't afford the petty troublesome adjustments that are found in the hand-made watch industry."[2]

To achieve this end, Timex simplified its designs to the extreme. Carping watchmakers noted that Timex cases were riveted, and hence could not be opened for repair. But Timex made no pretense at maintenance or repair. When a Timex stopped running, the owner simply threw it away and bought another—if he did not already own a spare or two.

In 1960, eleven years after United States Time brought out the first Timex, it made eight million of them. That was more than the total output of the Japanese watch industry and more than three times the entire United States production of jeweled watches. In 1962, one out of every three watches sold in the United States bore the name Timex. By that time the company was bringing out higher-priced lines of jeweled watches, which it sold like the others in such outlets as drugstores and airport shops, but which it also used to penetrate the jewelry trade. These watches were cheaper than comparable Swiss pieces; yet they gave equivalent service, were repairable at the plant, and were sold at conventional markups. Another ten years (1973), and Timex had 45 percent of the American market, 86 percent of domestic watch production.[3] An astonishing performance.

Appetite grows with the eating. All this success led Timex to look around for new worlds to conquer. It moved first into Canada and Britain, then into South Africa, perhaps because its marketing techniques were easier to introduce into English-speaking societies. By the mid-1970s, it was ready to move into France, setting up a plant at Besançon in the heart of French watch country to produce watches under the brand name "Kelton." Then it was Germany's turn: "They [German watch producers] were trying to sell West German customers on the idea that expensive wristwatches added to the owner's prestige. We set out to fill the vacuum in the lower price watch market and convince buyers that good watches don't have to be expensive."[4]

In each of these invasions of "hostile" territory, Timex adjusted its marketing to the special circumstances of local tastes and trade. In Germany, for example, it was crucial to gain the cooperation of jewelers and specialty watch stores, which accounted for more than three quarters of all watch sales in the country. Timex systematically wooed these outlets, offered an array of special inducements, including markups comparable to those on other watch lines and handling commissions on all repair jobs. It also spent millions of marks to test-market its line and prepare the ground for its usual publicity blitz. Within two or three years (1967–68) it was selling over half a million units, and the Swiss were complaining that Timex was compelling them to sell cheaper in the German market in order to hold their place. (A few years later, they were lamenting similar pressures in the East Asian market, where the surging Japanese industry was already pressing them on jeweled watches.) At that point (1972–73), the former United States Time Corporation, renamed in 1969 the Timex Corporation, was making 30 million pieces around the world, had sales of $200 million, and employed some 17,000 people in twenty plants. So potent had been its impact that the Fédération Horlogère made a point of studying its operations in minute detail and reported its findings to the Swiss industry in a monograph, "The Timex Formula."

This was the good old Swiss response to challenge: study it and learn from it. Nothing that Timex did was radically different from what had gone before; it was just that Timex had pushed modern techniques of production and marketing to their logical conclusion. Timex, in other words, was the extreme expression of older developments. The Swiss felt there was no reason why they should not or could not follow suit.

Just about this time, however, the Swiss watch industry found itself confronted by a very different kind of challenge, one that rested on a radical transformation of the technology of time measurement and resulted in the creation of what still looked like a watch but was in reality a new product. This was the quartz revolution, and the Swiss watch industry had far more difficulty coping with it than with the Timex formula.

The antecedents of this revolution go back to the turn of the century, when Pierre Curie observed the phenomenon known as piezoelectricity: certain crystals, among them quartz crystals, actually vibrate mechanically—change shape back and forth—when an alternating current of electricity passes through them. This oscillatory effect found one of its first and most important applications in radio broadcasting, where crystals were used as energy resonators and controllers; and this in turn led telecommunications engineers seeking reliable frequency standards to the realization that crystals could be made to produce stable vibrations—that is, could serve as a clock. All that was required was to divide frequencies of 100,000 and more cycles per second into slower, more usable beats. (See Figure A.7.) It was found, moreover, that quartz crystals showed almost no response to temperature variation (only two or three parts per million per one degree Celsius; or to atmospheric pressure (about one part per million per four inches of mercury). At the same time, their vibrations were isochronous for all amplitudes. Gone then were all the old problems of compensation and circular error. The first quartz clock proper was developed in the United States (Horton and Marrison, 1928); the first such clock to be installed at the Greenwich observatory dates from 1939 and tracked time to within two thousandths of a second per day.[5]

Large clocks powered by electric mains and installed in astronomical observatories and physical laboratories were one thing; battery-powered wristwatches, another. It took another thirty years before quartz clocks were miniaturized enough to be suitable for personal wear. Even then, they still suffered from serious disadvantages that made them acceptable only to a small avant-garde of novelty lovers and fashion pace-setters. They were expensive: the earliest models (1968–69) cost upward of a thousand dollars, more than twice that in today's money. They were bulky—crude in appearance alongside the slim, elegant forms of the best mechanical watches. They were inconvenient, because battery life was short and one could not afford to waste energy on a permanent display. The user had to press a button to read the time, not an easy thing to do when one is carrying a case in one hand and hurrying to catch a train. Finally, for all the vaunted accuracy of quartz, these early models were not so reliable as conventional mechanical timepieces. After about a year, the stability of the crystals would break down and the time readings would drift. For the price of a Timex one could throw away one's watch and start over. But not at $1,500 a copy!

With this new technology, watchmaking was really a new industry, and it was not altogether surprising to see the field invaded by firms that had never produced timekeepers before. Indeed, anyone at all could now buy the new circuits, have them cased by cheap, unskilled labor, perhaps in Third World countries, and sell them through any and every outlet that would take them. Experience in distribution and access to markets were more important, it seemed, than skill in manufacture, and an enterprise such as Gillette moved into the watch business on the assumption that it could sell timepieces like razors and blades.

Inevitably a great many of these early product offerings fizzled. The very rapidity of technical change made last year's model obsolete, and woe to the producer who could not dispose of stock fast enough to justify the long runs required. Durability seemed a secondary consideration (why build an obsolescent item to last?), and most makers thought it was simply not worth the effort to institute strict quality controls. As with dollar watches, it was cheaper to let the public screen the product and send back for replacement any pieces that failed.[6] This is just what buyers did, in record numbers: the Swiss in their *Schadenfreude* report rates of return of up to 40 percent, and as late as 1976 the J. C. Penney Company, a major American mass-market department store

chain, was shipping back 40 percent of the units sent by one supplier before they had put them out on the shelves.[7] Even experienced watch firms miscalculated, as did Seiko when it had to
withdraw its pioneer quartz model because of technical problems.
Gillette could afford to write off its misadventures; others were
not rich enough. Readers may remember the flood of unfamiliar
brand names that filled the newspaper advertisements as price
leaders and sale bait. The shake-out is still going on.

Under the circumstances, one can hardly blame the Swiss for
looking skeptically at these ugly ducklings and dismissing them as
faddish novelties. With our 20/20 hindsight, we know that this
was a serious mistake. But how were they to know that in 1968?

The answer, I think, is that they should have known it. There
had been a number of warnings that horological technology was
in the throes of a revolution and that the fundamental principle
of the oscillating controller was finding new forms of expression
vastly superior to the balance wheel and hairspring. Quartz timekeeping, as noted, went back to the 1920s, and atomic clocks to
the late 1940s and early 1950s. More to the point of watch manufacture, the introduction in 1960 of the tuning-fork controller in
Bulova's "Accutron" posed a direct challenge in the marketplace.
(See Figure A.8.) This technique, ironically, had been invented by
a Swiss engineer, Max Hetzel, and had been offered to major
Swiss makers, who were not interested; so Hetzel took his device
to the United States, where Bulova made such a success of it
that a number of Swiss companies swallowed their pride and
purchased the right to bring out their own models. Their contracts required them to note in their advertising that they were
using the tuning fork by courtesy of Bulova—which they did
grudgingly in type so small that most readers needed a loupe to
read it.

When Bulova marketed the Accutron, it issued an unprecedented guarantee: that the watch would not vary by more than a
minute a month (two seconds a day); what's more, that the watch
would keep to this standard of precision throughout its life.[8] Now,
it goes without saying that there are very few people who operate
within such limits. Even so, it has always been the rule that the
quality of instruments of time measurement is a function of their
precision. This was a point that had long been a feature of Swiss

watch advertising, with its emphasis on chronometer certification by Swiss observatories. In making this assertion, then, Bulova was saying in effect that it was offering a better product—better than all but the very best of mechanical timepieces.

The Swiss watch industry, as industry, took this challenge seriously enough to assign it as problem number one to the newly established collective research laboratory, the Centre Electronique Horloger (CEH) of Neuchâtel.[9] The task was to develop an even better chronometric technology, specifically quartz controllers in combination with integrated circuits. By 1967 the CEH was ready to submit its prototypes to the annual chronometry competitions. They arrived just in time to contribute to the demise of this almost sacrosanct institution. The contests could not survive the intrusion of this new and superior competitor: watches controlled by resonators vibrating tens of thousands of times a second could keep within limits of variation ten times smaller than even tuning-fork watches.

The story of this drama—to use Charles Thomann's word—is worth recalling, because it sums up the death of an old technology and birth of a new. In 1961 two quartz marine chronometers were submitted to the Observatory at Neuchâtel. They were voluminous and used far too much energy: the batteries had to be changed every week. But they kept far better time than any such instrument had ever recorded. One of them made a score of 1.2 points (0 would be perfect), as against 2.8 for the best mechanical chronometer. The *chronométriers* saluted and smiled. One year later, new instruments were submitted, ten times smaller. The best of them made a score of 0.8. The next year, this record was broken again, down to 0.13, and by 1967 it was reduced to 0.0099—two orders of magnitude, then, in six years. The best of the mechanical instruments could do no better than 2.3, a sensational result by the older standards, representing an enormous investment of skill, finesse, and obsessive dedication. The really bad news, though, was that the first quartz watches of small caliber were coming in, among them the products of the CEH. The best of these made 0.152, humbling an unbelievable record performance of 1.73 by the best of the mechanical watches.

Against that standard of accuracy, even the best *régleur de précision* was helpless. The old combat lost meaning, both for spectators and participants. Moreover it was quickly apparent that the very competitions that had consecrated the supremacy of the Swiss in the matter of horological precision might now serve as a vehicle for their overthrow. In particular, the Japanese announced their intention of competing in the production-line wristwatch class with quartz timekeepers that the Swiss were not yet ready to match, and that kind of humiliation was not to be risked or endured. In Thomann's words, "Swiss horology refused to envisage even the possibility of a defeat on its own ground." On April 26, 1968, the Council of State decreed that the competition in the wristwatch chronometer category was suspended. It was never revived, and with that gone, competition in the other categories lost much of its interest. A few years later, the suspension was made permanent. The *régleurs,* those athletes of chronometry, bitter at their premature retirement, were left to nurse their memories. They meet now once a year for dinner; also at funerals, for this was a specialty that demanded experience, and these are men of a certain age.[10]

So science had defeated art. Not that Switzerland lacked for science. Alongside the CEH, other Swiss laboratories and industrial enterprises were engaged in research into the new electronic technology and developing software and hardware applicable to horology. Among the most successful were the Institut Battelle in Geneva, which helped develop a quartz watch with analog dial for Omega in 1970, and FASELEC of Zurich, founded in 1967 by an international consortium (Philips, Brown Boveri, Fédération Horlogère, Landis & Gyr, et al.) to develop, among other things, low-energy microcircuits for horological use. As with the Accutron, moreover, whatever was not available in Switzerland was there for the buying abroad. In the late 1960s, for example, Ebauches S.A., building on personal contacts, got Hughes Aircraft (USA) to contribute its know-how to the manufacture of integrated watch circuits at Marin near Neuchâtel; and in 1972, Texas Instruments (USA) joined with Ebauches S.A. to furnish liquid crystal digital displays to Longines and ASUAG.[11]

In short, the Swiss watch industry had everything it needed to enter the new world. The only thing lacking was entrepreneurship: the manufacturers of watches were not interested. Those years of comfortable, sheltered monopoly rents had cost the industry what had once been its most precious characteristic—its *Neuerungsfreudigkeit,* its joy in innovation. A few *horlogers* were prepared to run off a few copies of prototypes—just by way of showing they could do it. But they were not prepared to go into mass production. Why sink a fortune into a fad? What did *électroniciens* know about watches anyway? They had never made one, never even measured time before. Besides, the new technology called for investments larger than the great majority of Swiss watch enterprises could afford. The only firms that had the means were the great holding companies such as ASUAG (which included Ebauches S.A.), and these did move to learn and use the new technology. But they made modules (the quartz equivalent of the mechanical movement) and components, not watches. If the Swiss *horlogers* were not going to buy, the combines would have to sell their output abroad and promote in effect the development of competitive industries.

The watch manufacturers changed their publicity to suit the circumstances. Patek Philippe, the world's most esteemed maker (not everyone would agree, but that is what auction prices tell us) no longer assured and reassured prospective buyers that, foolish though it might seem to spend so much money just to have a watch that kept time within a minute a month, this kind of accuracy did matter to people of discrimination and accomplishment. Instead, the new line went: "A Patek Philippe doesn't just tell you the time. It tells you something about yourself." (And tells other people, of course.)

In the meantime, neither Japanese nor American companies were idle. In Japan, Seiko brought out its first electronic model in 1968–69; and when this had to be withdrawn, it followed it quickly with another and better. In the United States, the Hamilton Watch Company took the lead, working with Electro-Data to bring out in 1969 the "Pulsar," the world's first all-electronic watch, offering a digital display (numbers) in place of the traditional analog dial (turning hands). The Japanese and American entry into the new quartz horology, then, though roughly contemporaneous with the Swiss, was qualitatively different. In Japan and the United States, the initiative came from individual enterprises, ready to bet their money on the new product and to produce it in large quantities (often too large). In Switzerland the first moves were collective, and there was still a critical gap to be crossed between research and development on the one hand and production and marketing on the other. Watch companies are like horses: you can lead them to know-how, but you can't make them use it.

The question remains why the Japanese and Americans were so quick to take up the new technique. The answer would seem to lie in the importance already attained in both economies by the manufacture of electronic devices (calculators, computer hardware, information and communication equipment, recording instruments) and their components, in particular the chips that hold the integrated circuits. The leading Japanese watchmakers, Seiko and Citizen, had already had some experience in this area, and they were soon joined from the other side by such electronics firms as Sharp, Ricoh, and Casio. In the United States a similar pattern emerged, with Hamilton followed by such horological innocents as Texas Instruments, Optel, Intel, Hughes, and Fairchild. Some of these went into the manufacture of timepieces or their components as a derivative activity: their production capacity exceeded demand for older uses, and they wanted to diversify. Others were invited in, as Electro-Data by Hamilton or Statek (USA) by Seiko, by watch firms needing access to the new technology. (Statek was the inventor in the early 1970s of a photochemical method for processing quartz crystals—a substantial advance over the mechanical technique used previously.)[12] In all this, national as well as professional lines blurred as knowledge was traded for profit across frontiers. Even the Japanese and Swiss exchanged know-how.

In 1970, after a couple of years of experience with the new quartz watches, the Fédération Horlogère in Bienne reassuringly noted that 98 percent of sales still consisted of mechanical timepieces and that the electronic, solid-state timekeeper, though accurate, did not constitute a sufficient advance to sweep the market. Yet what did they expect? Nothing so radically new simply sweeps the market, especially when it still costs a lot and the older product is giving adequate service for much less money. The Japanese were not moving much faster: as late as 1974, less than 6 percent of Japanese watches and movements used quartz controllers.[13]

What the Swiss makers did not take into account, though, was the future prospects of the new technology. Japanese manufacture of electronic watches bounded upward, increasing more than ten times in the four years 1974–1978 (1.8 million to 19.7 million units). This surge was fueled by, and promoted in turn, an extraordinarily high rate of technical improvement—much higher on electronic watches than on mechanical—yielding a steeply falling curve of costs (and prices) over time (the so-called learning curve). The Swiss had never seen such a curve: theirs had always been linear, with gentle negative slope; this was exponential, cascading downward.

In September 1979, horological professionals and amateurs from all over the world met in Geneva for the Tenth International Congress of Chronometry. The Japanese technicians, representing the Seiko and Citizen watch companies, brought with them the elaborate gestures of courtesy they had inherited from their ancestors and an array of charts showing the extraordinary progress of one decade of research and production.

The quartz watch used too much energy to allow permanent display? From 1973 to 1979, Seiko's digital quartz watches reduced current consumption by more than 60 percent while the number of liquid crystal display segments almost tripled. One model, with analog (conventional) dial and day-date display, used only one-tenth the current of their first quartz watch and was equipped with batteries that would last five years instead of one.[14]

The quartz watch was too thick? By 1978 the five-plus millimeters of the first Seiko model was down to less than one millimeter, and volume from four cubic centimeters to less than half a cubic centimeter.[15] All this, moreover, had been achieved without sacrificing economy, precision, or reliability—in contrast to mechanical timekeepers, where miniaturization does entail losses in all these areas.

The quartz watch was too expensive? Over this same decade, retail prices fell to perhaps 2 or 3 percent of what they had been at the start, and this for substantially better timepieces and in the face of gathering inflation.

This spectacular gain was the more impressive because the quartz watches embodied an increasing number of ancillary functions: calendar and chronograph work, simple and then multiple alarms, elementary calculators, double time readings (for use in two time zones), repetition, and *sonnerie*. What's more, all of these were obtained without significant increase in cost or size, or sacrifice of accuracy, whereas in mechanical watches every additional function imposes penalties. Today it is possible to buy a solid-state watch for about $100 that keeps better time than the finest detent pocket chronometer, repeats the minutes, sounds the hours, gives the day and date with due attention to the varying length of the months (including February in leap year), measures elapsed time to the tenth, even hundredth, of a second, serves as preset timer, and offers the possibility of setting several alarms to remind the wearer of a succession of engagements. A mechanical watch with such capability would not fit in any ordinary pocket, would weigh a pound, and would cost several hundreds of thousands of dollars—if one could find the workers to make it. They would spend years making it; and when they were through, they would swear never to take on such a task again. The history of clock- and watchmaking is dotted with stories of craftsmen who undertook excessively complicated projects and lost so much time and money in the doing that they were ruined.

The new technology also made possible performance that was simply beyond the reach of the old. In 1981 an American merchant house, Embassy Marketing of Northbrook, Illinois, advertised a "talking wristwatch"—a timepiece that combined the functions of a minute repeater, alarm timer, and clock-watch. Instead of announcing the time by chimes or buzzes or bells, however, the watch would speak to its owner in pleasant, clear tones. For the alarm it might say, "It's six forty-five A.M.," followed by fifteen seconds of Boccherini's "Minuet"; and if sleepyhead does not wake and shut off the watch, a polite reminder comes five minutes later: "It's six fifty. Please hurry." This, too, is followed by music, and if this is not enough, the sequence is repeated every five minutes. This marvelous device is the invention of Sharp, in spite of its name a Japanese manufacturer of pocket and desk calculators. When Sharp introduced it in the latter half of 1980, it was uncomfortably large and cumbersome. One year sufficed to get it down to wristwatch size, and Sharp licensed the logic to a manufacturer in Hong Kong, where labor is cheaper and exchange rates are favorable. The watch, which was sold with

Union Carbide batteries, is thus the product of a characteristic international combination of materials and factors. It sold in 1981 for $100, verily a trifle.[16] It is probably the first of a succession of supercomplicated portable voice and timing devices that will make the fantasies of science fiction a reality. Who can say how far solid-state physics will take us?

While Japanese and Americans made these extraordinary advances (if the Mercedes were an electronic watch, it would cost $300), the Swiss worked to make their mechanical timepieces better and cheaper. That is a universal characteristic of once-dominant technologies: they make some of their greatest improvements under sentence of obsolescence; the finest days of the sailing ship came after the advent of steam. Thus René Retornaz, director of the Fédération Horlogère, was able to point in 1974 to smarter, faster machines and higher standards of inspection. Under the pressure of competition, standardization proceeded apace: between 1970 and 1973, the number of calibers fell from 550–600 divided into 170–190 types, to 440–460 representing 100–110 types. (This was still far too many.) Production per head doubled, from 723 watches per year in 1966 to 1,305 in 1976. The possibilities of automation were explored as never before: I have seen one remarkable machine that makes Ingold's pantographs look like crude toys. It reads designs, tracks the outlines, and directs the cutting tools accordingly. With it the semiskilled worker can copy any part to any size and produce the most complicated movement desired; all the skill and labor is in the design, as it is in the production of integrated circuits, and in the machine. Executives of the firm that showed me the device, which is of German manufacture, said that they were thinking of reviving production of repeater watches and other multifunctional pieces. I have my doubts. Even with this machine, such watches would cost far more and do less than quartz "complicateds." Still, for those who can afford it, a mechanical repeater is much more fun.

The same for product innovation: Swiss watchmaking poured a disproportionate share of their modest research efforts into the improvement of their mechanical watches. Only about 2.5 percent of sales went to research and development, but if patent figures are any indication, about two-thirds of that sum were invested in the older technology.[17] Some of the results are impressive: for example, Fabriques d'Assortiments Réunies and Favre-Leuba brought out the fast-beat watch, with a frequency of 36,000 oscillations an hour as against the normal 18,000. This was the mechanical answer to the Accutron: the manufacturer claimed that "results can be obtained at least as accurate as an electronic watch with tuning fork."[18] Perhaps; though Richard Good, curator of clocks and watches in the British Museum, London, doubts whether any lever-escapement watch can hold its rate as long as a fork-controlled watch can. In any event, the introduction of quartz has made the point moot. Or has it? British horologist George Daniels tells me that he has invented an escapement that will revive the fortunes and reputation of the mechanical watch. The trouble is that he has not yet found a Swiss manufacturer ready to undertake commercial production. (There is apparently no question of getting a British firm to do this.) The Swiss are not prepared, perhaps, to listen to an Englishman. Or maybe they have now given up on mechanical watches. In any event, Daniels has not quit.

It is hard to love a quartz timepiece. As often as not the cases do not even open to expose the works. (The word "movement" would be inappropriate, for in the digital version nothing moves.) There is really nothing to see. But a movement! That is something else. A good movement, especially a complicated one, has art and grace and life. The parts sweep and swirl. Every edge is carefully beveled. The screws are sometimes heated to a deep electric blue. The bridges and plates are gilded or highly finished, and often decorated, chased, or damascened. The red jewels gleam, usually in a setting of gold. The balance wheel swings tirelessly to and fro; the hairspring breathes; the pallet arms move in and out like a living being. It is hard to believe that so much can be squeezed into so small a space. Even someone who knows nothing about the mechanics of timekeeping can admire so cunning and artful a device, and it is no accident that some of the best watches have been "skeletonized" or cased with transparent backs for the owner's enjoyment. This kind of watch is still a Swiss specialty, and such firms as Audemars-Piguet and Breguet (now owned by Chaumet of Paris) get thousands of dollars for tracery masterpieces under glass or, better yet, under artificial sapphire. For the professional, of course, these aesthetic considerations are enhanced by the efficiency and ingenuity of the mechanism. Even the tick of a good timepiece is a delight to those who can appreciate a strong, regular beat. Those of us who love and admire these wonderful mechanisms, then, wish those who would continue them all good luck. They will need it.

T HE SWISS HAVE PAID DEARLY for their slowness to adopt the new technology. As recently as 1978, their industry was still the largest: some 63 million pieces, out of 265 million for the entire world. Yet this already represented a sharp drop from peak output of 87 million (out of 227 million) in 1974, and Japan was rushing by. Japanese watch production tripled from 1970 to 1980, from 23.8 to 87.9 million units, while exports increased six times, from 11.4 to 68.3 million, or from 48 to 78 percent of output. By that time, Swiss export of watches and movements was down to 51 million pieces (45 million in 1981), of which an increasing share of components had been bought abroad, so that the designation "Swiss Made," once carefully reserved to objects almost entirely of home manufacture, now had to be redefined.[19] All of this was accompanied by a purge of the weaker firms: the more than two thousand watch houses of 1963 (peak number: 2,332 in 1956) were down to less than nine hundred in 1980 and still falling; while work force at the bench dropped from 55,320 to 26,228. (See Tables 10 and 11.)

Swiss performance has been weakest precisely in the area of most rapid growth: quartz complicateds—that is, watches with multiple indications and functions. These call for digital displays, which handle additional information at negligible marginal cost. The conventional analog dial, on the other hand, needs a wheel train to translate the electrical impulses into the usual visual signs, and such a train takes money, effort, and space. The Swiss have no special advantage in the digital area, but they do make the best wheel trains in the world. More to the point, an analog dial can be far more attractive than the utilitarian blinking numerals of a digital display, which leaves little or no room for style or elegance. Here Swiss horology still reigns supreme. So whereas Japanese watchmakers have gone over increasingly to digitals, the

Swiss have tried to marry old and new. Over 90 percent of Swiss quartz watches, I am told, are made with conventional dials. An economist would simply describe this as rational behavior. But what price rationality? The large and growing market for interesting and specialized pieces: the constituency composed of travelers, drivers, scientists, horse players, sports fans; the people who want to be awakened in the morning; the people who forget appointments and the people who remember appointments; the people who want a watch that can add, subtract, multiply, divide and the people who want a calculator that can also tell them the time; the people who want a lot of information for their money; young people; with-it people; people who love gadgets for their own sake—in short, most of the utilitarian demand and a good part of the romantic demand—is going elsewhere.

The one area of strength has been the manufacture of unfinished movements (including solid-state modules) and components. These upstream branches have always been more advanced technically, more mechanized and less labor-intensive, than the finishing and casing (*habillement*) sections of the industry, going back to Japy and the mass production of *ébauches* in the late eighteenth century. Now, in the context of the quartz revolution, it is Ebauches S.A., the big movements trust, that has been most active in acquiring the new technology, whether by Swiss research or by purchase from and collaboration with foreign firms. This strategy has yielded capability and capacity that, at least for now, surpass the demand of Swiss finishers. As a result, the long effort to hold back the growth of competitive industries by discouraging the sale of unassembled movements and parts has simply collapsed. More and more of Swiss exports have taken this form—1.8 million pieces in 1971; 32.7 million in 1980.[20]

From the national point of view, this is not a comfortable achievement. For one thing, no sector of the industry is so vulnerable to erosion by low-cost competitors: there is no loyalty to *chablons*. For another, this is a reversal of the traditional Swiss policy of reserving the higher-paid, skill-intensive tasks to Swiss while buying rough work outside as necessary. Yet it is understandable. The early stages of manufacture now embody some of the most knowledge-intensive processes in the industry—in particular, the design of circuitry and the photochemical preparation of crystals and modules—whereas the new technology has drastically simplified the tasks of assembly and casing: there is no *repassage* with digitals. There has been, in other words, an inversion of the traditional hierarchy of tasks within the industry, and the meek have inherited the horological earth. Under the circumstances, highly advanced industrial nations such as Switzerland may be better able to defend themselves on the battleground of anonymous components, which represent a small fraction of total cost and hence stand or fall on the quality of logic and performance (compare Britain's quasi-monopoly of crucible steel for almost a century), than of cost-conscious competition in finished timepieces.

None of this, of course, touches the production of the great "name" houses, whose reputation for quality and style and whose ability to confer prestige make cost considerations almost irrelevant. Here the old argument of Thorstein Veblen's *Theory of the Leisure Class* is to the point: at a high enough level, pricey prices may actually stimulate demand, for how better to announce one's exclusivity and success? In this area the Swiss are untouchable. No one comes close, and one can see Japanese buyers going to Lake Geneva to shop for the kind of watch that "tells you something about yourself" (Patek); that is a "precious and rare work of art" (Vacheron); an "exclusive creation for the discerning few" (Audemars); "the most expensive watch in the world" (Piaget). Readers of *The New York Times* in December 1980 were treated to one of those shopping articles on the Fifth Avenue watch trade that aim to tell well-off people what rich people are buying: self-winding, perpetual-calendar wristwatches by Patek and Audemars that retailed at $15,000; jewel-encrusted ladies' bracelet watches (no mere wristwatches, these) that sold for ten and, yes, one hundred times as much. The most extravagant of these, a "watch" that was simply a setting for large diamonds, was priced at over $4 million—a sum large enough to make even a doting tycoon hesitate. The seller, Vacheron-Constantin, sought to encourage the faint-hearted by promising a substantial capital gain of 20 percent a year. This was no watch; it was not even a piece of jewelry. It was an investment.[21] It was also a publicity coup for Vacheron, which found itself promoted above Patek and Piaget as maker of the world's costliest watches. Sniffed Patek: "You can take any movement and encrust it with diamonds."[22] (It is generally a good rule never to buy anything that promises so high a rate of return; for why should the seller want to exchange it for mere money?)

The Swiss here seem to be following the British precedent of the nineteenth century: as they are pressed harder in the lower and middle price ranges, they retreat into (or hold the fort in) that area where price is no longer a consideration. But this represents at best a small fraction of unit sales, though admittedly a much larger share of value. And as with the British earlier, the Swiss cannot afford to think of this bastion as inexpugnable. Seiko is moving into the manufacture of gold, high-fashion watches, with a television advertising campaign in support. A Japanese gold watch costs perhaps $500; a Swiss costs from two to twenty-five times as much. How long . . .?

Numbers are one thing; people another. While Geneva has sold diamonds, the Jura has shriveled. Job figures tell only part of the story, for many of the watch firms that have survived have done so only by cutting pay and hours or by introducing intermittent employment—which is a way of shifting part of the wage bill to the government. Jobs in the Jura sector were down in 1982 to half what they had been a decade earlier; in a region where watch-making accounted for over two-thirds of all industrial employ-ment, no one, in or out of the industry, escaped unscathed. In 1950 this was one of the most prosperous regions in the country, with income per head 42 percent higher than the national aver-age; by 1977 it was 25 percent below. To be sure, the Swiss mean is very high, and what passes for depression there might be an-other people's land of cockaigne. Aggregate income of the watch area—specifically the French-speaking core—went up after all 3.1 times from 1950 to 1977. Yet national income rose over that pe-riod by 7.4 times, and this difference implies a half-turn of the wheel of fortune.[23] The Jura watch districts were for two centuries a pole of attraction for job seekers. Now people are leaving. It is not the old people who go. They are fixed by habit, friendships, the waning of energy that comes with age. But the young are going to other, more secure jobs, to something more exciting than snow and the workbench and hours of repetitive motions. In a re-markable photographic and prose essay, *Quand nous étions horlogers* (When we were watchmakers), Simone Oppliger, daughter of a watchmaker father and seamstress mother, a twentieth-century Osterwald in her fashion, shows and recounts the dwindling of her parents' world: the population of her town, Renan (in the Vallon de Saint-Imier near La Chaux-de-Fonds), down in eight years (1970–1978) from 1,094 to 845 inhabitants; the first grade, which numbered twenty pupils in 1954, shrunk to six; the streets strangely still; the exteriors of the houses, particularly the community buildings, weathered and neglected; the interiors as Swiss-neat, comfortable, and respectable as ever.[24] (See Figure 30.)

Even the watchworkers' union is of little help in these circumstances. How do you fight structural change? Where once the union would have led the workers out into the streets, now it tells them to take what they can and let it go at that: make trouble, and you'll have nothing. To some, the union seems more concerned about shrinking membership than about wages and working conditions. Workers stopped going to meetings, so the union gave out small door prizes to reward attendance: glasses, clothes hangers, bottle openers, key cases. Like American movie theaters in the 1930s: for fifteen cents you could see three feature films and take home a plate. One disillusioned old-timer was not impressed: the hangers didn't really work—the clothes slid off.[25]

Simone Oppliger sees things more from below than above: "The bosses have never been part of my world."[26] A veteran of Third World *reportages* and friend of left-wing militants, she presents a cast of characters that may not be statistically representative. Socialist and Communist workers seem to hold a bigger place than their numbers would warrant. But articulate as they are, they probably make better copy than the mass of reconciled, even contented workers who, somewhat to Oppliger's surprise, rather like their small, rapid, repetitive tasks and are happy in their secure, familiar work station.[27] Besides, Renan has always been a foyer of "advanced" opinion; back in the early nineteenth century it was the first and often the last stop for political refugees from Prussian Neuchâtel. Oppliger has even found one worker who has been able to visit the Soviet Union three times, once in reward for his party loyalty, and who has never come away disappointed: "C'était magnifique."[28] That is the political version of the tenacity (obstinacy) that has kept many of these people at the bench for decades: the Jurassien is a dogged cuss. He has an irrepressible sense of his own merit and dignity; a sublime faith in the virtue and reward of labor; and a somber fatalism. With that kind of labor force, an industry can go down slowly, surely, and with a stiff upper lip. (See Figure 31.)

Are the employers better off? The leading manufacturer in Renan once seemed rich. He and his family had taken a business with two workers and built it up to two hundred fifty in the space of a generation. Now it is bankrupt. They can look back on ten years of struggle, with its ups and downs, its moments of hope, its bitter disappointments. They had bet on growth, kept going longer than they should have. They were not prepared to compete in a watch industry that had become a battleground. They got one huge order from the United States, but by the time they filled it, the client had changed his mind. Maybe he was in trouble, too. In any event, the sales contract did not stand up—something about a commitment to ship from Renan and the cases' having gone out from Zurich. In the best of circumstances, it is not easy to enforce a commercial obligation in another country. "We weren't big enough to face up to such people, and on their terrain at that. We weren't experienced enough in banking matters, in international commercial practice, because we had developed as a family enterprise with the conviction that all one had to do to succeed was work hard. We attached too much importance to moral values. That's not the language you have to use in the business jungle." Sour grapes? Self-pleading? Perhaps; but the fact is that the typical small Swiss watch enterprise—there were once about two thousand of them—was at a loss in the turmoil of technological change, plunging prices, rapid obsolescence, growing capital requirements. When our manufacturer from Renan turned for help to those much-vaunted Swiss banks, he found that they had already written him off: "It had already been decided in high places that we were beyond rescue."[29] The secret of profitable banking is to lend other people's money only to those who do not need it.

The Swiss watch industry has not given up—far from it. It continues to play its good cards: prestige, artistry, luxury, snobbery. But its best trump is gone: Switzerland's comparative advantage in skill and experience has been devalued overnight by the quartz revolution. Can a new watch industry be built on the legacy of the old? Can the Swiss do today what they did one hundred years ago in the face of the American challenge? And if they do, will they be doing it in Switzerland? A number of Swiss firms have already moved operations to Hong Kong and other low-cost centers, following the path of the *chablons* and uncased movements.[30] Labor in Hong Kong costs less than in Japan, works longer, is more docile. The place is filled with refugees who need a place to lay their head, and if they are good workers, the boss will let them sleep in the factory. Switzerland today is in the position of Britain two hundred years ago: how does one compete with a combination of low wages and up-to-date technology?

Meanwhile the Japanese cannot rest on their laurels. They had held first place for no more than three years when they were supplanted by Hong Kong, which is now the leading exporter of watches in the world. And Hong Kong in turn may not be able to hold its place. There are countries where labor is even cheaper; besides, the political future of the crown colony is in doubt. The new industry is caught up in swirling currents of technological change (the quartz watch is surely not the last word), and it is a rash person who would predict its future.

SOME OF MY READERS may remember the valedictory formula of those old Lowell Thomas travelogues: "As the sun sets on the snowy peaks and green valleys of the Swiss Jura, we bid farewell to . . ." Now we bid farewell to those master craftsmen who have brought us these wonders of the mechanical arts. Their time has come and probably gone. Their children are doing other things. Seven hundred years of science and skill, of one of man's greatest industrial achievements, are drawing to an end.

But if this is an end, it is also a beginning—like those college graduations that we in America call "commencements." We have entered a new era of time measurement, one that rests on a new science and new skills. Our instruments are degrees of magnitude better than those we had before. Meanwhile timekeeping devices become more widely used, imposing their relentless beat on populations that have never before given thought to minutes and hours. Is this good? Those of us who live under tight time discipline deplore it and flee it when we can. We seek vacations in places where we can put our watches away and let nature wake us and put us to sleep. For others, though, submission to time is the price of modernization, productivity, potential affluence. Who are we to deprecate what we live by and, living, have gotten rich by?

The motto of the National Association of Watch and Clock Collectors (USA) is *Tempus vitam regit:* Time rules life. That it does, like it or not. The mechanical clock and watch may go, like the clepsydra and sundial before them. The timekeeper remains.

Appendix A: Escapements

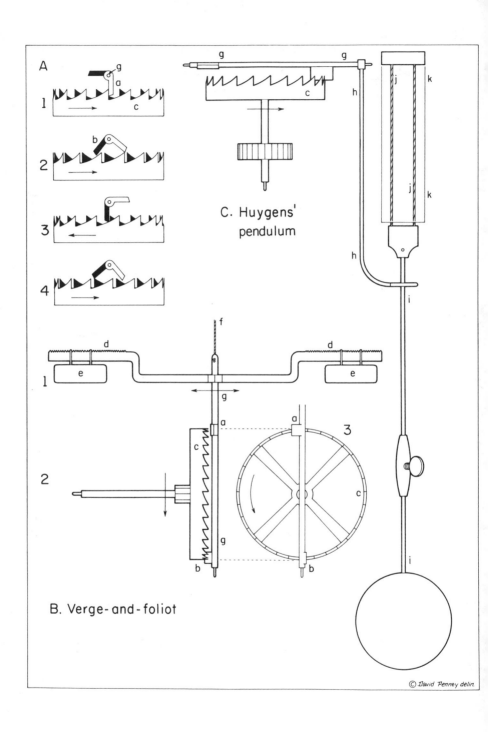

A

1

2

3

4

C. Huygens' pendulum

B. Verge-and-foliot

1

2

3

© David Penney delin.

Figure A.1. The standard verge escapement—on the lower left, with foliot regulator; on the right, with pendulum controller. The escapement device is the same in both: a staff (verge) with two pallets, usually set at a right angle to each other, which alternately block and release the scape wheel.

A. Upper left: the escapement sequence, seen from above: (1) Top pallet (white), *a*, holds wheel, *c*, which is turning in the direction of the arrow. (2) Wheel pushes top pallet aside and causes verge, *g*, to pivot, thereby bringing in the bottom pallet, *b*. (3) Bottom pallet (black) stops wheel and pushes it back (hence the name "recoil escapement"), as shown by arrow. (4) Wheel pushes bottom pallet aside, and resumes its forward rotation.

B. Lower left: side view of the verge-and-foliot. (1) At the top, the folio crossbar, *d–d*, which serves as an inertial brake on the turning verge. The weights, *e*, suspended on both sides can be moved to change (adjust) the rhythm of oscillation. The foliot is fixed rigidly to the verge, *g–g*, but the whole is suspended by a cord, *f*, to minimize friction. (2) The scape wheel, which is geared into the wheel train (to the left; not shown) by a contrate wheel (not shown). In the diagram, it is being held by the lower pallet on the verge, *b*. (3) The scape wheel and verge, rotated ninety degrees. The mechanism illustrated comes from the Dover Castle clock, now in the Science Museum, South Kensington, London. This clock was originally thought to date from the fourteenth century, but subsequent research revised this to around 1600. The technique, however, goes back to the dawn of mechanical timekeeping, in the thirteenth century.

C. Right side: the verge with pendulum regulator. At the top, the escapement mechanism, with scape wheel, *c*, and verge, *g–g*, horizontal. The wheel is being held at the moment by the right-hand pallet. In this instance the verge is controlled not by a foliot (or balance wheel), but by a long lever (the crutch), *h–h*, curved at the bottom and ending in a fork that loosely brackets the oscillating pendulum rod, *i–i*. It is this crutch that transmits impulse from the train to the pendulum to make up for the energy lost in each swing. But note that the pendulum hangs freely from cords, *j*, and contact between rod and crutch is reduced to a minimum; the aim is to limit perturbation of the regularity of oscillation. The diagram is taken from Christiaan Huygens' *Horologium* (1658). Huygens shows the clock as fitted with cycloidal cheeks, *k–k*, designed to alter the arc of swing and thus to eliminate circular error. They did not work very well.

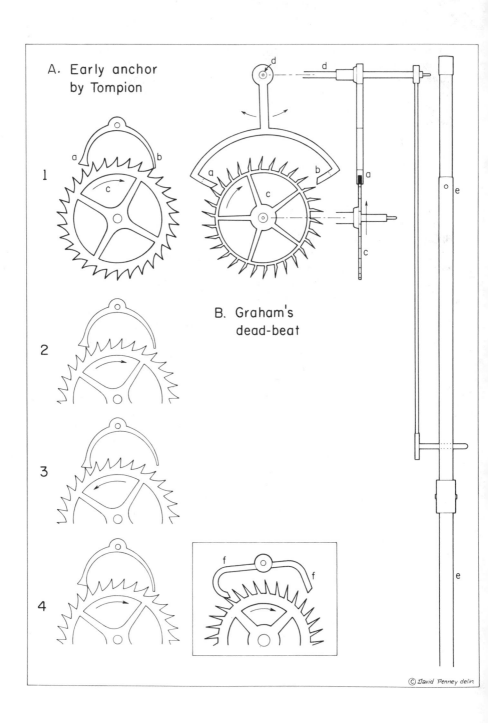

A. Early anchor by Tompion

1

2

3

4

B. Graham's dead-beat

© David Penney delin.

Figure A.2. Clock escapements of the anchor variety.

A. On the left, an early example of the anchor escapement, from Thomas Tompion's astrolabe clock in the Fitzwilliam Museum, Cambridge, c. 1676. The steps of the cycle are illustrated. (1) Tooth is about to leave exit pallet, *b*. (2) Tooth falls on face of entry pallet, *a*. (3) As entry pallet moves in, it pushes the scape wheel, *c*, back (recoil). (4) As entry pallet withdraws and releases tooth, the scape wheel, resuming forward motion, gives impulse through the tooth to the pendulum. The advantage of this escapement is that it needed a much smaller arc of swing than the verge and drastically reduced circular error.

B. Top center and on the right: the Graham version of the anchor, known as "dead-beat" because the pallets fall dead on the teeth of the scape wheel—no recoil, minimal friction. Impulse is given by a push from the just-released tooth of the scape wheel against the beveled end of the jeweled pallet, *a*. This diagram is taken from the mechanism of John Arnold's "Manheim" (an anglicized spelling) regulator, a long-case precision clock made in 1779 for the observatory of the elector palatine in Mannheim, Germany. It was noteworthy at the time for the use of diamond locking stones, *a* and *b*, in the anchor arms and ruby bearings for the pallet arbor, *d*. The pendulum, *e–e*, compensated for temperature changes by a composite rod of Arnold's invention. The clock was said to keep a rate within a few seconds a year.

Inset: the kind of anchor escapement used in shelf and long-case clocks mass-produced in the United States and Germany in the nineteenth century. The pallet arms are made from a single, bent steel strip, *f–f*.

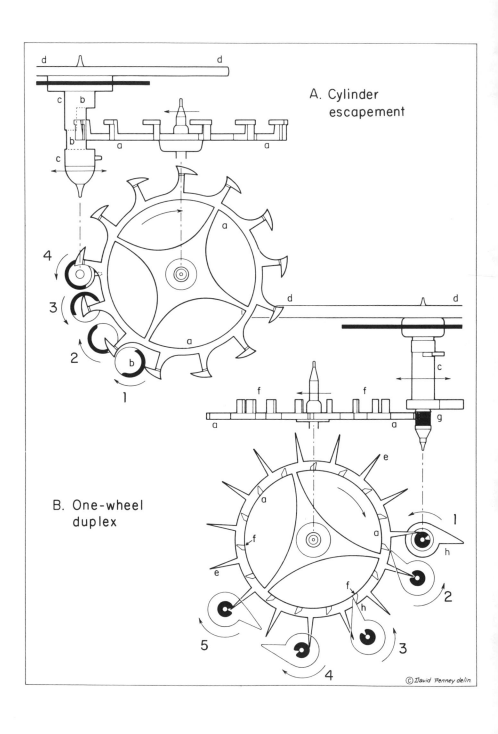

A. Cylinder escapement

B. One-wheel duplex

© David Penney delin.

Figure A.3. Frictional-rest escapements.

A. Upper left: Graham's cylinder escapement, taken from Thomas Mudge's No. 96, c. 1750. Above: a side view of the mechanism showing one of the teeth of the scape wheel, *a–a,* inside the cylinder, *b,* which is an integral part of the balance staff, *c–c.* Below: the escapement mechanism proper, viewed from above to show the successive steps: (1) Tooth at rest (frictional rest) on the outside of the cylinder, *b,* which is rotating clockwise (arrow). (2) Cylinder has rotated enough to let the tooth pass the lip and enter; in the course of entry, the curved outer side of the tooth gives impulse to the balance, *d–d.* (3) The balance swings back and with it the cylinder, holding the tooth at rest on its inner face. (4) The cylinder completes its return swing, which allows tooth to exit, once again while giving impulse—this time by a push to the exit lip of the cylinder. The advantage of this escapement is that it blocks the train from passing its irregularities on to the balance, which completes most of its to-and-fro swing while the scape wheel is locked in place. The disadvantage is the wear produced by rubbing of the point of the scape wheel teeth against the walls of the cylinder. Lubrication helped, but the oil deteriorated with time and the brass scape wheel worked like a file on the harder steel of the cylinder. Not until watchmakers learned to make steel cylinder scape wheels was durability much improved.

B. Lower right: the British one-wheel duplex escapement, taken from a watch by Vulliamy, London, No. szom, c. 1790. In this mechanism, the scape wheel, *a–a,* uses separate teeth for locking (long teeth in the plane of the wheel), *e,* and impulse (short, beveled teeth set at right angles to the plane of the wheel), *f.* Above: side view of the mechanism, showing one of the locking teeth entering a notch in the locking jewel, *g,* on the balance staff, *c,* just prior to release. Below: the escapement mechanism proper, viewed from above to show the successive steps: (1) *Locking* tooth held against the locking jewel on the balance staff as the balance rotates counterclockwise (arrow). (2) As the balance continues turn, the point of the *locking* tooth enters the notch on the locking jewel, which gives it room to escape. (3) This allows the following *impulse* tooth to catch and give a push to the impulse "roller," *h.* (4) The next *locking* tooth falls onto the locking jewel; the balance swings back in the other direction. (5) As the balance completes its return swing (arrow), it does *not* allow the locking tooth to escape. That will take place on the swing back (counterclockwise). The duplex watch became something of a British specialty in the first half of the nineteenth century, when it was seen as the next best thing to the pocket chronometer for portable precision timekeeping. It was used much less on the continent, and then almost invariably in a two-scape-wheel version, one for locking and one for impulse.

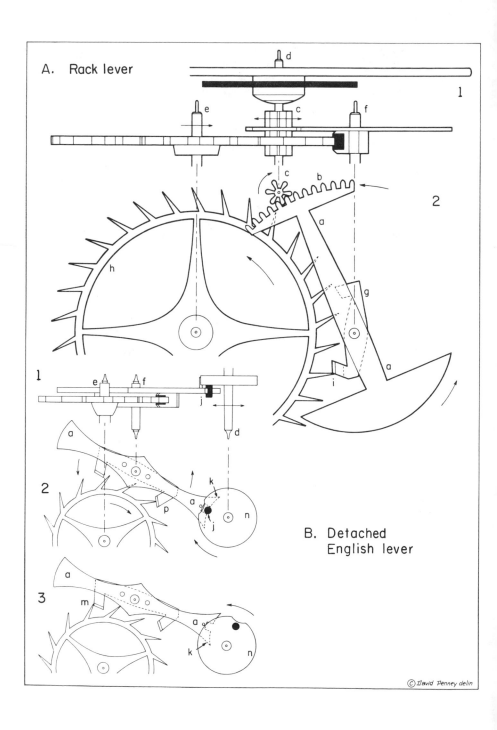

A. Rack lever

1

2

B. Detached
English lever

© David Penney delin

Figure A.4. The English lever.

A. Top: The rack lever, so called because the lever, *a–a*, ends in a rack, *b*, that meshes with and is moved back and forth by a pinion, *c*, on the balance staff. This example comes from a watch by Peter Litherland of Liverpool, No. 1336, c. 1800. Litherland invented (reinvented) the rack lever and patented it in 1792. (1) Top: side view of the mechanism, showing the balance staff, *d*, between the scape wheel arbor, *e*, and the pallet arbor, *f*. This arrangement is characteristic of a side lever, so called because the pallet frame, *g*, lies not between the balance and the center of the scape wheel, *h*, but off to the side. This "right-angle" arrangement would be characteristic of just about all subsequent British versions of the lever escapement. (2) View of the same mechanism from above, to show the action. Essentially this works the same as the Graham dead-beat, with pallets falling dead on the pointed teeth of the scape wheel, *i*. Impulse is given, as in the dead-beat, by the push of the tooth against the beveled end of the pallet jewel. The rack lever could not be expected to give performance comparable to that of the dead-beat, if only because the lever and balance are always in contact, so that the irregularities of the train are invariably communicated to the oscillating controller. But the watch gave good performance in everyday use, rarely stopped, and showed little wear by comparison with the cylinder. Because of its layout, moreover, it fairly invited transformation into the detached lever escapement shown below.

B. Detached English lever, single roller, from a watch by J. Sewill, London, No. 41205, c. 1880. This is in effect a modified rack lever, with pinion reduced to a single impulse pin or nib, *j*, and the rack cut down to two horns, *k*. In this way contact between train and balance is reduced to those moments when the impulse nib hits one or the other horn and knocks the lever back and forth. Otherwise the balance is free to accomplish its oscillation without perturbation by the train. From the top down: (1) Side view of the mechanism, with scape wheel arbor, *e*, pallet arbor, *f*, and balance arbor, *d*, from left to right in that order. (2) View from above, showing scape wheel giving impulse by pushing against the beveled edge of the pallet jewel on the right, *p;* impulse nib, *j*, in the fork, about to pivot the lever, *a–a*, counterclockwise (arrows). (3) Escapement wheel locked by pallet on left, *m;* balance roller, *n*, on the way back; this time the impulse nib, coming in from the right, will pivot the lever clockwise. Note that in both versions, rack lever and table-roller detached lever, the teeth of the scape wheel are ratchet-type (pointed)—typical of British watchmaking. The detached lever escapement proved in the long run the best for high-quality civilian performance, not only because it was capable of good precision (variance of, say, fifteen seconds a day), but also because it rarely set in the pocket. A good watch for everyday, knockabout use.

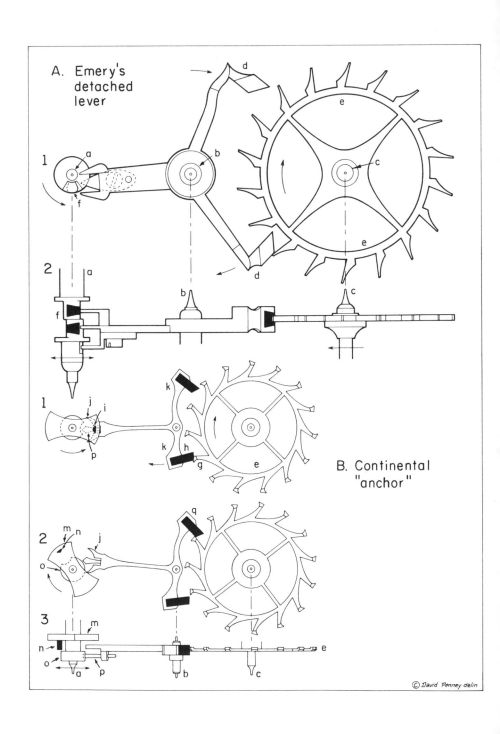

A. Emery's
   detached
   lever

B. Continental
   "anchor"

© David Penney delin

Figure A.5. The continental lever; called in French the "anchor escapement." This is sometimes called the "straight-line lever," because the pallet arbor, *b*, lies directly between the balance staff, *a*, and the center, *c*, of the scape wheel, *e*.

A. Top: the English prototype—Josiah Emery's version of Thomas Mudge's original lever, taken from Emery's No. 1089, c. 1785. From the Ilbert collection, now in the British Museum. (1) View from above, showing balance staff on the left, pallet arbor in center, claw-like pallet arms, *d*, and eighteen-tooth scape wheel. (2) Side view of the mechanism, showing two jeweled unlocking/impulse pins (cams), *f*, one above the other. This two-tier arrangement derives from Mudge, but Mudge did not use the straight-line layout. For Mudge, the lever was a one-of-a-kind *tour de force*. Emery was concerned to simplify and reduce costs. Even so, his lever watches were not cheap. He made about twenty of them, and those that have survived are much prized by collectors today. A few of them found their way to the continent, where they influenced the further development of this escapement.

B. Bottom half: three views of the mature continental lever (*échappement à ancre*), as used in a watch by Vacheron & Constantin of Geneva, c. 1900. From top to bottom: (1) View from above: scape wheel tooth, *g*, impulsing entry pallet, *h*; unlocking/impulse jewel, *i*, in the fork, *j*, rotating counterclockwise and pivoting (levering) anchor arms, *k*, clockwise. Here the horns of the fork are in the plane of the lever. (2) Scape wheel locked on exit pallet, *q*; balance now swinging back and bringing unlocking/impulse jewel around clockwise to where it will again enter the fork; but this time the lever pivots in the other direction, releasing the tooth, which will give impulse in its turn. Note the shape of the teeth: this club-tooth form (probably invented by A.-L. Breguet) is typically Swiss. (3) Side view, from left to right: balance arbor, *a*, pallet arbor, *b*, scape wheel arbor, *c*. In this advanced form, there are two rollers (hence the name "double roller"), one, *m*, carrying the unlocking/impulse jewel, *n*, the other, *o*, with notch (indentation) to receive a gold safety dart, *p*, whose function is to prevent the scape wheel from becoming unlocked except when the unlocking/impulse jewel is in the notch of the fork. This is the escapement that became by the last quarter of the nineteenth century the staple of mass-production watch manufacture, everywhere but in Britain.

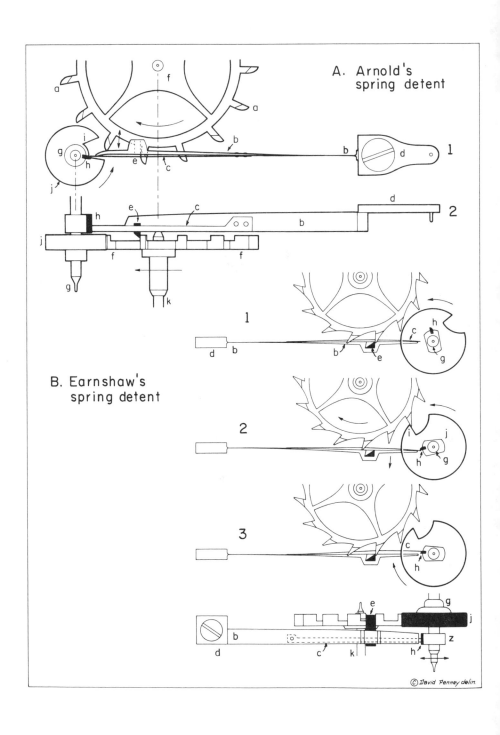

A. Arnold's
spring detent

1

2

B. Earnshaw's
spring detent

1

2

3

© David Penney delin.

Figure A.6. Detent ("chronometer") escapements.

A. John Arnold's version, from his watch (a pocket chronometer) No. 89/390, c. !785. (1) At top: view from above, showing escapement teeth, *a*, with epicycloidal profile; the double spring, the top one (detent spring), *b–b*, for locking, the lower, "longer" one, *c*, for passing. The detent spring is attached to the stud, *d*, at the right; the passing spring is pinned to the detent. The action shows the bottommost tooth of the scape wheel, *f*, locked by the detent jewel (dotted lines), *e*, which projects below the plane of the detent. Note that the detent is in tension—that is, that the force of the scape wheel is pulling the detent spring taut, away from the point of fixation. At this moment, the balance staff, *g*, is turning counterclockwise, and the unlocking jewel on the staff, *h*, is about to push the tip of the passing spring, *c*, toward the center of the scape wheel, carrying the detent, *b*, with it and thereby unlocking the wheel. All of that takes place in a flash: as soon as the unlocking jewel clears the passing spring, the detent springs back and locks the next tooth of the scape wheel. Meanwhile the balance swings back, but this time the soft passing spring allows the unlocking jewel to clear (it just bends out of the way) without disturbing the detent. Then the cycle is repeated. Impulse is given by the flying teeth of the scape wheel (in between lockings), catching and pushing the exit edge of the notch, *i*, in the impulse roller, *j*, on the balance staff. (2) Below: side view of the mechanism, with balance staff, *g*, to the left, scape wheel arbor, *k*, to the right. Note that the teeth rise above the plane of the scape wheel, *f–f*, to meet the locking stone, *e*, of the detent, *b*.

B. Thomas Earnshaw's version, from a chronometer by Howells of Kennington, London, No. 138, c. 1805. From top to bottom: (1) View from above. Scape wheel tooth held by locking jewel, *e*. The detent, *b*, is shown under the "longer" passing spring, *c*. Note that here the detent is in compression—that is, the force of the scape wheel pushes the locking stone toward the point of fixation, *d*. If this force were strong enough, it could cause the detent spring to buckle, and numerous horologists have fussed about this alleged defect of the Earnshaw design. In fact, it is not strong enough to do that, and the detent spring works perfectly well. (2) The balance staff, *g*, rotating counterclockwise, brings the unlocking jewel, *h*, around to catch the end of the passing spring and pull the detent back (away from the center of the scape wheel), which turns rapidly clockwise. The second tooth to the right of the one just released can be seen giving impulse as it catches the edge of the notch, *i*, in the impulse roller, *j*. (3) The detent spring has sprung back in time to catch the next tooth and lock the scape wheel. Meanwhile the balance, returning, can make its pass because the unlocking tooth, *h*, lifts the passing spring, *c*, out of the way without disturbing the detent. At bottom: a side view of the mechanism, showing the steel detent spring, *b*, a steel (often gold) passing spring (dotted lines), *c*, pinned alongside it and behind it; the scape wheel above, with teeth rising above the plane of the wheel; and balance staff, *g*, to the right, with solid sapphire impulse roller (a big stone), *j*, and the unlocking roller, *z*, with inset jewel nib, *h*.

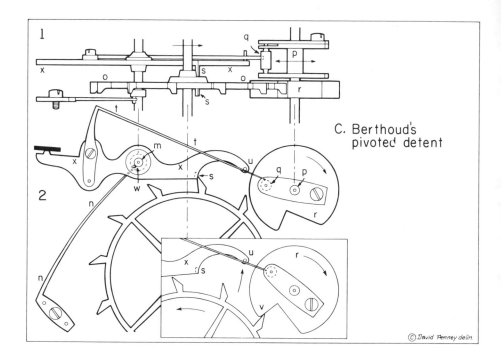

C. Berthoud's
pivoted detent

1

2

© David Penney delin

C. Louis Berthoud's version, taken from his marine chronometer No. 49, a small piece, one-day going, c. 1795; from the Ilbert collection, now in the British Museum. This is a pivoted rather than a spring detent—that is, the detent, *x–x*, pivots on an arbor, *m*, and is returned to the locking position by a separate return spring, *n–n*. In Berthoud's version this is a straight spring; in later versions, the return spring is a spiral coiled around the detent arbor. (1) Top: side view of the mechanism, detent, *x–x*, above on left, scape wheel below, *o–o*, balance staff, *p*, with unlocking roller, *q* (in plane of detent), and impulse roller, *r* (in plane of scape wheel), to the right. (2) Below: view from above showing scape wheel held by downward projecting locking pallet on corner of detent, *s* (dotted line). At this point, the balance, *p*, is swinging clockwise, bringing the unlocking roller, *q* (dotted circle), around to catch the passing spring, *t–t*, push it against the pin on the end of the detent, *u*, and thereby pivot the detent counterclockwise, releasing the scape wheel. The inset shows the detent clear, while the scape wheel gives impulse to the balance by catching the corner of the notch, *v*, in the impulse roller on the balance staff. As soon as the unlocking roller clears the end of the passing spring, the steel return spring, pressing against a small pin, *w*, alongside the detent pivot, *m*, brings the detent back into position to lock the next tooth.

The timekeeping accuracy of all the detent escapements lies in the detachment of balance from wheel train. Contact is limited to those two moments, exceedingly brief, when the unlocking roller pushes the detent aside and when the impulse roller receives its quick push from the scape wheel. The advantage of the spring detent is that there is no pivot to oil, for oil means trouble in the long run. On the other hand, the pivoted detent is less likely to set in the pocket. The one is better for boxed marine chronometers; the other, for pocket timepieces.

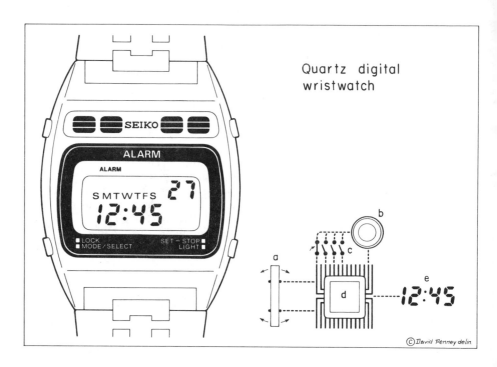

Quartz digital wristwatch

© David Penney delin.

Figure A.7. Quartz digital wristwatch, by Seiko, Japan. Continuous liquid crystal display of minutes, seconds, day, and date, with extra functions (by buttons on side of case) of month, alarm, time signal, and back light. Stainless steel case. These watches are guaranteed to keep their rates to within a few seconds a month. Such accuracy is due to the stability of the oscillation frequency of the quartz crystal controller, *a*, which gets energy from a small battery cell, *b*, and vibrates tens of thousands of times a second. These very rapid vibrations then have to be divided down—gathered together, as it were—so as to yield countable seconds. This is the task of a multistage binary circuit. Other circuits (setting and function switches, *c*) store and provide the information needed to perform the calendar, alarm, and other accessory functions. The signals from the time computer module (microchip, *d*) then pass to the display, where they energize the appropriate liquid crystal segments, *e*, to give a digital readout. (It is possible to combine the quartz controller with wheel-work transmission and analog display—that is, turning hands; but then it becomes very costly to add functions.)

This multiplication of functions has been made possible by the progressive miniaturization of electronic circuits. In 1960 such a circuit might have had ten transistors (in effect, gates—like the valves in the old radio sets) on a chip 25 mm. square; ten years later, at the dawn of quartz watch manufacture, it might have had a thousand transistors—thereby inaugurating the era of microprocessors. Another ten years, and the figure was between one and two hundred thousand—a gain in twenty years of more than four orders of magnitude. If it were necessary to build one of these highly complicated digital watches using the old vacuum tubes, the resulting device might well fill a small room, and the problem of heat disposal would be enormous. All of this seems very mysterious to someone habituated to the visible, moving works of a mechanical watch. In a quartz digital timepiece, nothing moves: it boggles the mind to think what a great artist/mechanician such as Graham or Mudge would have said of all this. Yet the principle is the same for both technologies; each comprises the same elements: (1) a source of energy (spring or battery); (2) an oscillating controller (balance or quartz crystal); (3) a counting device (escapement or solid-state circuit); (4) transmission (wheelwork or electric current); and (5) display (hands or liquid crystal segments). All clocks, save continuous timekeepers such as the sundial or clepsydra, have these five essential parts.

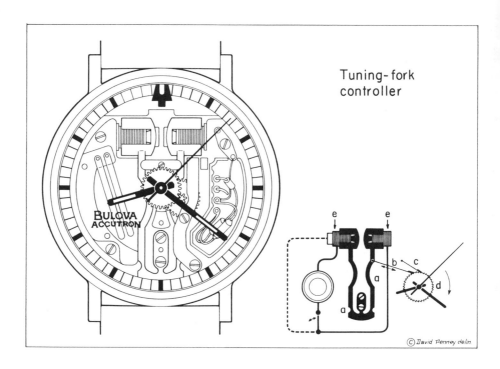

Tuning-fork controller

BULOVA ACCUTRON

© David Penney delin

Figure A.8. The Bulova "Accutron"—an electrically powered timepiece with tuning-fork controller, 1959. The see-through dial on the left shows the actual location of the components: the tuning fork, *a–a,* is vertical, centered above six o'clock, with drive coils, *e,* above. A circuit diagram is provided at the lower right. As shown, the vibrations of the tuning fork are used to drive the train directly by means of a tiny index pawl, *b.* A locking pawl, *c,* keeps the index wheel, *d,* from reversing. Since so small a tuning fork necessarily produces extremely rapid vibrations of very small amplitude, the parts have to be exceedingly small. The ratchet index wheel that drives the train, for example, is 0.09 in. in diameter with 300 teeth—distance between teeth, 0.001 in.; the jewels of the indexing and locking pawls are 0.010 in. square by 0.002 in. thick. Pushed by the extremely rapid but light strokes of the index pawl, the ratchet wheel (made of hardened beryllium copper) turns in just under a second, or about 38 million times a year. Even so, there is no wear, owing no doubt to the very small forces employed. The drive coils each have 8,000 turns of wire 0.015 mm. in diameter.

Thanks to the stability of the tuning-fork oscillator and the quality of the indexing and transmission mechanism, Bulova is able to guarantee the accuracy of the Accutron to less than a minute a month. Moreover, since there is no oil at the tuning fork, the Accutron will hold its rate far better than a lubricated mechanical timepiece. Repair, on the other hand, is not so simple, if only because of the need for microscopic magnification; the good old loupe will not do. Bulova provides microscope kits for this purpose.

Appendix B: Tables

Table 1. Watches passing through the Bureau de Garantie in Besançon, France, 1850–1871.

Year	Number	Year	Number
1850	59,863	1861	250,467
1851	67,876	1862	254,477
1852	76,471	1863	297,094
1853	94,897	1864	301,128
1854	106,076	1865	296,012
1855	141,934	1866	305,435
1856	160,165	1867	334,649
1857	177,555	1868	335,961
1858	190,013	1869	373,138
1859	191,876	1870	231,124
1860	211,811	1871	256,945

Source: Louis Martin, *Etude sur les transformations de l'industrie horlogère* (Besançon, 1900), pp. 13, 15.

Note: It is not entirely clear which, if any, of these figures include watches of foreign origin. This would seem to be the case in 1869; on p. 17 Martin gives the number of watches assayed in that year as: French, 310,849; foreign, 66,624. But the figures for 1850 and 1855 tally. See Martin, *Etude*, pp. 7–8.

Table 2. Distribution of employment in the Swiss watch industry, by canton, 1870–1920.

	1870	1888	1900	1910	1920
Neuchâtel	13,689	14,629	18,024	16,322	19,016
Bern	14,722	19,157	22,359	21,832	26,085
Solothurn	806	2,395	3,965	6,111	6,329
Geneva	3,234	2,416	2,202	2,435	3,376
Vaud	3,633	2,827	3,136	2,814	3,333
Basel-land	94	367	648	1,106	1,500
Six cantons, total	36,228	41,791	50,334	50,620	59,639
Proportion of industry as a whole (in percent)	92.1	94.7	95.4	95.1	94.9

Source: Albert Hauser, *Schweizerische Wirtschafts- und Sozialgeschichte* (Erlenbach-Zurich and Stuttgart: E. Rentsch, 1961), p. 225.

Table 3. Waltham watches: annual production (in thousands) approximated by first differences of serial numbers.

Year	Output	Year	Output	Year	Output
1857–58	13	1889–90	882	1921–22	200
1858–59	3	1890–91	118	1922–23	200
1859–60	3	1891–92	1,000	1923–24	200
1860–61	3	1892–93	400	1924–25	200
1861–62	11	1893–94	250	1925–26	200
1862–63	12	1894–95	350	1926–27	1,000
1863–64	72	1895–96	500	1927–28	500
1864–65	72	1896–97	500	1928–29	500
1865–66	72	1897–98	500	1929–30	200
1866–67	73	1898–99	500	1930–31	200
1867–68	75	1899–1900	550	1931–32	200
1868–69	60	1900–01	450	1932–33	200
1869–70	30	1901–02	1,000	1933–34	200
1870–71	50	1902–03	1,000	1934–35	500
1871–72	50	1903–04	1,000	1935–36	500
1872–73	91	1904–05	1,000	1936–37	300
1873–74	29	1905–06	500	1937–38	400
1874–75	80	1906–07	500	1938–39	300
1875–76	100	1907–08	2,000	1939–40	300
1876–77	100	1908–09	500	1940–41	400
1877–78	160	1909–10	500	1941–42	300
1878 79	191	1910–11	30	1942–43	370
1879–80	148	1911–12	30	1943–44	280
1880–81	176	1912–13	940	1944–45	350
1881–82	162	1913–14	1,000	1945–46	300
1882–83	163	1914–15	300	1946–47	400
1883–84	356	1915–16	400	1947–48	300
1884–85	294	1916–17	300	1948–49	449
1885–86	350	1917–18	1,000	1949–50	109
1886–87	300	1918–19	1,000	1950–51	42
1887–88	400	1919–20	500	1951–52	100
1888–89	300	1920–21	500	1952–53	130

Source: George E. Townsend, *Almost Everything You Wanted to Know about American Watches and Didn't Know Who to Ask* (privately printed, 1971), p. 48.

Table 4. Elgin watches: annual production (in thousands) approximated by first differences of serial numbers.

Year	Output	Year	Output	Year	Output
1867–68	30	1896–97	500	1925–26	1,000
1868–69	40	1897–98	500	1926–27	1,000
1869–70	30	1898–99	500	1927–28	2,000
1870–71	25	1899–1900	1,000	1928–29	1,000
1871–72	25	1900–01	300	1929–30	300
1872–73	25	1901–02	450	1930–31	200
1873–74	25	1902–03	250	1931–32	200
1874–75	100	1903–04	1,000	1932–33	300
1875–76	100	1904–05	1,000	1933–34	1,000
1876–77	100	1905–06	500	1934–35	500
1877–78	50	1906–07	500	1935–36	500
1878–79	50	1907–08	500	1936–37	1,000
1879–80	100	1908–09	500	1937–38	500
1880–81	100	1909–10	1,000	1938–39	500
1881–82	200	1910–11	1,000	1939–40	1,000
1882–83	400	1911–12	1,000	1940–41	1,000
1883–84	200	1912–13	500	1941–42	1,000
1884–85	200	1913–14	500	1942–43	1,000
1885–86	200	1914–15	500	1943–44	500
1886–87	500	1915–16	500	1944–45	500
1887–88	500	1916–17	1,000	1945–46	500
1888–89	500	1917–18	1,000	1946–47	500
1889–90	500	1918–19	1,000	1947–48	1,000
1890–91	440	1919–20	1,000	1948–49	1,000
1891–92	440	1920–21	1,000	1949–50	1,000
1892–93	120	1921–22	1,000	1950–51	1,000
1893–94	500	1922–23	1,000	1951–52	1,000
1894–95	500	1923–24	1,000	1952–53	1,000
1895–96	500	1924–25	1,000		

Source: George E. Townsend, *Almost Everything You Wanted to Know about American Watches and Didn't Know Who to Ask* (privately printed, 1971), p. 15.

Table 5. Swiss watch exports to the United States (money values in Swiss francs).

Year	Francs	Year	Francs
1864	8,477,192	1874	12,119,941
1865	11,301,954	1875	8,499,501
1866	13,093,408	1876	4,809,822
1867	10,362,418	1877	3,569,048
1868	10,469,728	1878	3,995,716
1869	13,222,578	1879	5,292,098
1870	16,512,162	1880	10,143,813
1871	17,105,752	1881	11,809,122
1872	18,312,511	1882	13,238,489
1873	13,054,147		

Source: U.S. Department of Commerce, *Commercial Relations*, vol. 1 (1882–1883), p. 497. These data must be used with caution: they were calculated by dividing the weight of exports by the average weight of a watch and then multiplying the result by the average price of a watch.

Table 6. Largest watchmaking enterprises in Switzerland in 1905, by number of employees.

Enterprise	Employed in the plant	Employed at home	Power (in horsepower)
1. Langendorf, S.A.[a]	1098	161	200
2. Longines, St.-Imier, S.C.A.	853	956	82
3. Omega, Bienne, S.A.[a]	724	156	155
4. Tavannes Watch, Tavannes, S.A.[a]	609	13	128
5. Billodes (Zenith), Le Locle, S.A.	574	38	100
6. Fontainemelon, S.A.[b]	558	34	66
7. Obrecht & Cie, Granges (Sol.), S.C.A.[a]	541	89	68

Source: M. Fallet-Scheurer, *Le travail à domicile dans l'horlogerie suisse et ses industries annexes* (Berne, 1912), p. 134.

a. Manufactured complete watches.

b. Manufactured *ébauches* and finished movements.

Note: S.A. = Société Anonyme = Inc. S.C.A. = Société en Commandite par Actions = Limited joint-stock partnership.

Table 7. Switzerland: horological exports, 1885–1913.

Year	Watches and watch movements (thousands of units)	Value (thousands of francs)[a]
1885	2,975	82,026
1886	3,107	82,796
1887	3,535	86,247
1888	3,697	83,939
1889	4,531	98,942
1890	4,789	104,067
1891	4,347	103,397
1892	4,027	93,703
1893	4,191	93,794
1894	4,373	90,664
1895	4,737	94,635
1896	5,346	103,508
1897	5,487	103,765
1898	5,792	109,208
1899	6,718	110,816
1900	7,314	120,193
1901	8,044	128,320
1902	7,335	118,679
1903	7,393	116,445
1904	8,005	121,016
1905	9,107	131,290
1906	9,990	150,402
1907	9,931	149,268
1908	8,576	129,297
1909	8,959	125,975
1910	10,417	147,017
1911	12,058	164,027
1912	12,545	173,773
1913	16,855	182,849

Source: F. Scheurer, *Les crises de l'industrie horlogère dans le canton de Neuchâtel* (Neuveville: Editions Beerstecher, 1914), pp. 136–137.

a. Includes clocks, watch cases, and components.

Table 8. United States: production of watches by classes, 1926–1955 (in thousands).

Year	Jeweled watches	Clock-watches		Watches containing imported movements
		Pocket watches	Wristwatches	
1926	2,089	8,270	1,130	3,870
1927	2,281	7,732	1,160	4,375
1928	1,745	8,010	990	3,843
1929	1,737	8,440	1,040	4,935
1930	1,330	5,720	780	2,663
1931	612	4,063	1,587	851
1932	488	2,887	1,778	433
1933	463	3,361	1,900	448
1934	950	5,404	2,672	920
1935	1,393	5,315	3,067	1,202
1936	1,702	7,537	3,541	2,222
1937	2,111	8,141	2,764	3,124
1938	1,042	5,709	1,472	2,386
1939	1,624	6,710	2,347	2,919
1940	1,912	7,030	2,814	3,537
1941	2,510	7,913	3,599	4,300
1942	2,070	3,051	1,367	5,293
1943	1,313	63	65	7,996
1944	1,014	21	82	6,915
1945	1,103	541	139	9,398
1946	1,720	2,931	2,000	9,657
1947	2,364	4,873	4,321	7,757
1948	3,018	6,779	4,523	9,046
1949	2,793	4,107	2,182	8,099
1950	2,480	4,504	2,761	9,508
1951	3,162	5,084	3,242	11,576
1952	2,433	3,295	2,758	11,637
1953	2,365	2,710	3,321	13,367
1954	1,716	2,676	2,975	10,485
1955	1,926	2,999	3,641	10,853

Source: U.S. Tariff Commission, as given in Jean-Jacques Bolli, *L'aspect horloger des relations commerciales américano-suisses de 1929 à 1950* (La Chaux-de-Fonds: Editions de la Suisse Horlogère, 1956), tables 8, 12, 13.

Table 9. World production of watches and watch movements, 1945–1972; selected countries, 1976–1978 (in thousands of units).

Year	Switzerland	Japan	United States	USSR	France	West Germany	East Germany	United Kingdom	Italy	Others	Total
1945	18,800	34	1,783	100	850	—	—	—	—	—	21,567
1946	20,666	140	6,651	150	1,500	—	—	26	—	—	29,133
1947	24,007	333	11,566	300	2,570	300	100	310	—	—	39,486
1948	24,358	592	14,321	700	2,600	730	400	717	—	200	44,618
1949	24,250	756	9,092	1,000	2,555	1,500	700	1,100	—	200	41,153
1950	25,000	694	9,784	2,150	3,200	3,500	1,710	1,385	—	300	47,723
1951	33,550	924	11,559	2,900	3,755	4,720	1,592	1,716	—	500	61,216
1952	34,300	1,217	8,554	3,635	3,680	4,830	2,139	2,053	300	800	61,508
1953	34,000	1,617	8,441	4,915	3,213	5,920	2,061	2,316	600	1,500	64,583
1954	32,000	2,002	7,396	5,450	3,360	6,416	2,408	2,931	700	1,500	64,163
1955	34,750	2,240	8,492	8,730	3,660	6,960	2,515	3,410	800	2,000	73,557
1956	40,900	2,686	9,449	11,145	4,680	7,830	2,460	2,835	800	2,300	85,085
1957	41,200	3,419	7,906	13,432	4,750	7,911	2,692	2,596	800	2,500	87,206
1958	34,300	4,309	9,503	15,315	4,390	7,715	2,833	2,328	1,000	2,500	84,193
1959	38,400	5,547	11,334	16,385	4,950	7,880	3,423	2,587	1,000	2,600	94,106
1960	42,200	7,147	9,555	16,531	5,300	8,120	3,580	2,952	1,200	2,800	99,385
1961	43,300	9,231	9,712	15,169	5,500	7,560	3,600	3,163	1,200	3,000	101,435
1962	46,000	10,774	11,939	18,000	5,700	6,951	2,113	3,055	1,300	3,000	108,832
1963	46,900	11,696	12,185	15,000	5,550	5,989	2,364	3,780	1,500	3,000	107,964
1964	49,200	13,214	12,383	14,800	5,900	6,738	1,998	3,950	1,800	3,000	112,983
1965	54,750	13,608	13,473	16,000	6,500	7,249	2,220	4,000	2,000	3,000	122,800
1966	62,400	15,531	15,200	16,500	7,600	7,220	2,400	4,300	2,100	3,000	136,251
1967	65,100	16,495	16,600	17,000	9,000	6,880	2,468	2,787	2,200	3,000	141,530
1968	68,700	17,300	17,110	20,000	9,000	8,223	3,060	3,131	2,300	3,000	151,824
1969	71,550	21,425	19,000	22,000	10,600	8,276	3,280	3,008	2,300	3,000	164,439
1970	73,600	23,922	19,400	22,000	10,900	8,214	3,310	3,500	2,400	3,500	176,746
1971	72,300	24,500	21,500	23,300	12,500	7,900	3,500	3,470	2,300	4,000	178,965
1972	70,000	25,600	22,500	24,500	14,100	9,227	3,409	5,382	2,300	4,500	196,983
1973	84,307	28,046	22,059	25,000	15,850	9,691	3,526	6,154	2,500a		215,850
1974	87,027	32,369	23,672	28,900	16,700	8,746	5,000a	8,300	2,500a		235,536
1975	70,000	30,239	26,443	28,600a	16,700	9,387	5,000a	8,169	2,500a		224,186
1976	64,000	34,001	28,030								
1977	68,000	44,737	30,804								
1978	63,000	49,192	27,139								

Source: Switzerland, Fédération Horlogère, Bienne.

a. Estimated.

Table 10. Switzerland: exports of watches, watch movements, and other horological products, 1926–1980.

| Year | Watches and finished movements | | Other products (value in thousands of francs)[a] |
	Units (in thousands)	Value (in thousands of francs)	
1926	17,185	232,143	26,118
1927	18,454	243,554	26,691
1928	20,131	260,315	40,122
1929	20,758	267,284	40,056
1930	16,248	201,930	31,523
1931	11,554	121,851	22,591
1932	8,206	70,980	15,324
1933	10,599	79,482	16,533
1934	12,534	90,498	18,583
1935	15,203	102,458	22,052
1936	17,740	125,957	25,614
1937	23,916	205,026	35,355
1938	21,850	206,139	35,179
1939	16,816	166,546	29,131
1940	15,229	179,108	35,068
1941	14,441	204,864	25,705
1942	13,957	260,389	24,252
1943	14,533	318,129	19,666
1944	11,838	281,897	21,463
1945	18,800	452,508	40,111
1946	20,665	543,931	61,239
1947	24,007	672,941	95,815
1948	24,357	655,169	88,204
1949	23,545	627,538	75,689
1950	24,226	656,685	73,482
1951	33,549	916,942	93,386
1952	33,323	990,311	92,231
1953	33,030	1,016,755	89,907
1954	31,088	953,070	86,845
1955	33,742	984,157	92,843
1956	39,676	1,134,525	100,000
1957	39,999	1,195,811	107,325
1958	33,314	1,026,818	91,289
1959	37,262	1,031,031	93,539
1960	40,981	1,146,326	112,922

(*continued*)

Table 10 (cont.)

| Year | Watches and finished movements | | Other products (value in thousands of francs)[a] |
	Units (in thousands)	Value (in thousands of francs)	
1961	42,021	1,186,371	126,756
1962	44,665	1,286,058	142,881
1963	45,532	1,345,084	152,764
1964	47,764	1,466,840	163,928
1965	53,164	1,616,199	182,274
1966	60,566	1,841,097	192,963
1967	63,213	1,966,153	204,899
1968	66,621	2,107,900	208,829
1969	69,469	2,241,245	237,452
1970	71,437	2,363,221	265,822
1971	70,178	2,389,575	261,800
1972	75,253	2,520,318	300,990
1973	81,778	2,861,377	373,685
1974	84,416	3,259,511	443,024
1975	65,798	2,720,297	420,984
1976	62,053	2,605,470	466,220
1977	65,916	2,855,771	522,985
1978	60,264	2,900,299	532,857
1979	48,954	2,726,770	532,912
1980	50,986	2,917,510	632,398

Source: Chambre Suisse d'Horlogerie, "Rapport," 1980.
a. Includes watch components (*chablons*).

Table 11. Japan: production and exports of watches and movements, 1951–1980 (in thousands of units).

Year	Output	Exports
1951	924	31
1952	1,217	15
1953	1,617	19
1954	2,002	8
1955	2,240	19
1956	2,686	176
1957	3,419	17
1958	4,309	60
1959	5,447	72
1960	7,147	145
1961	9,231	535
1962	10,574	1,050
1963	11,588	2,060
1964	13,408	2,980
1965	13,607	4,800
1966	15,490	7,346
1967	16,388	6,437
1968	17,601	7,262
1969	21,260	9,442
1970	23,778	11,399
1971	24,327	13,340
1972	25,464	15,335
1973	28,046	16,473
1974	32,369	18,664
1975	30,239	17,070
1976	34,001	21,411
1977	44,737 [a]	29.201 [a]
1978	49,192 [a]	30,986 [a]
1979	59,500 [b]	43,300 [b]
1980	87,889 [b]	68,300 [b]

Source: Switzerland, Fédération Horlogère, Bienne.

a. Including Japanese production outside Japan (mostly Hong Kong) and exports of such manufactures.

b. Including *chablons* (components).

Notes

Introduction

1. The word "tide" here has nothing to do with the ebb and flow of the oceans, but rather with time and season. Hence such derivative words as "Christmastide" and "eventide"; also "tidings." On the new atomic time standard, see P. Kartaschoff, *Frequency and Time* (New York: Academic Press, 1978), pp. 2–3.

2. Much of what follows is taken from Roberto L. Quercetani's excellent book *A World History of Track and Field Athletics 1864–1964* (London: Oxford, 1964), pp. xv–xvi and ch. 1.

3. That represents a lag of seventy years. Electric timers were used to clock sports events as early as 1892, perhaps earlier, and in 1902 the Japanese were already obtaining results to the hundredth of a second. Ibid., pp. 3, 5. See also J.-P. Bovay, *Omega Sports Timing: Photosprint History* (Bienne: Omega, 1980).

4. It is worth noting that in spite of the assurance with which the judges were ready to discriminate between these two contestants on so thin a basis, record times are officially rounded up to the nearest tenth of a second, though reported to the hundredth.

5. I owe this information to Norman F. Ramsey, Higgins Professor of Physics at Harvard University and one of the fathers of atomic timekeeping, whom I had the good fortune to sit next to at a dinner of the Society of Fellows. I want to thank Prof. Ramsey here for the offprints of his publications and for his informative letter of February 17, 1981.

6. Note that the temperature need not fall as low as freezing for a water clock to stop working properly. Viscosity changes with temperature, and this can make a big difference over the range from freezing to warm (say, 0 ° C. to 20 °) to the rate of a drip- or tubular-flow clepsydra (water clock). Moreover, the narrower the stream (as through a narrow tube), the more rapid the effects of falling temperature. On these matters, I have profited greatly from an exchange of correspondence with Prof. William F. Hughes of Carnegie-Mellon University. He tells me that a clepsydra that used as orifice a small, sharp-edged hole cut in the bottom would be essentially "independent of fluid viscosity and hence temperature" (letter of June 1, 1982). The references in the historical annals to temperature problems, however, would seem to indicate that this was not the structure employed, perhaps because it would have produced a relatively rapid flow and hence required an excessively large installation.

One other source of error in clepsydras should be mentioned: no natural water is free of impurities, and deposits of these will reduce over time the dimensions of the orifice, thereby altering the rate of flow. This is the kind of error (like friction in mechanical clocks) that has a way of "sneaking up on" the user. It would also make drip- or tubular-flow clepsydras that much more vulnerable to temperature variation.

7. I originally thought the notion that time is continuous was intuitively obvious and generally accepted. I have since learned that physicists have come to see time as part of a space-time continuum, and that some of them have hypothesized that time, like other apparently continuous phenomena such as light, is composed of quanta, that is, a stream of very tiny particles. A number of them have even tried to calculate the dimension of such a fundamental time unit: Werner Heisenberg estimated it at 10^{-26} seconds; Maurice Levy, at 10^{-24} seconds. This is such a short time that one second contains more of these time quanta than all the seconds since the birth of the earth. See François Le Lionnais, *Le Temps* (Paris: Delpire, 1959), p. 91. In our present state of technique, we have no way of marking and counting such small units of time, so that all of this is still in the realm of speculation; but even if the hypothesis were true, it would in no way affect the validity of time measurement as we know it.

For the purpose of this study, then, I shall treat time as continuous. By continuous, I mean that for any interval, however short, there is a difference between time before and after (for any ϵ, however small, $t_{a+\epsilon}-t_a>0$); that is, time is always passing. Or, to use the definition of the calculus:

$$\lim_{x \to a} t(x) = t(a),$$

where x and a are points in the domain of a function that assigns a time value (hours, minutes, seconds, subseconds) to every such point.

8. From a memorial of 1094 by Su Sung, official and scholar at the court of the Che Tsung emperor of the Northern Sung dynasty. The translation is from Joseph Needham, "The Missing Link in Horological History: A Chinese

Contribution," in Needham, *Clerks and Craftsmen in China and the West* (Cambridge: Cambridge University Press, 1970), p. 223. This essay was delivered in 1958 as the Wilkins Lecture to the Royal Society in London.

9. See William Markowitz, "Time Measurement Techniques in the Microsecond Region," *Engineer's Digest* 135 (July–August 1962): 9–18 (published by the U.S. Coast Guard).

10. On the water clock in China, the principal sources are: Joseph Needham, Wang Ling, and Derek J. de Solla Price, *Heavenly Clockwork: The Great Astronomical Clocks of Medieval China* (Cambridge: Cambridge University Press, 1960); and Joseph Needham, *Science and Civilisation in China,* vol. 4, pt. 2, *Mechanical Engineering* (Cambridge: Cambridge University Press, 1965). On the water clock in Islam, Needham notes that "arrangements logically equivalent" to those found in China may be inferred from a treatise of Razzaz al-Jazari, written in 1206 (Needham, *Science and Civilisation in China,* vol. 4, pt. 2, p. 534).

11. The first publication of this passage was by Lynn Thorndike, "Invention of the Mechanical Clock about 1271 A.D." *Speculum* 16 (April 1941), 242–243.

12. See O. Kurz, *European Clocks and Watches in the Near East,* Studies of the Warburg Institute, no. 34 (London and Leiden: Warburg Institute, 1975), p. 18; and Eric Bruton, *The History of Clocks and Watches* (New York: Rizzoli, 1979), p. 31. See also Johann Hügin, "Mechanische Uhrwerke vom Altertum bis zum Mittelalter," *Alte Uhren* 3 (July 1980): 215–216. Hügin uses *Hemmung* for both the mercury-flow wheel and the mechanical escapement.

13. Offhand, I can think of only six such book-length studies: A. Pfleghart, *Die schweizerische Uhrenindustrie, ihre geschichtliche Entwicklung und Organisation* (Leipzig: Duncker & Humblot, 1908); Antony Babel, *Les métiers dans l'ancienne Genève: histoire corporative de l'horlogerie, de l'orfèvrerie et des industries annexes* (Geneva: A. Kündig, 1916); Charles W. Moore. *Timing a Century: History of the Waltham Watch Company* (Cambridge, Mass.; Harvard University Press, 1945); Viviane Isambert-Jamati, *L'industrie horlogère dans la région de Besançon: étude sociologique* (Paris: Presses Universitaires, 1955); Carlo Cipolla, *Clocks and Culture, 1300–1700* (New York: Walker, 1967), François Jequier, *Une entreprise horlogère du Val-de-Travers: Fleurier Watch Co SA* (Neuchâtel: La Baconnière, 1972). There are surely others, but not many. There are, in addition, a number of German doctoral dissertations, more concerned with institutional arrangements than economic history. The literature on the technical history of horology is far more abundant, though much of it is antiquarian in character and, ironically, timeless in context.

14. These are the words of a committee of the Clockmakers' Company of London, in a petition for prevention of "the illicit introduction of foreign clocks and watches into this country." Great Britain, "Report of the Committee on the Petitions of Watchmakers of Coventry, etc.," *Parliamentary Papers,* 1817, vol. 6, no. 504, p. 63.

I. FINDING TIME

1. On the concept of time-givers, see Pierre Sansot et al., *Les donneurs de temps* (Albeuve: Editions Castella, 1981). On biological rhythms, see Edward S. Ayensu and Philip Whitfield, eds., *The Rhythms of Life* (New York: Crown, 1982); and Martin C. Moore-Ede, Frank M. Sulzman, and Charles A. Fuller, *The Clocks that Time Us: Physiology of the Circadian Timing System* (Cambridge, Mass.: Harvard University Press, 1982).

2. The passage comes down to us as a fragment from a play entitled *Boeotia* and was attributed by Roman authors to Plautus. The lines are quoted by Aulus Gellius in *Noctes Atticae* 3.3.5, a discussion of spurious and authentic Plautine works. The translation is by J. C. Rolfe in the Loeb Library edition (New York: Putnam's, 1927), p. 247. I have this information from Glen Bowersock, who writes further: "From the historical perspective, the introduction of sundials within the lifetime of the speaker is entirely plausible in a Plautine work. Pliny the Elder, *Nat. Hist.* 7.60, 213–215, discusses very interestingly the introduction of sundials in the early third century B.C. Since Plautus was writing in the later third century B.C., the novelty of these clocks would still have been a reasonable subject for jokes." Letter to me of June 11, 1981.

3. Lewis Mumford, *Technics and Civilization* (New York: Harcourt Brace & World, 1934), p. 15.

1. A Magnificent Dead End

1. He had written some fifteen years earlier, "no doubt with a number of assistants," what Joseph Needham calls "the best work of his time on pharmaceutical botany, zoology, and mineralogy." Joseph Needham, Wang Ling, and Derek J. de Solla Price, *Heavenly Clockwork: The Great Astronomical Clocks of Medieval China* (Cambridge: Cambridge University Press, 1960), p. 8.

2. For earlier Chinese astronomical clocks, see John H. Combridge, "The Astronomical Clocktowers of Chang Ssu-Hsun and His Successors, A.D. 976 to 1126," *Antiquarian Horology* 9 (June 1975): 288–301; idem, "Clocktower Millenary Reflections," ibid. 11 (Winter 1979); and Needham, *Science and Civilisation in China*, vol. 4, pt. 2, *Mechanical Engineering* (Cambridge: Cambridge University Press, 1965), pp. 466–474. Combridge's work represents an important advance in our knowledge of the technological sequence. The account in the clockwork chapter of the Su Sung monograph (first published in 1172, translated by Needham et al. in *Heavenly Clockwork*, pp. 28–47) shows "puzzling discrepancies" between and within text and illustrations. Combridge has sorted these out, showing that they were due to inclusion of material relating to the earlier clocks in the series, which material was incompletely updated and imperfectly conflated with second- and third-generation designs.

3. It did not attempt to automate the movements of the planets, which were of secondary interest in this connection—a "side issue," says Shigeru Nakayama. See Nakayama, *A History of Japanese Astronomy: Chinese Background*

and Western Impact (Cambridge, Mass.: Harvard University Press, 1969), p. 150. The positions of the planets were tracked by hand-adjusted beads on silk string.

4. On this mechanism, see above all John H. Combridge, "The Celestial Balance: A Practical Reconstruction," *Horological Journal* 104 (February 1962): 82–86; also the descriptions in Needham et al., *Heavenly Clockwork,* and Needham, *Science and Civilisation in China,* vol. 4, pt. 2. Needham and his associates originally thought that the buckets filled and the wheel turned every "quarter"—that is, every 14 minutes 24 seconds. On the basis of logic and the literature, however, Combridge calculates that the intervals were much shorter—about 24 seconds, in fact.

5. *Chin Shih* (History of the Chin [Jurchen] dynasty), cited in Needham et al., *Heavenly Clockwork,* p. 133.

6. *Sung Shih* (History of the Sung dynasty), cited ibid., pp. 127–128. In particular, the secret of the timekeeping water wheel seems to have been lost. The armillary sphere was copied by Kuo Shou-ching in about 1276 (since lost) and recopied in 1437 by Huangfu Chung-ho, whose instrument still exists and has been memorialized by a postage stamp of 1953. See ibid., p. 133, n. 2; Needham, *Science and Civilisation in China,* vol. 3, *Mathematics and the Sciences of the Heavens and the Earth* (Cambridge: Cambridge University Press, 1959), figs. 156, 163; Combridge, "The Astronomical Clocktowers," pp. 500–501.

7. Pasquale M. d'Elia, *Storia dell'introduzione del Cristianesimo in Cina, scritta da Matteo Ricci,* in d'Elia, ed., *Fonti Ricciane,* vol. 1 (Rome: Libreria dello Stato, 1942), p. 33 and n. 2. See also Carlo Cipolla, *Clocks and Culture, 1300–1700* (New York: Walker, 1967), p. 80.

8. These articles appeared in 1931. Text is as quoted in Needham, *Science and Civilisation in China,* vol. 4, pt. 2, p. 545.

9. Mathieu Planchon, *L'horloge: son histoire rétrospective, pittoresque et artistique* (Paris, 1898), p. 262.

10. Derek J. de Solla Price, "Joseph Needham and the Science of China," in Shigeru Nakayama and Nathan Sivin, eds., *Chinese Science* (Cambridge, Mass.: MIT Press, 1973), p. 15.

11. In many ways John Combridge is a latter-day Han Kung-lien. I am much beholden to him for his help in understanding the construction and operation of these models. (I should note, though, that he is not always satisfied with my understanding of these matters.) I am also grateful to William Andrewes, curator of the Time Museum in Rockford, Illinois, for information concerning the working of the museum's half-scale model of the Su Sung clock, also built to Combridge's drawings.

12. Needham, *Science and Civilisation in China,* vol. 4, pt. 2, p. 440; Needham, "Time and Knowledge in China and the West," in Julius T. Fraser, ed., *The Voices of Time* (New York: Braziller, 1966), p. 127. This stereotype may have been relevant in the 1950s, when Needham was sharply attacking it. But I am told that Chinese scholars no longer entertain such views. They may still be current, though, in less informed circles.

13. Needham et al., *Heavenly Clockwork*, p. 197.

14. "Anticipations or precedents": Willard J. Peterson, " 'Chinese Scientific Philosophy' and Some Chinese Attitudes towards Knowledge about the Realm of Heaven-and-Earth," *Past & Present* 87 (May 1980): 20–21. "The balance-sheet": Needham, *Science and Civilisation in China*, vol. 3, p. xli; "the whole march of humanity": ibid., p. xxxi.

15. "In person": Needham, *Science and Civilisation in China*, vol. 4, pt. 2, p. 545. "Owes its existence": ibid., p. 436. For the influence of Needham's hypothesis, see Ernst von Bassermann-Jordan, *Uhren: Ein Handbuch für Sammler und Liebhaber*, rev. and ed. Hans von Bertele (Braunschweig: Klinkhardt & Biermann, 1976), pl. 1, which describes Su Sung's clock as already containing the basic ideas of the mechanical escapement. This work goes back to 1961 at least, but the retention of such a statement in the eighth edition is a measure of the tenacity of the idea. See also Samuel A. Goudsmit et al., *Time* (New York: Time-Life Books, 1966), pp. 91–92. For Cipolla's views, see Cipolla, *Clocks and Culture*, p. 40.

16. The question is reminiscent of Tawney's paradox: "China ploughed with iron when Europe used wood and continued to plough with it when Europe used steel." Richard H. Tawney, *Land and Labor in China* (New York: Harcourt, Brace, 1932), cited by Anthony M. Tang, "China's Agricultural Legacy," *Economic Development and Cultural Change* 28 (October 1979): 5.

17. Kenneth Boulding, "Great Laws of Change," in Anthony M. Tang, Fred M. Westfield, and James S. Worley, eds., *Evolution, Welfare, and Time in Economics* (Lexington, Mass.: Lexington Books, 1976), p. 9. It may well be that Needham is now inclined to invert the question to: Why did Europe ever take this path?

18. For a chronology of inventions conveying the overwhelming technological priority of China over Europe, see the table entitled "Transmission of Certain Techniques from China to the West," in Charles Singer et al., eds., *A History of Technology*, vol. 2, *The Mediterranean Civilizations and the Middle Ages* (London: Oxford University Press, 1956), pp. 771–772. The table is based on Needham's researches, and hence is somewhat biased in favor of Chinese precedence. Even so, there can be no mistaking the early balance on technological account.

19. This device was credited to the duke of Chou, brother of the emperor Wu, founder of the Chou dynasty some eleven hundred years before our era. Planchon, *L'Horloge*, p. 246.

20. On the uneven performance of the sand clock, see Dàniel W. Hering, *The Lure of the Clock* (New York: New York University Press, 1932), p. 16, n. 2. On recourse to mercury: when Chang Ssu-hsun built his astronomical clock-tower in 976–979, he substituted mercury for water (it is not clear whether he did so at the time of construction or later), "and there were no more errors." Yet Su Sung tells us that after the death of Chang, "the cords and the mechanism went to rack and ruin" and there was no one to keep it going. Needham, *Science and Civilisation in China*, vol. 4, pt. 2, pp. 469–471. See also Combridge, "Clocktower Millenary Reflections," which offers a more detailed description

of the Chang machine: "the world's first astronomical clocktower" (p. 604).

21. The best source on Islamic horology is Donald R. Hill, *Arabic Water Clocks*, Sources and Studies in the History of Arabic-Islamic Science: History of Technology Series, no. 4. (Aleppo: University of Aleppo, 1981). See also idem, *On the Construction of Water Clocks* (London: Turner & Devereux, 1976); and idem, "A Treatise on Machines . . .," *Journal for the History of Arabic Science* 1 (May 1977): 33–46. The general survey of Eilhardt Wiedemann and F. Hauser, *Ueber die Uhren im Bereich der islamischen Kultur*, Abhandlungen der Kaiserlichen Leop.-Carol. Deutschen Akademie der Naturforscher 100 (Halle, 1915): 176–266, is still useful.

Anthony Turner has reminded me that there are some places in the classical Muslim world that know freezing temperatures—for example, the Iranian and Anatolian plateaus and some parts of the Spanish sierra. These do get better service from sundials, though, than Europe north of the Alps. It would be interesting to learn how they coped with low temperatures in their use of water clocks.

22. Needham, *Science and Civilisation in China*, vol. 4, pt. 2, p. 465.

23. This pattern of work still prevailed in China at the start of this century. See "A Manual for Apprentices in Trade" (1905), given *in extenso* in Lien-Sheng Yang, "Schedules of Work and Rest in Imperial China," *Harvard Journal of Asiatic Studies* 18 (December 1955): 301–325. Much of these paragraphs is derived from this article, which John Fairbank was good enough to call to my attention.

24. The Chinese divided the night into five equal watches, which obviously varied in length with the season. Usage also varied, however, in regard to the appropriate starting and ending points of the night, so that it is impossible to assign a mean-time designation to these units without knowing period and place (the Imperial Palace apparently had its own system). See the discussion in Needham et al., *Heavenly Clockwork*, pp. 203–205.

25. Ibid., p. 93, n. 1. The "clocks with weights" were not mechanical clocks, but *steelyard clepsydras*—that is, water clocks combined with a balance beam and adjustable weights. See John H. Combridge, "Chinese Steelyard Clepsydras," *Antiquarian Horology* 12 (Spring 1981): 530–535.

26. Etienne Balazs, *La bureaucratie céleste: recherches sur l'économie et la société de la Chine traditionnelle* (Paris: Gallimard, 1968), pp. 208–210.

27. On the merchant cities, yet the absence of bourgeois autonomy and freedom of enterprise: ibid., pp. 212–218, 200–205. Balazs cites with approval the aphorism of Max Weber: "City equals mandarin seat without self-government; village equals self-governing locality without mandarins." Weber, "Die Wirtschaftsethik der Weltreligionen, I: Konfuzianismus und Taoismus," in *Gesammelte Aufsätze zur Religionssoziologie*, vol. 1 (Tübingen: Mohr, 1922), p. 381, cited ibid., p. 210, n. 2.

28. See Alfred Chapuis, *Relations de l'horlogerie suisse avec la Chine: la montre "chinoise"* (Neuchâtel: Attinger Frères, n.d. [1919]), pp. 15–16.

29. Temporal hours are defined as equal fractions of daytime and nighttime. They therefore vary with the seasons, and daytime hours equal night-

time hours only at the equinoxes. On the theory of the sundial, see Edmond Guyot, *Histoire de la détermination de l'heure* (La Chaux-de-Fonds: La Chambre Suisse de l'Horlogerie, 1968), esp. pp. 11–14.

30. George N. Kates, *The Years that Were Fat: Peking, 1933–1940* (New York: Harper, 1952), pp. 253–254.

31. As cited by Needham et al., *Heavenly Clockwork*, p. 26. I have omitted Needham's parenthetical editorial insertions. Needham contends that Su Sung's statement is "quite logical and scientific" (ibid., n. 3) because an accurate calendar could be used, for example, to select the proper time for planting. This is, I fear, another instance of his excessive indulgence for Chinese "science." Chinese astronomical observations produced valuable records, but their application was essentially astrological and their effectiveness as a guide to action, while not unrelated to calendrical reality, was intrinsically coincidental.

32. There is no explicit statement in the sources that the jacks held in this way from quarter to quarter (displayed what one might call "jump quarters"), but this is what Combridge deduces from his reading of fifteenth-century Korean materials that he is now in the process of editing.

33. Needham, *Science and Civilisation in China*, vol. 4, pt. 2, pp. 482, 483, 494, 495.

34. Combridge tells me, in a letter of May 26, 1981, that Su Sung's astronomical middle chapter (not yet published in translation) reports the sun's position on the sidereally driven ring of the armillary sphere, at "nominal" dusk and dawn of solstices and equinoxes, to better than one-fourth of a Chinese celestial degree (equals 1/365.25 of a circle), or about one minute of time.

35. Joseph Needham, "The Missing Link in Horological History: A Chinese Contribution," in Needham, *Clerks and Craftsmen in China and the West* (Cambridge: Cambridge University Press, 1970), p. 236. What Needham gives here is a modified version of a chart by F.A.B. Ward of the growing precision of horological mechanisms over time. See Ward, "How Timekeeping Mechanisms Became Accurate," *Chartered Mechanical Engineer* 8 (1961): 604.

36. See the reference to I Hsing's clock, the first fully referenced example of these water-bucket mechanisms (eighth century), in Needham, *Science and Civilisation in China*, vol. 4, pt. 2, p. 499.

37. The museum recently placed it in a new environment and hopes for better. Letter of May 15, 1981, from William Andrewes.

38. Needham et al., *Heavenly Clockwork*, p. 16, n. 3.

39. Ibid., pp. 6–7. It was after this that the emperor ordered Su Sung to construct the great astronomical clock that was the peak of Chinese achievement in this domain and whose details have been recovered for us by Needham, Wang Ling, and Price, and by Combridge.

40. In fact, there is some evidence of astronomical instruments in private hands. Such instances, however, were few and no doubt exceptional, limited probably to "scholarly families connected with the bureaucracy." Needham et al., *Heavenly Clockwork*, p. 117, n. 1.

41. As recounted by naturalist Yang Hsiung (−53–+25). Ibid., p. 129.

42. Needham, *Science and Civilisation in China,* vol. 4, pt. 2, p. 504.

43. Ibid., p. 499. This from a book of 1111 by Wang Fu, politician-functionary-archaeologist of the late Sung dynasty (late eleventh–early twelfth centuries). Wang Fu is here comparing an astrarium model of the early twelfth century with I Hsing's clock of the eighth century. On the linguistic difficulties posed by these archival researches, see the passages from the *Sung Shih* (History of the Sung Dynasty) cited by Needham et al., *Heavenly Clockwork,* p. 122.

44. Cited in Needham et al., *Heavenly Clockwork,* pp. 18–19. I have incorporated in brackets one additional date and information from p. 19, n. 1.

45. Ibid., p. 136, n. 2.

46. Needham, *Science and Civilisation in China,* vol. 4, pt. 2, p. 500.

2. Why Are the Memorials Late?

1. Castiglione's *Il Cortegiano* appeared in 1528 and was already a classic by the time of Ricci's birth in 1552.

2. Pasquale M. d'Elia, *Storia dell'introduzione del Cristianesimo in Cina, scritta da Matteo Ricci,* in d'Elia, ed., *Fonti Ricciane,* vol. 1 (Rome: Libreria dello Stato, 1942), p. 16, n. 8.

3. See, for example, A. Koyré, "The Copernican Revolution," in René Taton, ed., *The Beginnings of Modern Science from 1450 to 1800* (London: Thames & Hudson, 1964), pp. 66–71.

4. Pasquale M. d'Elia, *Galileo in China: Relations through the Roman College between Galileo and the Jesuit Scientist Missionaries (1610–1640)* (Cambridge, Mass.: Harvard University Press, 1960), p. 5. Joseph Needham points out that in introducing the Chinese to a positional astronomy based on the ecliptic (the sun's perceived path around the earth) rather than on the celestial equator, the Jesuits were actually moving Chinese science backward. Further, they refrained from teaching the Chinese the new developments of European heliocentric astronomy over the next two centuries, and this for the worst of reasons: they would not propagate a theory condemned by the Church, and they feared that by correcting Ricci the astronomer they would cast doubt on the lessons of Ricci the preacher and theologian. "China, when it received Western science, received it in a form which was already antiquated" (J. J. L. Duyvendak, cited by J. Chesnaux and J. Needham, "Science in the Far East from the 16th to the 18th Century," in Taton, ed., *The Beginnings of Modern Science,* p. 590, n. 2.).

Chesnaux and Needham concede that the failure of the Chinese to "make a more original contribution to the progress of modern science" (ibid., p. 593) was not due exclusively to the defects of the Jesuit contribution. As good historical materialists, they also point to the "slow development of Chinese society." But surely there were cultural and intellectual impediments as well. Whatever the limitations of Jesuit astronomy, they did bring to China vastly superior instruments—above all, the telescope—and the Chinese did not lack for opportunity to make the same kind of observations that were generating in Europe substantially improved models of planetary and stellar movements. For a defense of the Jesuit contribution, showing among other things that Jesuit astronomers did try to communicate Galileo's observational discoveries to the Chinese, see d'Elia, *Galileo in China*, pp. 51–56. As for heliocentric theory, d'Elia argues that the greatest impediment was the Chinese long-standing belief in a geocentric universe: the Chinese were no different in this from the Catholic church.

5. D'Elia, *Galileo in China*, p. 6 (letter of May 12, 1605).

6. D'Elia, *Storia dell'introduzione*, p. 127.

7. Arnold H. Rowbotham, *Missionary and Mandarin: The Jesuits and the Court of China* (Berkeley: University of California Press, 1942), p. 61.

8. See the work of Paul Bairoch, "Ecarts internationaux des niveaux de vie avant la Révolution industrielle," *Annales: économies, sociétés, civilisations* 34 (January–February 1979): 164. On the comparison of Chinese and European population figures, see Fernand Braudel, *Civilisation matérielle, économie et capitalisme*, vol. 1 (Paris: A. Colin, 1979), pp. 23–29.

9. Cited in Alfred Chapuis, *Relations de l'horlogerie suisse avec la Chine: La montre "chinoise"* (Neuchâtel: Attinger Frères, n.d. [1919]), p. 15. The ambassadors made much of the fact that Chinese timekeepers provided only approximate readings, whereas European mechanical clocks could be read to the nearest minute.

10. Written c. 1705. Cited by Jonathan D. Spence, *Emperor of China: Self-Portrait of K'ang Hsi* (New York: Vintage, 1975), p. 63 (unpaginated).

11. Ibid., pp. 67–68. This seems to be a free translation. See "Instructions sublimes et familières de Chang-Tzu-Quogen-Hoang-Ti [the temple name of K'ang Hsi]," in *Mémoires concernant l'histoire, les sciences . . . des Chinois par les missionnaires de Pékin*, vol. 9 (Paris, 1783), pp. 179–180, where it is put somewhat differently: "Toward the end of the Ming dynasty, the Europeans having entered China and made for the first time one or two sundials, the Ming emperors took them for a valuable treasure"; K'ang Hsi then goes on to speak of his success in copying Western clocks. I have conflated the French and Italian texts.

12. Silvio A. Bedini, "Chinese Mechanical Clocks," *Bulletin of the National Association of Watch and Clock Collectors* 7 (June 1956): 220. See also Georges Bonnant, "The Introduction of Western Horology in China," *La Suisse horlogère* (International Edition, in English) 74, no. 3 (1959): 28–38. One of the few early Chinese watches to have survived and been recorded is illustrated in Chapuis, *Relations de l'horlogerie*, opp. p. 42. Though made about 1730, it bears stamps for the K'ang Hsi reign (a common practice) and resembles a European enamel watch of the late seventeenth century. The work, as Edouard Gélis describes it, is rather crude: "One would think it the work of an apprentice, done without the knowledge of the master, because the master would not have accepted such a piece." Gélis describes the decoration as beautiful, which is hard to verify or gainsay on the basis of half-tone illustrations. But to my eye it seems far short of the finesse and grace of European dials and cases at that period. For Gélis and his collection, see Edouard Gélis, *Horlogerie ancienne: Histoire, décor et technique* (Paris: Grund, 1949).

13. Bedini, "Chinese Mechanical Clocks," p. 214.

14. Cited in Carlo Cipolla, *Clocks and Culture, 1300–1700* (New York: Walker, 1967), p. 86.

15. This description of the rape of the Imperial Palace is taken from the admirably ironic treatment by Christopher Hibbert, *The Dragon Wakes: China and the West, 1793–1911* (London: Longmans, 1970), ch. 16: "The Burning of the Palace."

16. Simon Harcourt-Smith, *A Catalogue of Various Clocks, Watches, Automata . . . in the Palace Museum and the Wu Ying Tien, Peiping* (Peking: Palace Museum, 1933), p. 1.

17. Cipolla, *Clocks and Culture,* pp. 155, 89.

18. Letter to Colbert, no date (1675), in Onno Klopp, ed., *Die Werke von Leibniz . . . in der Königlichen Bibliothek zu Hannover,* vol. 3 (Hannover, 1864), pp. 212–213.

19. On Chinese cartography and the reactions to Ricci's efforts: Henri Bernard, *Aux portes de la Chine: les missionaires du seizième siècle, 1514–1588* (Tientsin: Hautes Etudes, 1933), pp. 232–234. See the attempted reconstruction of Ricci's map in the article by Henry Yule in the *Encyclopedia Britannica,* 11th ed., s.v. "Ricci, Matteo."

20. Cited in Carlo M. Cipolla, *Guns, Sails and Empires: Technological Innovation and the Early Phases of European Expansion, 1400–1700* (New York: Pantheon, 1965), p. 119, n. 3.

21. Joseph Needham, *Science and Civilisation in China,* vol. 4, pt. 2, *Mechanical Engineering* (Cambridge: Cambridge University Press, 1965), pp. 524–526 and 525 note d. There are comparable examples of defensive denial in the Islamic world: for example, that of Prince Tewfik of Egypt in the 1860s, who was convinced "that the Arab writers described long ago the steam engine, railroad, etc., etc." David S. Landes, *Bankers and Pashas: International Finance and Economic Imperialism in Egypt* (Cambridge, Mass.: Harvard University Press, 1958), p. 325.

22. Wei Jun obviously did not understand the function of the fusee wheel as a device to equalize the force of the unwinding mainspring. On this, see Chapter 5.

23. Jacques Gernet, *Chine et christianisme: action et réaction* (Paris: Gallimard, 1982), pp. 88–89.

24. A. E. van Braam, *An Authentic Account of the Embassy of the Dutch East India Company,* vol. 2 (London, 1798), pp. 47–48, cited in Cipolla, *Clocks and Culture,* pp. 89–90.

25. It is very difficult to translate unspecified ducats (presumably Venetian ducats) into today's money. The gold ducat weighed about 3.5 grams and was equivalent to about half an English pound. The silver ducat was worth about half that. Five hundred ducats, whether gold or silver, was a lot of money. I am reminded of the $500 that Soviet troops were paying for Mickey Mouse watches in Berlin after the city's capture in World War II. To be sure, the Soviets were paying in military scrip, guaranteed by the U.S. Treasury— that is, paying with other people's money. But so were the mandarins who were paying ridiculous sums for these toys.

26. Van Braam, *An Authentic Account*, vol. 1, pp. 242–243, cited in Cipolla, *Clocks and Culture*, p. 156.

27. As translated in Ssu-yu Teng and John K. Fairbank, eds., *China's Response to the West: A Documentary Survey, 1839–1928* (Cambridge, Mass.: Harvard University Press, 1979), p. 19.

28. Hibbert, *The Dragon Wakes*, p. 179.

29. These watches, it would seem, did not have a minute hand, which probably means that they did not have a balance spring, hence varied anywhere from fifteen minutes to an hour a day. (On the consequences of the balance spring for timekeeping accuracy, see Chapter 7.) Since Chao I was writing in the eighteenth century, some three generations after the introduction of the balance spring in the West (c. 1675), one may infer that watches exported to China were technically obsolete—or that Chao I was ignorant of watches. Probably both were true. This conforms, by the way, to the usual pattern of exports to less sophisticated markets: watchmakers, then as now, reserved precision pieces to those users who demanded (valued) them. The Chinese, like the Indians and Turks, were more concerned with other features: appearance, and such complications as moon-phase dials, automata, *sonnerie,* and repetition.

30. As translated and quoted in Beatrice Sturgis Bartlett, "The Vermilion Brush: The Grand Council Communications System and Central Government Decision Making in Mid Ch'ing China" (diss., Yale University, 1980), pp. 214–216. I am grateful to Dr. Bartlett for bringing this material to my attention; also for a critical hearing of my views on Chinese horology.

31. Ray Huang, *1587, a Year of No Significance: The Ming Dynasty in Decline* (New Haven: Yale University Press, 1981), pp. 7–8.

32. Cited in Spence, *Emperor of China,* p. 63 (unpaginated). Beatrice Bartlett tells me that she does not think this verse means what it seems to say—that it is not talking about time but about K'ang Hsi's diligence and devotion to duty. She also remarks that it is inconceivable that the emperor would ever have permitted the memorials to be delivered late to his desk—certainly not more than once. I am sure she is right so far as that goes. But in view of K'ang Hsi's well-documented interest in clocks and clockmaking, I think he is also writing in praise of timekeepers and in criticism of what he perceives to be the unpunctuality (in the literal sense) of those around him. The cautionary tale of Fu-heng's misplaced reliance, written about a century later, would seem to justify K'ang Hsi's pessimism on this point.

3. Are You Sleeping, Brother John?

1. On the Norwich clock: J. D. North, "Monasticism and the First Mechanical Clocks," in J. T. Fraser and N. Lawrence, eds., *The Study of Time II: Proceedings of the Second Conference of the International Society for the Study of Time, Lake Yamanaka-Japan* (New York: Springer, 1975), p. 385. There is a summary of the expenses of the Norwich clock in an article signed by A. W. [Way] entitled, "Original Documents: Agreement between the Dean and Chapter of St.

Paul's, London, and Walter the Orgoner . . . Communicated by Sir Frederic Madden, K. H.," *Archaeological Journal* 12 (1855): 173–177. (I owe this reference to Mr. Anthony Turner.)

On Richard of Wallingford, see the definitive edition of his work by John D. North, *Richard of Wallingford*, 3 vols. (Oxford: Clarendon, 1974). See also Robert W. T. Gunther, *Early Science in Oxford*, vol. 2, *Astronomy* (Oxford: Oxford Historical Society, 1923), pp. 349–370; and Derek J. de Solla Price, *The Equatorie of the Planetis* (Cambridge: Cambridge University Press, 1955), pp. 127–130. On the reconstruction of the Wallingford mechanism and its implications, see E. Watson, "The St. Alban's Clock," *Antiquarian Horology* 11 (Winter 1979): 576–584. On Dondi, see H. Alan Lloyd, "Giovanni de Dondi's Horological Masterpiece," *La Suisse horlogère* (International Edition, in English) 70 (July 1955): 49–71; and Silvio A. Bedini and Francis R. Maddison, *Mechanical Universe: The Astrarium of Giovanni de' Dondi*, in *Transactions of the American Philosophical Society*, n.s., vol. 56, pt. 5 (Philadelphia, 1966).

2. Richer [Richerus], *Histoire de France (888–995)* ed. Robert Latouche, in Louis Halphen, ed., *Les classiques de l'histoire de France au moyen age* vol. 2, *954–995* (Paris: Les Belles Lettres, 1964), pp. 57–63. See also Pierre Dubois, *Collection archéologique du Prince Pierre Soltykoff: Horlogerie* (Paris, 1858), pp. 161–168.

3. See F. Picavet, *Gerbert, un pape philosophe, d'après l'histoire et d'après la légende* (Paris, 1897), ch. 6: "La légende de Gerbert."

4. Derek J. de Solla Price, "Clockwork before the Clock," *Horological Journal* 97 (December 1955): 810, 814. This is a paper he presented to the British Horological Institute and the Antiquarian Horological Society in October of that year.

5. On Price's view of the clock as a "degenerate branch from the main stem of mechanized astronomical devices," see Price, "On the Origin of Clockwork, Perpetual Motion Devices, and the Compass," *U.S. National Museum Bulletin*, no. 218, Contributions from the Museum of History and Technology, paper 6 (Washington, D.C.: Smithsonian Institution, 1959), p. 86. The fallen angel metaphor first appeared in Price, "Clockwork before the Clock," p. 814.

6. Price, "Clockwork before the Clock," pp. 814, 810.

7. Derek J. de Solla Price, "Clockwork before the Clock and Timekeepers before Timekeeping," *Bulletin of the National Association of Watch and Clock Collectors, Inc.* 18 (1976): 399. This position leads Price to make an astonishing statement: "A consequence of the belief that Man had always 'kept time' is the erroneous supposition that prior to the mechanical timekeepers, which dominated so much of the medieval and later practice, there must have been various alternative non-mechanical timekeepers such as sundials, water and sand clocks and similar devices" (ibid., p. 398). But this was exactly so, and Price not only knows it but speaks of these other kinds of timekeepers in other places.

8. "Historically speaking": Price, "Clockwork before the Clock," p. 814. To be sure, development from complexity to simplicity may occur in the limited, meliorative sense that early versions of a device or technique may well be

inefficient and cumbersome—complicated along Rube Goldberg lines—by comparison with later, more refined versions. But these latter are anything but fallen angels.

9. Yet Price's thesis has found its way into the general literature on the history of technology. See D. S. L. Cardwell, *Turning Points in Western Technology* (New York: Science History Publications, 1972), p. 16. There is something to be said for an air of emphatic conviction.

10. See Chapter 12 below.

11. See his response in October 1969 at a meeting of the Antiquarian Horological Society, where it was suggested that these early *horologia* may well have been water clocks. *Antiquarian Horology* 6 (June 1970): 414. Cf. North, "Monasticism and the First Mechanical Clocks," p. 384.

12. G. H. Baillie, trans., *The Planetarium of Giovanni de Dondi, Citizen of Padua* (along with additional material from another Dondi manuscript translated by H. Alan Lloyd), ed. F. A. B. Ward (London: Antiquarian Horological Society, 1974), p. 14. Also Bedini and Maddison, *Mechanical Universe,* pp. 11, 14; and Anthony J. Turner, "The Tragicall History of Giovanni de' Dondi," *Journal of the History of Astronomy* 6 (1975): 126–131. The most reliable published version of the manuscript is Giovanni Dondi dall'Orologio, *Tractatus astrarii . . . (Biblioteca capitolare de Padova, Cod. D 39),* ed. Antonio Barzon, Enrico Morpurgo, et al., with photo reproduction of codex, "Codices ex ecclesiasticis Italiae bibliothecis selectis . . .," no. 9 (Vatican City: The Vatican, 1960).

Price suggests that "this [dismissal of the "common clock"] may be bravado to quite a large degree" ("On the Origin of Clockwork," p. 85, n. 7). Whose?

13. Jacques Le Goff, "Labor Time in the 'Crisis' of the Fourteenth Century: From Medieval Time to Modern Time," in Le Goff, *Time, Work, and Culture in the Middle Ages* (Chicago: University of Chicago Press, 1980), p. 44.

14. Most Jews, including many who recite the prayer, are not aware of its meaning. The blessing has been "improved" in translation in non-Orthodox prayer books. The rooster has been expunged and some lofty concept substituted, such as: "Blessed art Thou . . . who hast given the mind understanding to distinguish between day and night." For rabbinical commentary on the prayer, see Philip Birnbaum, ed., *Daily Prayer Book* (New York: Hebrew Publishing Co., 1949), p. 16n and the references cited.

15. These were the so-called *temporal* or unequal hours inherited from the ancients. Day and night were divided respectively into twelve equal parts, which differed from each other except at the equinoxes and varied with the changing seasons. Gustav Bilfinger, in his classic study *Die mittelalterlichen Horen und die modernen Stunden* (Stuttgart, 1892), calls them *Horen* in opposition to our equal hours or *Stunden.*

16. From his *De Oratione* and his *De Jejunio,* as cited in F. Claeys Bouuaert, "Heures canoniques," in R. Naz, ed., *Dictionnaire de droit canonique,* vol. 5 (Paris: Letouzey et Ané, 1953), pp. 1116–17. See also C. W. Dugmore, "Canonical Hours," in John G. Davies, ed., *A Dictionary of Liturgy and Worship* (New York: Macmillan, 1972). These hours were also hallowed for Christians

by the timing of events connected with the crucifixion of Jesus.

17. See Emile Bertaud, "Horloges spirituelles," in M. Viller et al., eds., *Dictionnaire de spiritualité ascétique et mystique, doctrine et histoire,* vol. 7, pt. 1 (Paris: Beauchesne, 1969), p. 749.

18. Cited in Bouuaert, "Heures canoniques," p. 1115, col. 1.

19. Ibid., p. 1118.

20. Ibid.

21. The matter still requires investigation, but such works as I have been directed to point in this direction. See Arthur Vööbus, *History of Asceticism in the Syrian Orient: A Contribution to the History of Culture in the Near East,* vol. 2, *Early Monasticism in Mesopotamia and Syria* [Catholic University of America and Catholic University of Louvain, "Corpus Scriptorum Christianorum Orientalium," vol. 197; "Subsidia," vol. 17] (Louvain, 1960), pp. 286–288, who notes that certain moments were considered especially propitious: the noon hour, for example, and sunset, "just before the doors of heaven are closed and darkness overcomes light." But he also notes some communities that fixed hours by the clock: third, sixth, ninth (Tertullian's public time breaks). Here, as with Muslims later, sundials and water clocks presumably sufficed.

22. One result of this double meaning was to blur the horary sequence. Since the offices (the "canonical hours") went on at some length, the terms *tierce, sext,* and *none* came to be applied not so much to points of time as to bands. And these in turn coincided with the four daylight "hours" of the ancients. The Romans called these *prima, tertia, sexta,* and *nona.* Medieval Christians used the term *tierce* from sunup to midmorning; *sext* from then to midday; *none* from midday to midafternoon; vespers from then until nightfall. (I use verbal rather than numerical bounds because the ecclesiastical clock kept temporal—that is, unequal—hours, that varied with the season.) See the chart in Antonio Simoni, *Orologi italiani dal cinquecento all'ottocento* (n.p.: Antonio Vallardi Editore, 1965), pl. 6, opp. p. 33. It may be that it was this demarcation of the canonical hours by bands of time that led in the end to the shifting of *none* (the end of the ninth hour) from midafternoon to midday, giving us our English *noon.* See Cardinal Gasquet, trans., *The Rule of Saint Benedict* (London: Medieval Library, 1925), ch. 48, "Of Daily Manual Labour": "Let None [the office] be said somewhat before the time, about the middle of the eighth hour." This was the summer schedule, in effect from Easter to the first of October. In the latitude of central Italy and during that period, the middle of the eighth hour would have come at the latest, at the summer solstice, a little before 2 P.M. mean time; at the earliest, near the equinox, a little after 1:30 P.M.—not at 3:20, as Gasquet thought (p. 129, n. 186). (This very misreading testifies to the ease with which the horary designations lent themselves to slippage. Compare the similar problems we have in counting birthdays and centuries.) In other words, by the sixth century, *none* had already moved forward half the way to what we now know as noon.

For some alternative explanations of this shift, see Le Goff, *Time, Work, and Culture,* pp. 44–45. One suggests that the monks could not wait to eat, and so moved the *none* office forward; the other, that urban workers took their pause at *none* and were also too impatient to wait until midafternoon. I have difficulty with both of these. The second is simply anachronistic: the slippage was, as noted above, already well under way before there was an urban working class to take into account. The first is problematic. There is some evidence that hunger pangs could influence the *horarium*. At the Cistercian abbey of Villers (near Brussels), for example, a semifast (one meal a day, after *none*) was observed during the period preceding Ash Wednesday, and *none* was recited around 11 or 11:30 A.M. Albert d'Haenens, "La clepsydre de Villers (1267): comment on mesurait et vivait le temps dans une abbaye cistercienne au XIIIe siècle," in Oesterreichische Akademie der Wissenschaften, Philosophisch-Historische Klasse, Sitzungsberichte, vol. 367: *Klösterliche Sachkultur des Spätmittelalters: Internationaler Kongress Krems an der Donau 18. bis 21. September 1978*, Veröffentlichunger des Instituts für Mittelalterliche Realienkunde Oesterreichs, no. 3 (Vienna: Verlag der Oesterreichischen Akademie der Wissenschaften, 1980), p. 341.

On the other hand, it is easy enough to move an activity or pause without changing the clock; witness what has happened to our dinner time. And in fact, the Benedictine summer *horarium* provided for taking the main meal after *sext*, that is, around midday; after which the brethren took a nap (called by the Spanish *siesta*, after the sixth hour) until the heat went down (*ad calorem declinandum*). B. van Haeften, *S. Benedictus Illustratus sive Disquisitionum Monasticarum Libri XII* (Antwerp, 1644), p. 776, col. 1. (I am grateful to my colleague Giles Constable for making a photocopy of this exceptionally rare tome available to me.) In the same way, afternoon did not begin at Cluny in summer, regardless of the solar hour, until all morning prayers (that is, *sext*) had been sung and the midday meal finished. Ulrich, *Consuetudines Cluniacenses*, vol. 1, p. 18, as cited by Jean Leclercq, "Experience and Interpretation of Time in the Early Middle Ages," in John R. Sommerfeldt et al., eds., *Studies in Medieval Culture*, vol. 5 (Kalamazoo: Medieval Institute, Western Michigan University, 1975), p. 12.

It may well be that custom sought sanction in nomenclature, and that the term *none* came to be applied semi-ironically to this midday pause in those northern lands where climatic conditions did not require it. Van Haeften speaks of the usage as a Flemish abuse. See also Cuthbert Butler, *Benedictine Monachism*, 2nd ed. (London: Longmans, 1924; rpt. Cambridge: Speculum Historiae, 1961), p. 282. (Note that Butler's conversions of medieval hours to mean time have to be used very cautiously, since he equates the twelfth temporal hour, *duodecima hora*, with sunrise and sunset Italian standard time and reckons from there. What he should have done is equate the end of the sixth hour, day and night, with 12 M. and 12 P.M. Italian true—solar—time and worked his conversions from there.)

23. Gasquet, trans., *The Rule of Saint Benedict*, ch. 43.

24. Ibid., ch. 11: "How Matins, or the Night Watches, Are to Be Celebrated on Sundays."

25. Cited in C. B. Drover, "A Medieval Monastic Water-Clock," *Antiquarian Horology* 1 (December 1954): 56. Further on the hours of prayer in medieval Christianity, see Ernest L. Edwardes, *Weight-Driven Chamber Clocks of the Middle Ages and Renaissance* (Altrincham: J. Sharrat, 1965), pp. 4–7; Gasquet,

trans., *The Rule of Saint Benedict,* chs. 8–16; Guy de Valous, *Le monachisme cluni-sien des origines au XVe siècle,* vol. 1, in *Archives de la France monastique,* vol. 39 (Paris: A. Picard, 1935), pp. 148–149, 292–293, 330–333.

26. Gordon Moyer, "The Gregorian Calendar," *Scientific American* 246 (May 1982): 146. A fascinating article.

27. This material on the computus and its link to time measurement is based largely on research by Richard A. Landes, and it is he who called my attention to the material cited in the text and notes that follow. For these manuscripts, see C. W. Jones, *Bedae Opera de Temporibus* (Ithaca: Cornell University Press, 1943), pp. 144–161; idem, *Beda: Opera Didiscalia,* pt. 2 (Corpus Christianorum Series Latina, vol. 123 B), pp. 242–256; A. Cordoliani, "Contribution à la littérature du comput au Moyen Age," *Studi Medievali,* ser. 3, vol. 1 (1960): 107–138; vol. 2 (1961): 169–208 (with bibliography).

28. Harriet P. Lattin, ed., *The Letters of Gerbert* (New York: Columbia University Press, 1961), pp. 189–191 (letter no. 161).

29. Paris, Bibliothèque Nationale, Latin 5239, saec. X, St-Martial de Limoges, ff. 205^1–211. The manuscript has been mentioned in Lynn Thorndike, *A History of Magic and Experimental Science,* vol. 1 (New York: Columbia University Press, 1923), p. 692, and Jones, *Bedae Opera de temporibus,* p. 155; also idem, *Bedae Pseudepigrapha: Scientific Writings Falsely Attributed to Bede* (Ithaca: Cornell University Press, 1939), p. 128. These time units, which reflected real and potentially measurable differences in duration, should be distinguished from hypothetical fractional units proposed by systematic scholastics. Thus Honorius Augustodunensis (of Autun), twelfth century, divided the canonical hour into 4 points, 10 minutes, 15 parts, 40 movements, 60 marks, and 22,500 atoms. Kazimierz Piesowicz, "Lebensrhythmus und Zeitrechnung in der vorindustriellen und in der industriellen Gesellschaft," *Geschichte in Wissenschaft und Unterricht,* no. 8 (1980): 484, n. 23, citing A. J. Gurevich, "Predstavleniia o vremeni v srednevekovei Evrope" (Concepts of time in medieval Europe), in *Istoriia i psikhologiia* (History and Psychology) (Moscow, 1971), p. 170. (I have not been able to read the Gurevich work.)

30. Raoul Glaber, *Les cinq livres de ses histoires,* ed. Maurice Prou (Paris, 1886), bk. 5, ch. 1, p. 114–115.

31. Ibid., p. 117.

32. Joan Evans, *Monastic Life at Cluny, 910–1157* (Oxford: Clarendon, 1931), p. 80–81.

33. Interestingly, translations of this rhyme have deprived it of its sense: in both English and German, Brother John (Bruder Jakob) is no longer responsible for ringing the bells; he is sleeping through them.

4. The Greatest Necessity for Every Rank of Men

1. Cardinal Gasquet, trans., *The Rule of Saint Benedict* (London: Medieval Library, 1925), ch. 48.

2. John Drummond Robertson, *The Evolution of Clockwork* (London: Cas-

sell, 1931), pp. 15–18, offers a hypothetical reconstruction of the adaptation of an alarm mechanism to timekeeping. Compare J. D. North, "Monasticism and the First Mechanical Clocks," in Julius T. Fraser and N. Lawrence, eds., *The Study of Time II* (New York: Springer Verlag, 1975), p. 393: "It is not inconceivable that such an oscillatory striking device, triggered at suitably chosen intervals by a hydraulic clock, pointed the way to the first mechanical escapement proper." And Derek Price has written: "Alternatively [to Needham's China connection], it is possible that the Western escapement is quite independent and an adaptation of the bell-ringing machinery which accompanied the early rotating water-clocks." But Price tells me that he does not think so. See Derek J. de Solla Price, "Clockwork before the Clock and Timekeepers before Timekeeping," *Bulletin of the National Association of Watch and Clock Collectors, Inc.* 18 (1976): 402. On this sequence of development, see also idem, *Weight-driven Chamber Clocks of the Middle Ages and Renaissance* (Altrincham: J. Sherratt, 1965), ch. 1; John D. North, "Opus quarundam rotarum mirabilium," *Physis* 8 (1966): 337–372; and idem, *Richard of Wallingford*, vol. 2 (Oxford: Clarendon, 1974), p. 362.

3. Robertson, *Evolution*, pp. 18–19, suggests that the scape wheel in its earliest form was probably a pinwheel; and that it must have taken some learning to make a crown wheel, so-called because of its carefully curved, pointed teeth at a right angle to the plane of the wheel. The "strob" version of the verge escapement employed by Richard of Wallingford around 1330 used a double pinwheel. See the illustration of a reconstruction in Ernest L. Edwardes, *The Story of the Pendulum Clock* (Altrincham: J. Sherratt, 1977), pl. 1 (Figure 5, below). Contrast the verge-and-foliot, ibid., pp. 227–228.

4. The older word has left one small trace of its presence in the word *tocsin*. On the replacement of *sain* or *sein* by *cloche*, see Claude Duneton, *La puce à l'oreille: anthologie des expressions populaires avec leur origine* (Paris: Stock, 1978), pp. 259–260. The standard etymological dictionaries give divergent dates for the introduction of *cloche,* ranging from the eleventh century on. Duneton opts for the end of the thirteenth.

5. Albert d'Haenens, "La quotidienneté monastique au moyen âge: pour un modèle d'analyse et d'interprétation," in Oesterreichische Akademie der Wissenschaften, Philosophisch-Historische Klasse, Sitzungsberichte, vol. 367: *Klösterliche Sachkultur des Spätmittelalters: Internationaler Kongress Krems an der Donau 18. bis 21. September 1978,* Veröffentlichungen des Instituts für Mittelalterliche Realienkunde Oesterreichs, no. 3 (Vienna: Verlag des Oesterreichischen Akademie der Wissenschaften, 1980), p. 39.

6. The classic and most eloquent statement of the link between monastic activity and the invention of the clock is to be found in Lewis Mumford, *Technics and Civilization* (New York: Harcourt, Brace, 1934), ch. 2, pt. 2: "The Monastery and the Clock." But Mumford had already made use of Abbot P. Usher, *A History of Mechanical Inventions* (New York: McGraw-Hill, 1929), which refers to this connection and takes it back to Charles du Cange, historian and lexicographer of the seventeenth century. For a more recent analysis along these lines, see the excellent pages by H. E. Hallam, "The Medieval

Picture," in Eugene Kamenka and R. S. Neale, eds., *Feudalism, Capitalism and Beyond* (Canberra: Australian National University Press, 1975), pp. 32–35. (I owe this reference to Joel Mokyr, historian of the Low Countries and Ireland.) My own introduction to the Benedictine connection came from histories of the clock—I can no longer remember which ones.

7. On William of Hirsau, the Cistercian Rule, and medieval Cologne, see North, "Monasticism and the First Mechanical Clocks," pp. 382–383. A word of caution: the evidence of increased recourse to water clocks, many with accessory dial and alarm arrangements, does not necessarily imply high levels of accuracy and reliable performance. The one such water clock for which we have some detailed instructions was installed in the abbey at Villers in 1267—that is, toward the very end of the era of hydraulic timekeepers. We know about it because directions for use were apparently inscribed on loose, uncut pieces of slate, preserved by the accident of having been used as fill some centuries later. That in itself strikes me as passing strange: Why consign something that has to be consulted frequently to so inconvenient and unmanageable an object? In any event, the instructions would seem to indicate that this clepsydra, equipped though it was with dial, had no provision for maintaining stable water pressure, so that the rate slowed as the vessel emptied. The sacristan was directed to pour in varying amounts at different times. The results cannot have been accurate, and it was necessary to reset the clock whenever the sun allowed. For this purpose the abbey itself served as a clock, and the points of reference were given in such terms as, "in the middle of the first window," "in the second corner of the first window," and so on. On cloudy days, one used an arbitrary setting—after the Mass, ten o'clock. All of this was a far cry from what had been achieved in China and Islam. On the Villers clock the fundamental source is Paul Sheridan, "Les inscriptions sur ardoises de l'abbaye de Villers," *Annales de la Société archéologique de Bruxelles* 9 (1895): 359–362, 454–459; 10 (1896): 203–215, 404–451. More accessible is the article by Albert d'Haenens, "La clepsydre de Villers (1267)," in Oesterreichische Akademie der Wissenschaften, *Klösterliche Sachkultur des Spätmittelalters*, pp. 321–342.

8. Both citations from Jacques Le Goff, "Labor Time in the 'Crisis' of the Fourteenth Century," in Le Goff, *Time, Work, and Culture in the Middle Ages* (Chicago: University of Chicago Press), pp. 45–46.

9. Ibid., p. 46.

10. Toujours draps de soie tisserons,
Et n'en serons pas mieux vetues.
Toujours serons pauvres et nues,
Et toujours faim et soif aurons.
Jamais tant gagner ne saurons
Que mieux en ayons a manger.
Du pain avons a partager,
Au matin peu et au soir moins . . .
Et nous sommes en grand'misere,
Mais s'enrichit de nos salaires
Celui pour qui nous travaillons.

Des nuits grand partie nous veillons
Et tout le jour pour y gagner . . .

These verses are from *Li romans dou chevalier au Lyon* by Chrétien de Troyes, troubadour at the court of Champagne, cited in Jean Sablière, *Courir après l'ombre: Petite histoire anecdotique et pittoresque de la mesure du temps* (n.p., n.d. [1965]), p. 101. The verses were written about 1168, more than a hundred years before the appearance of the mechanical clock.

11. See Gustav Bilfinger, *Die mittelalterliche Horen und die modernen Stunden* (Stuttgart, 1892), pp. 163–166; Alphonse Wins, *L'horloge à travers les âges* (Mons: Léon Dequesne, 1924), pp. 100–104.

12. These are cited in W. Rothwell, "The Hours of the Day in Medieval French," *French Studies* 13 (July 1959): 243. The statutes for the tanners are from 1324.

13. On Japanese timekeepers, the best discussion is still that of Robertson, *Evolution*, pp. 190–287. See also H. H. N. Mody, *Japanese Clocks* (Rutland, Vt.: Tuttle, 1967). The Japanese earlier produced clocks with two controllers (foliots), one for the day rate, the other for the night. They did not go over to constant hours until the program of economic and scientific development under Meiji (reigned 1868 on) made retention of the old system counterproductive.

14. C. F. C. Beeson, *English Church Clocks, 1280–1850* (London: Antiquarian Horological Society, 1971), p. 16. Bilfinger, in *Die mittelalterliche Horen,* states that equal hours did not come in until the latter part of the fourteenth century. This may well be true for most places, though I suspect that wherever a chiming tower clock was built, equal hours soon followed—no doubt partially at first. Bilfinger was handicapped in his dating by his spotty data on the history of the early mechanical clocks and, for want of direct information, was ready to infer the introduction of the new timekeepers from the adoption of the new time standard.

Rothwell, "The Hours of the Day," p. 242, argues along similar lines on the basis of French literary texts: the vernacular literature continues to give times in canonical hours until the end of the fourteenth century. Yet one can also find such examples at the end of the fifteenth century, when clocks were long familiar, at least in urban areas. Clearly the two systems of time measurement went on side by side, with equal hours steadily gaining at the expense of temporal.

15. The best source is the highly suggestive book of Alexander Murray, *Reason and Society in the Middle Ages* (Oxford: Clarendon, 1978), chs. 6–8.

16. The text assumes that although the *temporal* hours of the medieval church were equal to one another, before the invention of the mechanical clock the church bells rang the *canonical* hours. These were not equally spaced, and hence did not lend themselves to arithmetical reckoning. Surprisingly, the Murray book says nothing about the contribution that timekeeping had to make to numeracy. *Et pourtant . . .*

17. See the article of R. S. Lopez on the fiscal consequences of cost overruns for communes caught up in overambitious cathedral projects: "Economie et architecture médiévales," *Annales: Economies, sociétés, civilisations* 7 (Octo-

ber–December 1952): 433–438.

18. Antonio Simoni, *Orologi italiani dal cinquecento all'ottocento* (n.p.: Antonio Vallardi Editore, 1965), pp. 16–17.

19. My translation. See also the prose of Charles Dahlberg, trans., *The Romance of the Rose* (Princeton: Princeton University Press, 1971), p. 343.

20. Some of these were Jews, who were also skilled in the art of making navigational and astronomical (the record speaks at one point of astrological) instruments. See Jeanne Vielliard, "Horloges et horlogers catalans à la fin du moyen âge," *Bulletin hispanique* of the *Annales de la Faculté des Lettres de Bordeaux* 63 (July–December 1961): 161–168; available in English translation by Charles K. Aked, "Catalan Clocks and Clockmakers to the End of the Middle Ages," *Antiquarian Horology* 10 (Spring 1978): 722–727.

21. Wins, *L'Horloge*, p. 168; Edwardes, *Weight-driven Chamber Clocks*, pp. 24–25; Giuseppe Brusa, *L'arte dell' orologeria in Europa: sette secoli di orologi meccanici* (Busto Arsizio: Bramante Editrice, 1978), Addenda & Corrigenda (1982), p. 21 and n. 18.

22. On this clock, see Bilfinger, *Die mittelalterliche Horen*, p. 176; North, "Monasticism," p. 385 and n. 36. This was once thought to be the earliest such clock; but now see Beeson, *English Church Clocks*, p. 16.

23. Wins, *L'Horloge*, p. 77; also A. Lecocq, "Notice historique et archéologique sur les horloges de l'église Notre-Dame de Chartres," in *Mémoires de la Société Archéologique d'Eure-et-Loire*, vol. 4 (1865), p. 295. The clock was signed by its maker *"Ma faicte Beaumont"*— in itself a sign of pride in artistry.

24. Dante, *Paradiso*, canto 10.

25. The medieval French is much better:
> L'Orloge est, au vray considerer,
> Un instrument tres bel et tres notable,
> Et s'est aussy plaisant et pourfitable,
> Car nuict et iour les heures nous aprent
> Par la soubtilite qu'elle comprent
> En l'absence meisme dou soleil:
> Dont on doit mieuls prisier son appareil,
> Ce que les autre instruments ne font pas,
> Tant soient faits part art et par compas:
> Dont celi tiens pour vaillant et pour sage
> Qui en treuva primierement l'usage
> Quant par son sens il commenca et fit
> Chose si noble et de si grant proufit . . .

II. KEEPING TIME

1. As Derek de Solla Price has shown. Price, "Gears from the Greeks: The Antikythera Mechanism—a Calendar Computer from ca. 80 B.C.," *Transactions of the American Philosophical Society*, n.s., 64 (1974): 1–70.

2. E. Watson, "The St. Alban's Clock," *Antiquarian Horology* 11 (Winter 1979): 581.

5. My Time Is My Time

1. These chains were not the usual link chains, but very much like the bicycle chains of today. (See Figure 9.) As clocks became smaller, the chains became smaller—little masterpieces of miniaturization. This was painfully fine and repetitious work, and it was one of the rare branches of clock manufacture that were open to women. Chains came in at least as early as the fourth decade of the sixteenth century. See a bill of 1539 charging for converting a fusee from gut to chain drive: Klaus Maurice, *Die deutsche Räderuhr: Zur Kunst und Technik des mechanischen Zeitmessers im deutschen Sprachraum*, vol. 1 (Munich: Beck, 1976), pp. 84–85 and figs. 26 and 27.

2. The authenticity and originality of this clock were long debated, but were eventually established by internal evidence and contemporary documentation. The case for authenticity was initially made by Ernst von Bassermann-Jordan, *Die Standuhr Philipps des Guten von Burgund* (Leipzig, 1927). There is an excellent brief summary of the controversy in Maurice, *Die deutsche Räderuhr,* vol. 1, pp. 85–86.

Another solution to the problem of diminishing force was the *stackfreed,* a spring brake working on a cam so indented as to reduce the action of the brake as the mainspring winds down. The stackfreed, a far less accurate compensator than the fusee, found application only in Germany, and then only in iron clock and watch movements. It was cheaper and demanded substantially less skill to make than did a fusee. (See Figure 10.)

3. On the exceptional technical, intellectual, and economic advantages of Nuremberg, see Wolfgang von Stromer, "Haec opera fient in oppido Nuremberga Germaniae ductu Ioannis de Monteregio: Regiomontan und Nürnberg, 1471–1475," in *500 Jahre Regiomontan, 500 Jahre Astronomie: Ausstellung der Stadt Nürnberg und des Kuratoriums "Der Mensch und der Weltraum e.V."* . . . *2.10.1976–2.1.1977* (Nuremberg: Germanisches Nationalmuseum, 1977).

4. Sometimes these signatures are concealed in remote parts of the mechanism or reduced to initials, with the aim of conveying a certain verisimilitude: if this were forgery, surely the signature would not be hidden that way; and maybe the seller hasn't noticed it. See Maurice, *Die deutsche Räderuhr,* vol. 1, pp. 87–89.

5. See the Italian watch pictured in Ernst von Bassermann-Jordan, *Uhren: Ein Handbuch für Sammler and Liebhaber,* 8th ed., rev. and ed. Hans von Bertele (Braunschweig: Klinkhardt und Biermann, 1976), p. 61. This watch is simply a drum (tambour) clock with a ring soldered on so that it can be slung around the neck. The author dates it as perhaps as early as 1500.

6. See Giuseppe Brusa, *L'arte dell'orologeria in Europa: sette secoli di orologi meccanici* (Busto Arsizio: Bramante Editrice, 1978), pp. 41–42; Enrico Morpurgo, *L'origine dell'orologio tascabile* (*The Origin of the Watch*) (Rome: Edizioni

"La Clessidra," 1954). As the title indicates, this monograph is presented in the original Italian and in English translation. See also Fulgido Pomella, *L'orologio da portare addosso* (Ivrea: Priuli & Verlucca, 1978), pp. 30–34.

7. The time museum in La Chaux-de-Fonds has a tiny stackfreed watch with iron movement that measures no more than twelve millimeters. *Collections du Musée International d'Horlogerie, La Chaux-de-Fonds, Suisse* (La Chaux-de-Fonds, 1974), p. 22. The catalogue dates the watch as c. 1648; I would put it a generation earlier, as does Jürgen Abeler, *Meister der Uhrmacherkunst* (Wuppertal: privately published, 1977), pp. 588–589. Abeler gives the movement diameter as ten millimeters.

8. William Bouwsma, "Anxiety and the Formation of Early Modern Culture," in Barbara C. Malament, ed., *After the Reformation: Essays in Honor of J. H. Hexter* (Philadelphia: University of Pennsylvania Press, 1980), pp. 215–246.

9. Ibid., p. 225.

10. Iris Origo, *The Merchant of Prato: Francesco di Marco Datini, 1335–1410* (New York: Knopf, 1957), p. 187.

11. Ibid., p. 376.

12. Bouwsma, "Anxiety," p. 220.

13. Ibid., p. 240.

14. Ibid., p. 221.

15. Ibid., p. 219.

16. From *I Libri della famiglia,* cited in epigraph to Giuseppe Brusa, ed., *Gli Orologi,* Cataloghi del Museo Poldi Pezzoli, no. 1 (Milan: Museo Poldi Pezzoli, 1974). The passage in question is to be found in book 3: "On Management." The bookseller and biographer Vespasiano di Bisticci (Florence, late fifteenth century) tells of one of his subjects that "he placed great store by time and never lost an hour." S. Stelling-Michaud quotes this with the remark that it would not even have made sense a century or two earlier. "Quelques aspects du problème du temps au moyen âge," *Schweizer Beiträge zur allgemeinen Geschichte* 17 (1959): 26. Interestingly, the equivalent expression today would be, "He doesn't lose a minute." Our clocks are better.

17. The *locus classicus* is Max Weber, *The Protestant Ethic and the Spirit of Capitalism,* trans. and ed. Talcott Parsons (London: Allen & Unwin, 1930), first published in *Archiv für Sozialwissenschaft und Sozialpolitik,* vols. 20–21 (1904–5).

18. Eva Groiss, "Das Augsburger Uhrmacher-Handwerk," in Klaus Maurice and Otto Mayr, eds., *Die Welt als Uhr: Deutsche Uhren und Automaten, 1550–1650* (Munich: Bayerisches Nationalmuseum, 1980), p. 85.

19. See also Guy Thuillier, *Pour une histoire du quotidien au XIXe siècle en Nivernais* (Paris and The Hague: Mouton, 1977), p. 213, n. 11. I owe this reference to Patrice Higonnet.

Natalie Davis, who is currently preparing a monograph on early modern Lyons, writes (January 23, 1983): "My sense is that *all* the clock-making ateliers at Lyon in the crucial decades of the Reformation 1550–1575 were Protestant. Once the persecution and persuasion of the Counter-Reformation set in, why, the picture changes. But even the figures from Vial suggest what a significant proportion of the trade was Protestant in the seventeenth century." The reference is to Eugène Vial and Claudius Côte, *Les horlogers lyonnais de 1550 à 1650* (privately printed at Macon: Protat Frères, 1927).

20. The government tried to keep some of these artists at home by buying conversions; for example, Louis Duhamel (not every Protestant had a biblical name) received the piddling sum of six livres and the more substantial patent of *horloger du roi*. Tardy, *Dictionnaire des horlogers français*, vol. 1 (Paris: Editions Tardy, n.d.), p. 202.

21. The assertion in the text goes against the thesis of Warren Scoville, *The Persecution of Huguenots and French Economic Development, 1680–1720* (Berkeley: University of California Press, 1960). Scoville argues that economic effects of the Revocation have been much exaggerated and that the capital and talent lost by the exodus of Protestants were rapidly made up by Catholic entrepreneurs and craftsmen. I have never been persuaded by this argument, which seems to me to fail in both logic and evidence. It does not hold for the watch industry.

22. Sigrid and Klaus Maurice, "Stundenangaben im Gemeinwesen des 16. and 17. Jahrhunderts," in Maurice and Mayr, eds., *Die Welt als Uhr*, p. 156.

23. On regional and local differences in counting the hours, see Bassermann-Jordan, *Uhren*, ch. 1, sect. C; and Karl Fischer, "Die Uhrmacher in Böhmen und Mähren zur Zeit der Gotik und Renaissance," *Bohemia: Jahrbuch des Collegium Carolinum* 7 (1966): 44–46. These two sources contradict each other on some of the popular designations of these time systems; also on the Italian day, which Bassermann-Jordan says started half an hour after sundown, and Fischer says began at midnight.

24. Traveling clocks may well be older—though I have never seen one, whether in the metal or in illustration, from before the seventeenth century. Morpurgo, *L'Origine*, p. 33, speaks of Bernardo Caravaggio of Cremona, who "is said to have built the first coach watches [*i primi orologi da viaggio*]" during a stay in Paris in 1480. (The English translation of this passage, on p. 70, is inexact.)

25. On army size and problems of command, see William H. McNeill, *The Pursuit of Power: Technology, Armed Force and Society since A.D. 1000* (Chicago: University of Chicago Press, 1982), ch. 4 and the sources cited there.

26. *Richard III,* act 5, scene 3.

27. Morpurgo, *L'Origine*, p. 33.

28. Viret, Protestant of Lausanne, sings the praises of these army cocks in his *Exposition chrestienne*, vol. 2 (Geneva, 1564), cited in Lucien Febvre, *Le problème de l'incroyance au XVIᵉ siècle* (Paris: Albin Michel, 1942; rpt. 1968), p. 365.

29. Julien Le Roy, "Mémoire pour servir à l'histoire de l'horlogerie, depuis 1715 jusqu'en 1729," in Henry Sully, *Règle artificielle du temps,* new ed., rev. and enl. by Julien Le Roy (Paris: Grégoire Dupuis, 1737), p. 384. This

campaign was part of the so-called War of the Spanish Succession. Sully repaired watches not only for his employers but for their friends and fellow officers.

30. M. Fallet-Scheurer, *Le travail à domicile dans l'horlogerie suisse et ses industries annexes* (Berne, 1912), p. 213.

31. Eugene Buffat, "La montre Roskopf," *Horlogerie ancienne* 10 (1981): 74. Compare the history of the marine chronometer and the deck watch. A generation and more passed after their invention before navies and shipowners began to buy them for their vessels. In the meantime, many a captain and mate bought one out of pocket for his own use.

6. Of Toys and Ornaments and Serious Things

1. See Gottfried Mraz, "Die Rolle der Uhrwerke in der kaiserlichen Türkenverehrung im 16. Jahrhundert," in Klaus Maurice and Otto Mayr, eds., *Die Welt als Uhr: Deutsche Uhren und Automaten, 1550–1650* (Munich: Bayerisches Nationalmuseum, 1980), pp. 39–54. This is the catalogue of an exhibition held first in Munich and then at the Museum of Technology (Smithsonian Institution) in Washington, D.C. In scholarship and illustration it is a model of its kind.

2. On these extraordinary German clocks of the sixteenth to eighteenth centuries, the best sources are Mraz, "Die Rolle der Uhrwerke"; and Klaus Maurice, *Die deutsche Räderuhr: Zur Kunst und Technik des mechanischen Zeitmessers im deutschen Sprachraum,* 2 vols. (Munich: Beck, 1976).

3. On the growing exclusiveness of the German clock- and watchmakers' guilds, see the long-forgotten article of Carl Friedrich, "Die Augsburger Uhrmacherei während des 18. Jahrhunderts," *Allgemeines Journal der Uhrmacherkunst* (Leipzig) 11 (1886): nos. 24 et seq. Also the valuable article of Eva Groiss, "Das Augsburger Uhrmacher-Handwerk," in Maurice and Mayr, *Die Welt als Uhr,* pp. 63–89, which offers statistics on the distribution of masters by mode of admission (inheritance, marriage, or purchase).

4. On the Pinchbecks: Rita Shenton, *Christopher Pinchbeck and His Family* (Ashford, Kent: Brant Wright, 1976). On Jaquet-Droz: Charles Perregaux and F.-Louis Perrot, *Les Jaquet-Droz et Leschot* (Neuchâtel: Attinger Frères, 1916), chs. 28–29. On the Chinese market in the eighteenth century: Alfred Chapuis, *Relations de l'horlogerie suisse avec la Chine: La montre "chinoise"* (Neuchâtel: Attinger Frères, n.d.), pp. 28–30; Joseph Needham et al., *Heavenly Clockwork: The Great Astronomical Clocks of Medieval China* (Cambridge: Cambridge University Press, 1960), p. 152; H. Alan Lloyd, *The Collector's Dictionary of Clocks* (New York: A. S. Barnes, 1964), s.v. "Chinese Market."

5. On French enamel watches of the seventeenth century, see especially E. Develle, *Les horlogers blésois au XVIᵉ et au XVIIᵉ siècle,* 3rd ed., followed by *Peintres en émail de Blois et de Châteaudun au XVIIᵉ siècle* (Nogent-le-Roi: Librairie des Arts et Métiers, 1978); and Edouard Gélis, *L'horlogerie ancienne: histoire, décor*

et technique (Paris: Grund, 1949). Also Alfred Leiter and Alma Helfrich Dorner, *Email-Malerei: Kostbarkeiten unter den Taschenuhren: Zur Technik und Geschichte der Email-Uhren* (Kornwestheim: Werbe-Verlag, 1977); and W. Deonna, *Les arts à Genève des origines à la fin du XVIII*^e *siècle* (Geneva: Musée d'Art et d'Histoire, n.d. [1942]), pp. 407–408.

6. For a brief discussion of the technique, as applied to another medium, see Rosalind Savill, "Six Enamelled Snuff-Boxes in the Wallace Collection," *Apollo* 111 (April 1980): 305. She relies primarily on the description in Diderot and d'Alembert, *Encyclopédie*, vol. 5 (Paris, 1755), s.v. "Email." For an earlier description, with emphasis on the problems of obtaining the colors desired, see Jacques-Philippes Ferrand, *L'art du feu ou de peindre en émail* (Paris, 1721). Ideally the technique called for starting with the color with the highest melting point, then lowering the temperature with each additional color—a very difficult task with a nonelectric furnace. In practice, I imagine, similar effects were obtained by shortening the exposure to heat, keeping careful watch to withdraw the painted *boîte* or *plaque* in time.

7. On "paint factories" in Geneva, see France, Archives Nationales, F¹² 1325B: "Observations générales sur la Manufacture d'horlogerie, en réponse à la lettre du Comité de Salut Public du 22 Frimaire an 3ᶜ de la République, f. 11.

8. But see T. P. Camerer Cuss, *The Camerer Cuss Book of Antique Watches*, rev. ed. by T. A. Camerer Cuss (Woodbridge, Suffolk: Antique Collectors' Club, 1976), pp. 54–55. Cuss shows a French watch of about 1600 by P. Chapelle that conforms in every way to the simple, so-called puritan standard. The author describes this as "not untypical of the period," but I find it exceptional. The same book (pp. 56, 63–65) shows three English watches in the puritan style: one, c. 1625, by David Ramsay; a second, described as such, by John Snow, c. 1640; and a third by David Bouquet, French émigré, almost surely Huguenot, again c. 1640.

9. Maurice, *Die deutsche Räderuhr*, vol. 1, p. 142.

10. On the observatory in Kassel, see Ludolf von Mackensen (with contributions by Hans von Bertele and John H. Leopold), *Die erste Sternwarte Europas mit ihren Instrumenten und Uhren: 400 Jahre Jost Bürgi in Kassel* (Munich: Callwey, 1979), pp. 12–21. This book, exemplary in its scholarship, was published originally as the catalogue of an exhibition, "Staatlichen Kunstsammlungen Kassel für das Astronomisch-Physikalische Kabinett," at the Hessisches Landesmuseum.

11. Ernst Zinner, *Deutsche und niederländische astronomische Instrumente des 11.–18. Jahrhunderts*, 2nd ed. (Munich: Beck, 1967; rpt. 1979), pp. 21–22.

12. L. Defossez, *Les savants du XVII*^e *siècle et la mesure du temps* (Lausanne: Editions du Journal Suisse d'Horlogerie et de Bijouterie, 1946), p. 58. For more on Bürgi (in addition to Mackensen, *Die erste Sternwarte*): Hans von Bertele, "Precision Timekeeping in the Pre-Huygens Era," *Horological Journal* 95 (December 1953): 794–816; idem, "Nouveaux documents sur l'oeuvre d'un horloger suisse de génie: Jost Bürgi, le second Archimède," *Journal suisse d'horlogerie* (1956–57), nos. 11, 12; Klaus Maurice, "Jost Bürgi oder über die Inno-

vation," in Maurice and Mayr, *Die Welt als Uhr,* pp. 90–104; J. H. Leopold and K. Pechstein, *Der kleine Himmelsglobus 1594 von Jost Bürgi* (Lucerne: J. Fremersdorf, 1977).

13. On this technique, see Peter Padfield, *Tide of Empires: Decisive Naval Campaigns in the Rise of the West,* vol. 1, *1481–1654* (London: Routledge & Kegan Paul, 1979), p. 31.

14. On navigation and latitudes, see J. B. Hewson, *A History of the Practice of Navigation* (Glasgow: Brown, Son & Ferguson, 1951), pt. 1, ch. 3, and pt. 2, chs. 1 and 2; W. E. May, *A History of Marine Navigation* (New York: Norton, 1973), ch. 4; E. G. R. Taylor, *The Haven-Finding Art* (New York: Elsevier, 1971), pt. 4. (Taylor notes, on p. 159, that sailors accustomed to navigating in the Mediterranean found it difficult at first to think in terms of latitude. So the scale of quadrants was marked with place names—say, a given cape or river-mouth—and the seaman oriented himself with reference to a specific landmark.) One of the best sources on the history of seamanship, beautifully written and based on the author's personal knowledge of "the way of a ship in the midst of the sea," is Samuel Eliot Morison's chapter on the navigation of Christopher Columbus, in Morison, *Admiral of the Ocean Sea: A Life of Christopher Columbus* (Boston: Little, Brown, 1942), pp. 183–196. See also the important corpus of work by John H. Parry—for example, *The Age of Reconnaissance* (London: Weidenfeld and Nicolson, 1964), pp. 53–113; and *The Discovery of the Sea* (New York: Dial, 1974), pp. 31–68.

15. Morison, *Admiral of the Ocean Sea,* pp. 186–187.

16. David Waters, *The Art of Navigation in England in Elizabethan and Early Stuart Times,* 2nd ed., vol. 1 (Greenwich: National Maritime Museum, 1978), p. 122.

17. William Lytle Schurz, *The Manila Galleon* (New York: Dutton, 1931; pb. rpt. 1959), p. 253, citing the *Giro del mondo* of the "much traveled" Gemelli Careri (end of the seventeenth century). Also Casimiro Diaz, who called this "the longest, most tedious, and most dangerous voyage in all the seas." Schurz's book is surely one of the great classics of maritime history and a delight to read, but the absence of footnote references in so scholarly a work, which rests on extensive research in manuscript archives and contemporary writings, is exasperating.

I have "improved" the old English translation used by Schurz by conflating it with the French version in Jean-Pierre Berthe, ed., *Le Mexique à la fin du XVIIe siècle vu par un voyageur italien, Gemelli Careri* (Paris: Calmann-Lévy, 1968), p. 31. Berthe's introduction is invaluable. See especially pp. 21–23 on the reliability of Careri as reporter.

18. Schurz, *The Manila Galleon,* p. 265, citing the buccaneer Raveneau de Lussan.

19. Ibid., p. 256. For a lively account of the perils of Pacific navigation, see Oliver E. Allen et al., *The Seafarers: The Pacific Navigators* (Alexandria: Time-Life Books, 1980).

20. On English scorn for Spanish navigation, see Peter Earle, *The Wreck of the Almiranta: Sir William Phips and the Search for the Hispaniola Treasure* (London:

Macmillan, 1979), pp. 30–31. I owe my knowledge of this delightful book to Peter Hertner.

21. An early alternative to dead reckoning was calculation of longitude from compass variation: the deviation of the compass needle from true north (as taken from a reading of the North Star) changes with movement east or west and should in theory enable the sailor to work out his longitude. This was the system Columbus used for part of his first voyage. The problem, though, is that this variation is not even along a given latitude, while the magnetic pole itself shifts over time. As a result, seamen preferred to rely on dead reckoning, for all its shortcomings. See Rupert T. Gould, *The Marine Chronometer: Its History and Development* (London: J. D. Potter, 1923; rpt. 1960), p. 4; Taylor, *The Haven-Finding Art*, pp. 246–247; Lloyd A. Brown, *The Story of Maps* (Boston: Little, Brown, 1949), p. 211.

22. F. Marguet, *Histoire de la longitude à la mer au XVIIIe siècle, en France* (Paris: Auguste Challamel, 1917), p. 19. The sentence runs on as loosely as the sea.

23. Morison, *Admiral of the Ocean Sea*. p. 196.

24. Ibid., p. 195.

25. Richard Hakluyt, from his *Principal Navigations*, 2nd ed., vol. 1 (1598), cited in Waters, *The Art of Navigation in England in Elizabethan and Early Stuart Times*, 2nd ed., vol. 3, p. 553.

26. Morison, *Admiral of the Ocean Sea*, p. 186.

27. For an eloquent lament on the shortcomings of mapmaking, see Charles Pierre Claret d'Eveux de Fleurieu (usually catalogued under the last name), *Voyage fait par ordre du Roi en 1768 et 1769* . . . (Paris, 1773), pp. xlviii et seq. Fleurieu was incensed by the dependence of masters on the pilots responsible for dead reckoning, whom he characterized as a prejudiced, superstitious, and obstinate lot: they "think they have accounted for their errors when they have blamed them on the effect of currents and secret causes" p. xxxvi).

28. Cited in Brown, *The Story of Maps*, p. 208. Eden is best known as author and editor of a widely read collection of travel accounts: *The Decades of the Newe World, or West India* (1555).

7. My Time Is Your Time

1. The earliest reference I have found to a minute hand is in the so-called Almanus Manuscript of a certain Brother Paul, who stayed a year in Rome in 1475–76 and spent much of his time repairing clocks for prelates and other dignitaries. See Klaus Maurice, *Die deutsche Räderuhr: Zur Kunst und Technik des mechanischen Zeitmessers im deutschen Sprachraum*, vol. 1 (Munich: Beck, 1976), p. 7; and John H. Leopold, *The Almanus Manuscript* (London: William Heinemann, 1971), p. 59.

The earliest known spring-driven timepiece with a second hand—perhaps the earliest timepiece of any kind—is an unsigned Orpheus clock in the Fremersdorf collection. The date is estimated at between 1560 and 1570. See V. Himmelein and J. H. Leopold, *Prunkuhren des 16. Jahrhunderts: Sammlung Joseph Fremersdorf* (Stuttgart: Württembergisches Landesmuseum, 1974), pp. 36–41; and P. G. Coole and E. Neumann, *The Orpheus Clocks* (London: Hutchinson Educational Ltd., 1972), pp. 82–84. This clock has other features of technical interest, and the beauty of case and dial matches the cunning of the movement. To be sure, Johann Wenzel, "Warum haben Uhren Sekundenzeiger?" *Alte Uhren* 3 (January 1980): 38, notes that this clock is so imprecise that the function of the second hand is really just to show that the clock is moving. Yet that is what most second hands are for, even today.

2. For reasons that probably have to do with a mistaken feel for the French language, this word has come into English as *remontoire*. Those who find it awkward may prefer the more accurate though longer term, *constant-force device*.

3. L. Defossez, *Les savants du XVIIᵉ siècle et la mesure du temps* (Lausanne: Editions du Journal Suisse d'Horlogerie et de Bijouterie, 1946), p. 58; Ernst von Bassermann-Jordan, *Uhren: Ein Handbuch für Sammler und Liebhaber,* rev. and ed. by Hans von Bertele (Braunschweig: Klinkhardt and Biermann, 1976), p. 370. At the time Defossez wrote his study, the character of Bürgi's inventions was still largely unknown. Defossez warned that one must take care in inferring invention from the features of such clocks of Bürgi's as had survived: "they have perhaps been retouched, modified after the death of the author" (p. 56). And referring to the marvelous clock in the museum in Vienna, he wrote: "It is said that the Vienna clock is fitted with an escapement that has less recoil than a verge—that is, a kind of recoil anchor escapement." *"On dit"*—because at that time the Vienna clock had not yet been opened for study. Defossez's confusion on this score is not surprising: within a century of Bürgi's death, his work had passed into myth. In 1738, for example, one writer credited him with the invention of the pendulum clock—an attribution not so fanciful as one might think when one recollects the swinging arms of his cross-beat escapement. See his biography in *Allgemeine deutsche Biographie,* vol. 3 (Berlin: Duncker & Humblot, 1967, rpt. of 1876 ed.), p. 604.

We owe our subsequent knowledge of Bürgi's work largely to Bertele. See Hans von Bertele, "Precision Timekeeping in the Pre-Huygens Era," *Horological Journal* 95 (December 1953): 794–816. Also the other sources cited in Chapter 6, notes 10 and 12, above.

4. On Galileo's clock (Figure 14, below), see Christiaan Huygens, *Oeuvres complètes,* vol. 3, *Correspondance, 1660–1661* (The Hague: Martinus Nijhoff, 1890), p. 21: letter to Ismael Boulliau of February 12, 1660.

5. See the minute of the Accademia of August 11, 1662, cited in G. H. Baillie, *Clocks and Watches: An Historical Bibliography,* vol. 1 (London: N. A. G. Press, 1951; rpt. London: Holland Press, 1978), p. 62.

6. Huygens, *Oeuvres complètes,* vol. 2, p. 185: letter of June 21, 1658.

7. Ibid., vol. 3, p. 12: letter to Boulliau of January 22, 1660.

8. Ibid., vol. 2, pp. 430–431, 441–442. At that time, Huygens was convinced that such tiny impulses would produce swings of equal amplitude. They did not, indeed could not; but they came close enough to yield a re-

markably even performance.

    9. Ibid., vol. 3, p. 12: letter to Boulliau of January 22, 1660.

    10. Leopold's letter of May 22, 1673, is given in Fabroni, *Lettere inedite di uomini illustri* (Florence, 1773–1775), cited in Baillie, *Clocks and Watches*, p. 87.

    On Galileo and the pendulum clock, the best discussion I know—marked by characteristic concision, balance, and good judgment—is in Giuseppe Brusa, *L'Arte dell'orologeria in Europa: Sette secoli di orologi meccanici* (Busto Arsizio: Bramante, 1978), pp. 115–116. Also on the controversy, see John Drummond Robertson, *The Evolution of Clockwork* (London: Cassell, 1931), chs. 6 and 7.

    This does not exhaust the Italian claims to the great invention. Regarding the hypothesis that Galileo's conception was the inspiration for pendulum clocks by Johann Philipp Treffler (a German maker employed by Prince Leopold of Tuscany) that predated Huygens' first models, see Silvio A. Bedini, "Johann Philipp Treffler, Clockmaker of Augsburg," *Bulletin of the National Association of Watch and Clock Collectors, Inc.* 7 (December 1956): 361–372; 7 (February 1957): 415–427; 7 (April 1957): 481–492; 7 (June 1957): 541–547. Brusa is dubious and notes that a copy of Huygens' pendulum clock reached Florence that very first year (1657); and that the early Italian versions that have survived are of the Huygens variety.

    Another version of the Italian claim to priority has been put forward on behalf of a clockmaker named Camerini, of Turin, maker of a small bracket clock dated 1656 which is now in the collection of the Science Museum in London. See Enrico Morpurgo, *Precious Watches from the 16th to the 19th Century, with a Historical Survey of Italian Watchmaking* (n.p.: Organizzazione Omega Italiana, n.d.), p. 19; also H. Alan Lloyd, *The Collector's Dictionary of Clocks* (New York: A. S. Barnes, 1964), pp. 52–53, who says that Camerini's clock "shows no sign of conversion from balance to pendulum." In fact, its pendulum is a later alteration, at least in the opinion of the Science Museum. See F. A. B. Ward, *Science Museum: Descriptive Catalogue of the Collection Illustrating Time Measurement* (London: H.M.S.O., 1966), p. 37, no. 153. On that point, see also the illustrations and description of a Camerini clock dated 1662, likewise with later conversion to pendulum, in Brusa, *L'Arte dell'orologeria*, plates 290–291, and description, p. 418. It is most unlikely that if Camerini had in fact built a pendulum clock in 1656, he would ever have gone back to a balance regulator in 1662. Brusa (p. 116, n. 139) dismisses the Camerini claim. Incidentally, the Camerini clocks, like the one designed by Galileo, have rigid pendulum suspension and fixed linkage to the escapement.

    There has also recently appeared a flurry of publications arguing for the invention of some version or feature of the pendulum clock by John Fromanteel, the Dutch-English maker who went to work for Coster in the Hague in 1658 and who has always been credited with bringing the pendulum clock from Holland to England. See R. D. Dobson, "Huygens, the Secret in the Coster-Fromanteel 'Contract': The Thirty-Hour Clock," *Antiquarian Horology* 12 (Summer 1980): 192–196; Ernest L. Edwardes, "The Suspended Foliot and New Light on Early Pendulum Clocks," *Antiquarian Horology* 12 (June 1981): 614–626, 634; Dobson, "The Development of the Pendulum Clock," *Antiquar-*

ian Horology 13 (March 1982): 270–279. The plausibility of this hypothesis aside, the articles show the continuing excitement (passion) of questions of priority—even when more than three hundred years old.

11. The circular pendulum swings slower: (1) because, for a given radius (length of rod), the arc traversed by a cycloidal pendulum is shorter (in geometric terms, the base of the cycloid is shorter than the chord subtending the circular arc of swing, and the cycloidal curve lies entirely within the segment defined by that chord and arc); and (2) because the common cycloid is brachistochronous (from *brachistos,* meaning the shortest, and *chronos,* time)—that is, it is the fastest path for an object falling from a higher point to a lower one not in the same vertical line. Here again the result goes against intuition, which would expect the fastest path to be the straight line connecting the two points, if only because it is the shortest distance. This time the greater velocity imparted by the steeper fall at the start of the drop more than makes up for the longer traverse, indeed minimizes the time required. Anything shallower would not move the falling or rolling object fast enough; anything deeper would make it cover too long a path. In practical-fantastic terms: if a contestant in the soap-box auto derby could somehow persuade the organizers to let him run his car in a cycloidal chute, he would beat all the rest—assuming that his vehicle was no worse than the others.

The proof of the brachistochronism of the cycloid was found in 1696, in response to a challenge by Johann Bernoulli to "the sharpest mathematicians in the world." Bernoulli gave the contestants six months to solve the problem. Leibniz solved it the day he received it. What is more, he correctly predicted that only five persons would solve it and named them: himself, Newton, the two brothers Bernoulli, and Guillaume de l'Hospital, who needed some help from Bernoulli in the process. (Is there any other field in which the order of excellence [power] is so accurately known as in mathematics?) Jakob Bernoulli's solution, though not so ingenious and elegant as that of his younger brother and rival, anticipated the development of a new branch of mathematics, the calculus of variations. Charles C. Gillispie, ed., *Dictionary of Scientific Biography,* vol. 2 (New York: Scribner's, 1970), s.v. "Johann Bernoulli," p. 53.

12. An appropriate formula for circular error K is: $K \simeq 1.65\alpha^2$, where α is the semiarc of swing in degrees. It holds for any length of pendulum. Here is a table of selected values of circular error:

Semiarc	Additional Seconds per Day
0° 30′	0.41
1° 00′	1.65
1° 30′	3.71
2° 00′	6.60
3° 00′	14.85
5° 00′	41.12
6° 00′	59.21
9° 00′	133.22
10° 00′	164.46
12° 00′	236.84

On all this, see the discussion in A. L. Rawlings, *The Science of Clocks and Watches* (Luling, Texas: Caldwell Industries, 1974, rpt. of 2nd ed. of 1948), pp. 50–52.

13. See his letters of February 5, 1659, to Pascal and January 22, 1660, to Boulliau, in Huygens, *Oeuvres complètes,* vol. 2, p. 340, and vol. 3, pp. 12–13.

14. A smaller arc of about twelve degrees was later achieved (late seventeenth, early eighteenth centuries) by the use of especially long pallets falling shallowly on the teeth of the scape wheel. Such escapements were used with seconds-beating pendulums on the continent but did not give as good results as the anchor (see below). Ernest L. Edwardes, *The Story of the Pendulum Clock* (Altrincham: J. Sherratt, 1977), p. 98.

15. They were built in 1676 by Thomas Tompion at the request of Sir Jonas Moore, Surveyor General of the Ordnance and instigator of the Royal Observatory, and John Flamsteed, his young protégé, newly appointed Astronomical Observator to the Crown. The best discussion is Derek Howse, "The Tompion Clocks at Greenwich and the Dead-Beat Escapement," *Antiquarian Horology* 7 (December 1970): 18–34; and 7 (March 1971): 114–133. See also idem, *Greenwich Time and the Discovery of the Longitude* (Oxford: Oxford University Press, 1980), pp. 33–37. For an even longer forty-foot pendulum c. 1750, see John R. Millburn, "Some Horological Extracts from Stukeley's Diaries," *Antiquarian Horology* 6 (September 1969): 209.

16. On the invention of the anchor escapement, see especially Edwardes, *The Story of the Pendulum Clock,* ch. 3, who inclines to Hooke. On Hooke's temperament, see Defossez, *Les Savants,* p. 225.

17. Bellair to Huygens, July 16, 1659, in Huygens, *Oeuvres complètes,* vol. 2, p. 440.

18. Ibid.

19. Hans von Bertele, "The Development of Equation Clocks: A Phase in the History of Hand-Setting Procedure," *La Suisse horlogère* (International ed.), 74 (1959), no. 3, p. 44, fig. 6; Bassermann-Jordan, *Uhren,* p. 115 and figs. 199–201.

20. The best source on these complicated machines is Bertele, "The Development of Equation Clocks," which appeared in *La Suisse horlogère* (International ed.) in five installments: 74 (1959), no. 3, pp. 39–46; 74 (1959), no. 4, pp. 15–24; 75 (1960), no. 1, pp. 17–27; 75 (1960), no. 4, pp. 37–48; and 76 (1961), pp. 25–36. See also J. Wenzel, "Equation Clocks," *Antiquarian Horology* 13 (September 1981): 24–43; and Tom Robinson, *The Longcase Clock* (n.p.: Antique Collectors' Club, 1981), pp. 171–184. On the Mercator-Fromanteel and the Hooke-Tompion equation clocks, see Robert W. Symonds, *Thomas Tompion: His Life and Work* (London: Batsford, 1951; rpt. Spring Books, 1969), pp. 128–129.

21. Paris did not go over to a mean-time standard until 1830. London had probably accomplished this shift some time before 1800. This contrast between British and French practice was first called to my attention by Sebastian Whitestone of Bobinet Ltd., clock and watch dealers in London. It is astonishing that no one, at least to my knowledge, has remarked the difference before. But see a letter from F. A. B. W. (Ward) to the *Antiquarian Journal* 13

(December 1981): 205, regarding a chiming longcase equation clock by Robin of Paris that strikes apparent (solar) time rather than mean time. The interpretation of this difference in practice is mine.

22. On Huygens' experiments with a pendulum marine chronometer, see Defossez, *Les Savants,* pp. 173–180.

23. For examples of a pendulum pocket watch, see Bassermann-Jordan, *Uhren,* pp. 270–271. Some pocket watches were subsequently fitted with a false or mock pendulum, consisting in a simple disk attached to the balance wheel and oscillating with it. Sector windows would be cut in the dial or the watch cock (the protective cover of the balance), so that the user could see the to-and-fro movement. The purpose may have been to arrogate some of the prestige of the true pendulum, just as some pocket watches were called pendulum watches by way of asserting their accuracy. (The expression was used of watches without as well as with false pendulum.) My own sense, however, is that the false pendulum's primary purpose was the same as that of the small seconds bit today—simply to tell the user that the watch was running. For illustrations of these false-pendulum watches, see Reinhard Meis, *Taschenuhren: Von der Halsuhr zum Tourbillon* (Munich: Callwey, 1979), figs. 116, 117, 123, 128, 130; T. P. Camerer Cuss, *The Camerer Cuss Book of Antique Watches,* rev. and enl. by T. A. Camerer Cuss (Woodbridge, Suffolk: Antique Collectors' Club, 1976), pp. 95, 96, 98; Eugène Jaquet and Alfred Chapuis, *Technique and History of the Swiss Watch* (London: Hamlyn, 1970), pl. 44.

24. On the above sequence of events, see Baillie, *Clocks and Watches,* pp. 93–102, which provides invaluable extracts and summaries of journals and correspondence concerning the invention of the balance spring. This is Baillie at his best.

25. Huygens, *Oeuvres complètes,* vol. 5, *Correspondance, 1664–1665,* p. 427.

26. A. R. Hall, "Robert Hooke and Horology," *Notes and Records of the Royal Society of London* 8 (1950): 167–177.

27. Unfortunately, the manuscript does not include the sketch of the mechanism that Hooke prepared at the time (as we may infer from the use of letters to denote parts in his description). Hall has prepared a drawing that he calls "only a very crude attempt to give some meaning to Hooke's description" (Hall, "Robert Hooke and Horology," p. 173). I marvel at his temerity: the device, as inferred by Hall, has all the complication and more of Le Roy's later effort.

28. Hall, "Robert Hooke and Horology," p. 177. Hooke's version of events is somewhat different. In a postscript to one of his Cutlerian lectures at Oxford, published in 1675 (but dated 1676), Hooke reviewed the history of his efforts to find the longitude by means of an accurate clock or watch. He says that in 1660 he agreed with "several Persons of Honour" to sponsor a device of his invention; that these gentlemen consented to pay him "several thousand" pounds; but that he then broke off because his patrons insisted on a clause that if anyone should find a way of improving on Hooke's principles during the lifetime of such patent as might be granted, that person, and not Hooke, should have the benefit of the patent. Hooke says he was then warned that he might as well disclose all, since someone else was sure to come up with his in-

vention within six months. To which he replied, "I would try them one [time] seven years." It took fifteen. Cited in Baillie, *Clocks and Watches: An Historical Bibliography,* p. 98.

29. Letter to R. Moray of September 18, 1665, in *Oeuvres complètes,* vol. 5, p. 486.

30. Some technicians have succeeded in rating early balance-spring verges to a variance as small as one to two minutes a day, but that kind of performance owes something to the advantage of hindsight and test conditions. There is reason to believe, however, that balance-spring verges continued to improve, to the point where Pierre Le Roy, citing his father, Julien, described a well-regulated watch as one gaining or losing less than a minute a day. *Etrennes chronométriques, ou calendrier pour l'année bissextile M.DCC.LX* (Paris, 1760), p. 66. That strikes me as a stunning performance, but then Julien and Pierre Le Roy were not the average watchmaker.

31. See the strictures of Henry Sully's *Règle artificielle du tems* (1717), cited in Jaques Savary des Bruslons, *Dictionnaire universel du commerce . . .,* new ed. (Copenhagen, 1761), vol. 3, pp. 991–992, s.v. "Montre."

32. Jaquet and Chapuis, *Technique and History of the Swiss Watch,* p. 87. This and related devices, such as the *secondes d'un coup* (jump seconds) were difficult and costly to make and found little use until the second quarter of the nineteenth century, when the Swiss made a specialty of what came to be known as chronographs—that is, watches with independent, stop fractional seconds. These watches, though, could not return the stop-seconds hand to zero, so that the user had to note the starting position each time before setting the hand in motion—a serious inconvenience. This difficulty was remedied in turn by the invention by Adolphe Nicole of the fly-back chronograph. The date usually given is 1862, yet the essentials of the design were already registered in 1844 as British patent number 10,348, including the heart-shaped cam that is the key to the fly-back arrangement. See fig. 5 of the patent specification. The earliest surviving examples, though, do date from the 1860s. There may have been a couple of decades' slippage between design and execution. See ibid. pp. 164–165, where the inventor of the fly-back chronograph is given as Henri-Féréol Piguet, who worked for Nicole in London and is said to have communicated his design to his employer in 1861. On these seconds-beating and fractional-seconds watches in general, see the valuable article of Adolphe Chapiro, "Les montres à 'secondes mortes,'" *Association Nationale des Collectionneurs et Amateurs d'Horlogerie Ancienne* (generally known as ANCAHA) 20 (December 1977): 27–47.

33. Alexander Pope, *Essay on Criticism* (1711).

34. Savary des Bruslons, *Dictionnaire universel du commerce,* vol. 3, p. 992.

8. Approaching the Asymptote

1. I have used the translation in Ernest L. Edwardes, *The Story of the Pendulum Clock* (Altrincham: J. Sherratt, 1977), p. 95. This gives photographs of the Latin text and the translation *in extenso.* There is a French translation in

Christiaan Huygens, *Oeuvres complètes,* vol. 17 (The Hague: Martinus Nijhoff, 1932). The Edwardes translation first appeared in the *Horological Journal,* July and August 1954, and was reprinted in *Antiquarian Horology* 7 (December 1970): 35–55.

2. On all this, see L. Defossez, *Les savants du XVII* ͤ *siècle et la mesure du temps* (Lausanne: Editions du Journal Suisse d'Horlogerie et de Bijouterie, 1946), pp. 258–262.

3. On the development of such devices, see Ferdinand Berthoud, *Histoire de la mesure du temps par les horloges,* vol. 2 (Paris, 1802; facsimile edition, Paris: Berger-Levrault, 1976), ch. 3; Rupert T. Gould, *The Marine Chronometer: Its History and Development* (London: J. D. Potter, 1923; rpt. Holland Press, 1960), figs. 29 and 32, and pp. 110ff.; Vaudrey Mercer, *John Arnold and Son, Chronometer Makers, 1762–1843* (London: Antiquarian Horological Society, 1972), esp. pp. ix and x and pl. 21.

4. Charles Bellair to Huygens, July 16, 1659, in Huygens, *Oeuvres complètes,* vol. 2, p. 440.

5. The best source is Defossez, *Les savants,* pp. 269–277. These concepts found more explicit application in algebraic form in C. E. L. Camus, "Sur la figure des dents des roues et des ailes de pignons pour rendre les horloges plus parfaites," in *Histoire et Mémoires de l'Académie Royale des Sciences,* Année 1733 (Paris, 1735), pp. 117–140; and then in two papers by Leonhard Euler for the Imperial Academy of Sciences in St. Petersburg (published 1760 and 1767). Neither geometric nor algebraic demonstrations could have much direct influence on horological practice, for few watchmakers could follow them; indeed, Euler's work was in Latin. But the ideas did work their way into the craft via example and such manuals of method as J.-A. Lepaute, *Traité d'horlogerie* (Paris, 1755), which includes as ch. 19 a "Traité des engrenages" by J.-J. Lefrançois de Lalande; and Ferdinand Berthoud, *Essai sur l'horlogerie,* 1st ed., 2 vols. (Paris, 1763), 2nd ed., 2 vols. (Paris, 1786; facsimile rpt., Paris: Berger-Levrault, 1978). I have used the second edition. Berthoud (in vol. 2, ch. 5) provides "cookbook" instructions for tracing appropriate profiles and refers the reader to Camus's *Cours de mathématique.*

6. Manuscript by John Carte, Bodleian Library, Oxford, as quoted by Alan Smith, ed., *A Catalogue of Tools for Watch and Clock Makers by John Wyke of Liverpool* (Charlottesville: University of Virginia Press, 1978), p. 1. This catalogue dates from the third quarter of the eighteenth century and represents a further stage in the equipping and industrialization of watch manufacture: the machines and tools were now standard items of trade.

7. Berthoud, *Essai sur l'horlogerie,* vol. 1 (1786), devotes a number of chapters to these engines and tools (chs. 25–29), beginning with the wheel-cutting machine. "This machine," he writes, "is so useful, its exactness is so essential for timekeepers, and it is so imaginatively conceived, that I could not dispense with giving its design" (vol. 1, p. 141).

8. Derek Howse, "The Tompion Clocks at Greenwich and the Dead-beat Escapement," *Antiquarian Horology* 7 (December 1970): 18–34; and 7 (March 1971): 114–133.

9. Defossez, *Les savants,* p. 283.

10. The diamond did prove suitable, however, for use unpierced as an end stone or cap jewel—that is, as a butt to keep the pivot from moving up and down in the bearing hole.

11. *Reasons of the English Clock and Watchmakers against the Bill to Confirm the Pretended New Invention of using Precious and Common Stones about Watches, Clocks, and other Engines* (broadside; n.p. [London, 1705?]), p. 2. See G. H. Baillie, *Clocks and Watches: An Historical Bibliography*, vol. 1 (London: N.A.G. Press, 1951; rpt. London: Holland Press, 1978), p. 132, which gives an inexact summary of the contents.

12. The operative word may be "legislature." In 1687 Edward Barlow had applied to the crown for a patent for his repeater mechanism, but was refused by the king for want of priority, by reason of Daniel Quare's invention of an alternative device. Baillie, *Clocks and Watches*, p. 115. And Hooke had certainly had monopoly in mind when he explored the possibilities of a chronometer with balance spring.

13. The history of watch jeweling remains to be written; there is almost nothing on the subject. One of the few pieces available is Gerd Ahrens, "Frühe Lochsteine aus Rubin," *Chronométrophilia* (La Chaux-de-Fonds) 6 (May 1979): 53–56, which has the merit of being based on an actual examination of watches from the early eighteenth century. This is crucial because Huggeford's use of a ruby as ornament was to find many imitators in later years, especially once jeweling was perceived as the sign of a fine watch. On Breguet's efforts to domesticate the technique of jeweling, see George Daniels, *The Art of Breguet* (New York and London: Sotheby Parke Bernet, 1975), pp. 34–35. See also Gerd Ahrens, "Frühe tragbare Ankeruhren in Deutschland," *Schriften der 'Freunde Alter Uhren'* 16 (1977): 80–85, which describes a well-jeweled German watch of the 1790s, perhaps slightly later. This was an experimental copy of a Josiah Emery lever, hence of an English model.

14. See H. Alan Lloyd, "Mechanical Timekeepers," in Charles Singer et al., eds., *A History of Technology*, vol. 3, *From the Renaissance to the Industrial Revolution* (Oxford: Clarendon, 1957), pp. 670–671; also Cecil Clutton, in the Introduction to the 1978 reprint of Berthoud, *Essai sur l'horlogerie*, vol. 1, p. xvii.

15. See his letter of January 28, 1804, cited in the *Encyclopaedia Londinensis* 10 (1811), quoted in Cedric Jagger, *Paul Philip Barraud* (London: Antiquarian Horological Society, 1968), p. 37.

16. The claim to priority for Tompion is advanced by Derek Howse, "The Tompion Clocks at Greenwich and the Dead-beat Escapement."

17. It may well be that the inspiration for this escapement also came from Thomas Tompion, Graham's master and teacher. There are reports of a new escapement by Tompion that has sounded to some like a rudimentary cylinder or an anticipation of a cylinder, and these have led some simply to credit him with the invention. The only trouble is that no such watch by Tompion has survived, and such reasonably contemporary descriptions as I have read of this device sound to me more like a double-hook or mini-anchor escapement. On the other hand, the cylinder may be seen from one point of view as just a special, circular form of anchor.

18. Writing in 1786, Ferdinand Berthoud, who was himself thoroughly

familiar with the cylinder, chose the verge escapement for his ideal watch. Berthoud, *Essai sur l'horlogerie,* vol. 2 (1786), ch. 47: "De la construction & de l'exécution d'une Montre dans laquelle on réunit tout ce qui peut contribuer à sa justesse." His main points were that verges were easy to adjust and rate, and that little friction made for little wear. He also noted (par. 2364) that a verge "can be made and repaired by ordinary workers." The skills required for a cylinder, then, were in scarce supply; but when commercial considerations made a shift to the cylinder advisable, continental craftsmen learned soon enough to make them. Compare the similar argument of André Hessen, *Mémoire sur l'horlogerie* (London and Paris, 1785), p. 26. I owe my knowledge of this informative brochure to Jean-Claude Sabrier.

19. On the story of Queen Charlotte's watch, see especially Paul M. Chamberlain, *It's About Time* (New York: Richard R. Smith, 1941; rpt. London, 1964), ch. 1; also Cecil Clutton and H. Quill, eds., *Pioneers of Precision Time-keeping,* Antiquarian Horological Society, monograph no. 3 (n.p., n.d.), pp. 8–16; also R. Good, "The First Lever Watch Made by Thomas Mudge," ibid., pp. 31–45.

20. On Litherland, see R. Vaudrey Mercer, "Peter Litherland & Co.," *Antiquarian Horology* 3 (June 1962): 316–323.

21. See Cecil Clutton, "Two Early Lever Watches," *Antiquarian Horology* 3 (December 1960): 129; also Alan Treherne, *The Massey Family: Clock, Watch, Chronometer and Nautical Instrument Makers* (Newcastle-under-Lyme: Borough Museum [1977]), p. 14.

22. The continental form of lever was different from that adopted in Britain. It derived from Mudge via the work of Josiah Emery, a Swiss maker established in London, as copied and modified by Robert Robin in Paris beginning about 1785. (See Figure A.5.) Another early maker of lever watches on the continent was Breguet, who brought it to a level of performance (variations of fifteen seconds a day) almost as good as it would ever reach. But Breguet's watches were made without regard to cost, and his highly original technical solutions were not easily adapted to commercial production. On Robin and the early continental lever, see J. D. Sabrier, "Histoire de l'échappement à ancre: les montres à ancre françaises de la première génération," *Association Nationale des Collectionneurs et Amateurs d'Horlogerie Ancienne* 23 (December 1978): 45–60. On Breguet's work, see George Daniels, *The Art of Breguet* (London: Sotheby Parke Bernet, 1975), pp. 304–313 and passim. The best concise history of the lever excapement is to be found in Cecil Clutton and George Daniels, *Watches,* 2nd ed. (London: Sotheby Parke Bernet, 1979), pp. 119–129.

23. G. H. Baillie, *Watchmakers and Clockmakers of the World* (London: Methuen, 1929; rpt. of 3rd ed., London: N.A.G. Press, 1972), s.v. Le Roy, Pierre.

9. The Man Who Stayed to Dinner

1. Rupert T. Gould, *The Marine Chronometer: Its History and Development* (London: J. D. Potter, 1923; rpt. Holland Press, 1960), pp. 10–11. On Digby,

see the biographical article in the *Encyclopedia Britannica,* 11th ed. Aside from the crass credulity of this proposal, it suffered from a serious defect of logic; for even if one granted its premises, the very application of Digby's powder would presumably have healed the dog. And then what? Repeated mutilation over long voyages?

2. Newton to Burchett, October 1721, in Alfred Rupert Hall and Laura Tilling, eds., *The Correspondence of Isaac Newton,* vol. 7, *1718–1727* (Cambridge: Cambridge University Press, 1977), pp. 172–173. See also Newton to the Admiralty, August 26, 1725, ibid., pp. 330–332.

3. The variation of Riefler's clocks averaged a hundredth of a second a day or less. F. A. B. Ward, *Time Measurement: Historical Review,* 4th ed. (London: Science Museum, 1970), p. 21. See also Klaus Erbrich, *Präzisionspendeluhren von Graham bis Riefler* (Munich: Callwey, 1978), pp. 44–46. At this level of precision, verification had to wait on intervals of at least a week.

4. Compare their reaction a little earlier to Henry Sully's petition to be allowed to practice his art in Paris, though a foreigner. Even Sully's friend Julien Le Roy, great maker though he was, joined in opposing and defeating the request. Julien Le Roy, "Mémoire pour servir à l'histoire de l'horlogerie, depuis 1715 jusqu'en 1729," in Henry Sully, *Règle artificielle du temps* (Paris, 1737), pp. 388–389.

5. Rupert T. Gould, the man who restored Harrison's long-neglected timekeepers to running order in the 1920s, in one of the finest lectures on horological history ever delivered, had some interesting things to say about the innovations and workmanship of Harrison's No. 3. Among other things, the remontoir "is the only *mechanically-perfect* remontoire I have ever met with—and I have studied more than a hundred devices of the kind. It gives an *absolutely constant* torque at the escape wheel, whether its driving springs are fully-wound, half run-down, almost due for rewinding, or actually being rewound." Also No. 3 "embodies, among its minor refinements, two sets of roller-bearings—steel rollers pivoted into a revolving cage and running on a steel race—which, although made about 1750, look as if they had been taken out of a modern car." Gould, *John Harrison and His Timekeepers,* 4th ed. (Greenwich: National Maritime Museum, 1978), p. 9 (rpt. from *Mariner's Mirror* 21 [April 1935]: 115–139).

6. Ibid.

7. Ibid., p. 11. Harrison wrote this in 1763.

8. Thanks to this arrangement, the train had little power over the motions of the balance. In Harrison's words: "In this my Time-keeper, the Wheels have only about *One-eightieth* Part of the Power over the Balance that the Balance-spring has; and, it must be allowed, the less the Wheels have to do with the Balance the better." For a somewhat different analysis of the escapement, clearer in some ways than Gould's, see G. H. Baillie, *Clocks and Watches: An Historical Bibliography,* vol. 1 (London: N.A.G. Press, 1951; rpt. Holland Press, 1978), pp. 272–274.

9. Parallax, to quote Webster's 3rd ed., is "the apparent displacement (or the difference in apparent direction) of an object, as seen from two different points." It is a phenomenon thoroughly familiar to any photographer who

has worked with a view-finder camera (as opposed to a single-lens reflex). In astronomy, parallax is the difference in direction of a heavenly body as seen from some point on the earth's surface, and from such other conventional point as the center of the earth. In calculations of lunar distances, all readings had to be reduced for purposes of comparability to what they would have been if made from the earth's center.

10. This was why the astronomical technique preferred for longitude measurements on land, namely observations of the occultation of Jupiter's moons, was simply not feasible at sea. Such observations required the use of a long telescope—"fourteen feet long at most," according to the still hopeful Astronomer-Royal Flamsteed—and there was simply no way to keep such a tube steady on shipboard. They even tried putting the observer in a special sling-chair, as one would put a clock in gimbals, without success. Charles H. Cotter, *A History of Nautical Astronomy* (Toronto: Hollis & Carter, 1968), pp. 187–188. This is, to my mind, the best technical history of navigation. By comparison, the others are exercises in *petite histoire*.

On the Davis quadrant and Hadley's octant, see especially Deborah Jean Warner, "Astronomers, Artisans and Longitude," in Tekniska Museet Symposium, *Transport Technology and Social Change* (Stockholm: Tekniska Museet, 1980), pp. 133–140.

11. Letter of August 26, 1725, to the Admiralty, in Hall and Tilling, eds., *The Correspondence of Isaac Newton,* vol. 7, p. 331. Newton made the same remarks in a letter of 1721 (ibid., p. 172): "A good Watch may serve to keep a recconing at Sea for some days & to know the time of a celestial Observ[a]tion: & for this end a good Jewel watch may suffice till a better sort of Watch can be found out."

12. Ibid., p. 199. On Newton's lunar tables, see E. Guyou (a member of the French Bureau des Longitudes), "La méthode des distances lunaires," *Revue maritime* 153 (June 1902): 948.

13. Alexis Clairaut's best effort in this area appeared as the *Théorie de la lune déduite d'un seul principe de l'attraction réciproquement proportionnelle aux quarrés des distances* (St. Petersburg, 1752). This was submitted for and won a prize offered by the Russian Academy of Science for a contribution to lunar theory. Euler, who had been invited to Russia by Catherine I in 1727 and succeeded Daniel Bernoulli in the chair of mathematics at St. Petersburg, and who had then moved in 1741 to Berlin, published a series of studies dealing with the movements of celestial bodies: *Theoria motuum planetarum et cometarum* (Berlin, 1744); *Novae et correctae tabulae ad loca lunae computanda* (1746); *Tabulae astronomicae solis et lunae* (1746); *Scientia navalis, seu tractatus de construendis ac dirigendis navibus,* 2 vols. (St. Petersburg, 1749); *Theoria motus lunae* (Berlin, 1753). On Clairaut and the history of astronomical navigation, see E. Guyou, "La méthode des distances lunaires," pp. 943–963.

It is an interesting sidelight on the science of the day that Russia, still a backward country politically, socially, and economically, was able to invite and hold some of the leading scientific and intellectual luminaries of western Europe and could become a significant center of study and research.

14. The intellectual link between the two men is well conveyed by their correspondence: Eric G. Forbes, ed., *The Euler-Mayer Correspondence: 1751–1755* (New York: American Elsevier, 1971). See also Forbes, "Tobias Mayer (1723–1762), a Case of Forgotten Genius," *British Journal for the History of Science* 5 (1970): 1–20.

Ironically, Mayer did not base his lunar equations on those of Euler, but rather on a technique the latter had developed to analyze the motions of Jupiter and Saturn. Forbes, *The Birth of Navigation: The Solving in the 18th Century of the Problem of Finding Longitude at Sea* [cover title: *The Birth of Navigational Science: Tobias Mayer*] Maritime Monographs and Reports, no. 10 (Greenwich: National Maritime Museum, 1974), p. 20, n. 56.

15. Cotter, *History of Nautical Astronomy,* p. 204, citing Guyou, "La méthode."

16. On Maskelyne's urging, the Board of Longitude awarded Mayer, or rather Mayer's widow, £3,000 for his contribution to the solution of the longitude problem; and £300 to Euler for his mathematical theorems. Cotter, *History of Nautical Astronomy,* p. 205.

17. For more on Mayer and lunars, see: D. H. Sadler, *Man Is Not Lost: A Record of Two Hundred Years of Astronomical Navigation with the Nautical Almanac, 1767–1967* (London: National Maritime Museum and Royal Greenwich Observatory, 1968); and Eric G. Forbes, "The Foundation and Early Development of *The Nautical Almanac,*" *Journal of the Institute of Navigation* 18 (1965): 391–401.

18. Both Gould and Baillie, following Harrison, judge Maskelyne's test and rating procedures as unsuitable and unfair. Gould, *The Marine Chronometer,* pp. 63–64; Baillie, *Clocks and Watches,* pp. 271–272. See the exchange between the two adversaries: Nevil Maskelyne, *An Account of the Going of Mr. Harrison's Watch . . .* (London, 1767); and John Harrison, *Remarks on a Pamphlet Lately Published by the Rev. Mr. Maskelyne under the Authority of the Board of Longitude* (London, 1767).

19. On Harrison, see Humphrey Quill, *John Harrison: The Man Who Found Longitude* (London: John Baker, 1966), p. 117; on Maskelyne, see Gould, *The Marine Chronometer,* p. 77. In 1774 Maskelyne had just helped draft a law for a new longitude prize, 14 Geo. III 66, which set such difficult requirements for a timekeeping device that the prize was never won. There, Maskelyne remarked, he had "given the Mechanics a bone that would crack their teeth."

20. This was the variance after correction for predicted rate and temperature variance. Gould, *The Marine Chronometer,* pp. 59–60; Quill, *John Harrison,* p. 137; Eric G. Forbes, "The Origin and Development of the Marine Chronometer," *Annals of Science* 22 (March 1966): 7–9.

21. Gould, *John Harrison and His Timekeepers,* p. 13.

22. On Harrison's continuing sense of deprivation, see H. Alan Lloyd, "John Harrison," *La Suisse horlogère* (international edition, in English; offprint, n.d.), pp. 15–16.

10. The French Connection

1. There is an excellent discussion of the new research mode in Jean Mayer, *Les Européens et les autres: de Cortés à Washington* (Paris: A. Colin, 1975), pp. 90–100.

2. Humphrey Quill, *John Harrison: The Man Who Found Longitude* (London: John Baker, 1966), pp. 157–159. Berthoud must have thought at one point that he had a deal, because on March 14, 1766, he wrote Paris describing the works of Harrison's watch and recommending payment of £500.

3. A summary of this letter, which was printed in the *Revue chronométrique*, 8 (1874–75): 19–23, is given in G. H. Baillie, *Clocks and Watches: An Historical Bibliography*, vol. 1 (London: N.A.G. Press, 1951; rpt. Holland Press, 1978), p. 269.

4. Berthoud's chronometers are described in detail in Conservatoire National des Arts et Métiers, *Catalogue du Musée*, sect. JB, *Horlogerie* (Paris, 1949), pp. 253–262. Pierre Le Roy's clock is described on pp. 247–250. This catalogue, exceptional in its technical scholarship, has long been out of print, but with the appointment of the new curator, Catherine Cardinal, plans were made for preparation of a new one incorporating the latest scholarship. Unfortunately the promised funds have not materialized, and the project has had to be shelved.

5. Two moments of contact are indispensable: (1) when the controller lifts the locking piece and releases the scape wheel; and (2) when the scape wheel gives some of its energy to the controller to replace what has been lost to friction and air resistance. Without that impulse, the controller would eventually run down and stop.

6. Pierre Le Roy, *Précis des recherches faites en France depuis l'année 1730 pour la détermination des longitudes en mer par la mesure artificielle du temps* (Amsterdam, 1773), p. 23. See Paul Ditisheim, Roger Lallier, L. Reverchon, Le Commandant Vivielle, *Pierre Le Roy et la chronométrie* (Paris: Editions Tardy, 1940), pp. 68–69. This collective work, printed during wartime in a limited edition of 450 copies, remains the best source on the work and life of France's greatest horologist.

7. He gives his reasons in his *Mémoire sur la meilleure manière de mesurer le tems en mer . . . contenant la description de la montre à longitudes présentée à Sa Majesté le 5 août 1766* (n.p., n.d. [Le Roy says 1770]), pp. 34–35. See also Le Roy, *Précis des recherches* (1773), pp. 26–27.

8. Commandant Vivielle, "Les chronomètres de Pierre Le Roy et la marine française: la conservation de l'heure et le problème des longitudes; les épreuves à la mer des chronomètres de Le Roy," in Ditisheim et al., *Pierre Le Roy et la chronométrie*, pp. 19–21. Pingré reported on these tests in a volume entitled *Journal du voyage de M. le Marquis de Courtanvaux, sur la frégate l'Aurore, . . .* (Paris, 1768), sometimes catalogued under Courtanvaux.

9. Ibid., p. 21. See also [Jacques-Dominique] Cassini, fils, *Voyage fait par ordre du Roi en 1768, pour éprouver les montres marines inventées par M. Le Roy* (Paris, 1770).

10. Rupert T. Gould, *The Marine Chronometer* (London: J. D. Potter, 1923; rpt. Holland Press, 1960), pp. 138ff., avoids the use of the term "detent escapement," preferring to speak of the "chronometer escapement." But the former term has passed into general use, beginning with its introduction in France by Pierre Le Roy as *échappement à détente*. G.-A. Berner, ed., *Dictionnaire professionnel illustré de l'horlogerie* (La Chaux-de-Fonds, n.d.), pp. 316–317, gives it as the standard term in both French and English, with *Chronometer-hemmung* as the German equivalent. For its use in a standard English work, see Cecil Clutton and George Daniels, *Watches* (New York: Viking, 1965), pp. 24ff.

11. Gould, *The Marine Chronometer,* p. 89. In saying this, Gould was clearly interpreting "detent" in the broad sense of a detaining or locking device, as had Abraham Rees a century before. Ibid., p. 84n. (On Rees, see *Rees's Clocks, Watches and Chronometers* [1819–20; rpt. Newton Abbot: David & Charles, 1970], p. 206.) For a distinguished example of the general sense (mis-sense) of the matter see Ernst von Bassermann-Jordan, *Uhren,* rev. and ed. Hans von Bertele (Braunschweig: Klinkhardt & Biermann, 1976), p. 384, which states that in 1769 "Pierre Le Roy perfects his first sea clock, whose chronometer escapement he had already published in 1748." Compare Eric Bruton, *The History of Clocks and Watches* (London: Orbis, 1979), p. 99, which captions an illustration of the clock as follows: "The most fundamental invention incorporated was a detached detent escapement."

Part of the difficulty may perhaps have arisen from Gould's having written (*The Marine Chronometer,* p. 87) that Le Roy was the author of *"the first detached chronometer escapement."* By this he presumably meant the first detached escapement used in a marine chronometer; but many readers have taken it to mean the first example of what we have come to know as a chronometer escapement—that is, a spring or spring-loaded detent escapement.

12. Le Roy, *Précis des recherches,* p. 24. Le Roy used the term in the sense that British horologists today would use it—to apply to an escapement whose locking piece locks by itself, separately from the balance regulator.

13. Cecil Clutton and George Daniels, *Watches,* 2nd ed. (London: Sotheby Parke Bernet, 1979), pp. 45–46, speak of the locking piece as a pivoted lever, and point out that this is in effect the escapement later reinvented by Robin and generally known as the Robin escapement.

14. Rupert T. Gould, *John Harrison and His Timekeepers,* 4th ed. (Greenwich: National Maritime Museum, 1978), p. 14 (rpt. from *Mariner's Mirror* 21 [April 1935]: 115–139).

15. It is clear that this was the feature of his clock that Le Roy had least confidence in, for good reason. The glass tubing required introduced an element of fragility and complexity into the mechanism that was at variance with its purpose and Le Roy's goal of simplicity. In his *Précis des recherches,* pp. 27–28, he writes that a better way to compensate for temperature change would be to hold temperature constant by surrounding the clock with lamps and adjusting the heat by changing the number of lamps or altering their wicks, then placing the whole in a basket wrapped in blankets—in short, by putting the clock in an incubator (*étuve*). He says that the idea was first put

forward by Massi in 1720, but that somehow he missed it. (That was a bit of luck.)

As for the mercury compensation he did employ, the idea probably came from Graham's mercury pendulum; but Le Roy, who was extremely jealous of his originality, writes that he is persuaded that impartial persons will find the link "very indirect and distant." Ibid., p. 32.

16. The *Exposé succinct des travaux de MM. Harrison et Le Roy, dans la recherche des longitudes en mer, et des épreuves faites de leurs ouvrages* (Paris, 1768) was translated into English and published in London in the very year of its appearance in Paris. See *A Succinct Account of the Attempts of Messrs. Harrison and Le Roy for Finding the Longitude at Sea, and of the Proofs Made of Their Works*, trans. by a Fellow of the Royal Society (London: F. Newberry, 1768).

17. Le Roy, *Précis des recherches*, p. 10, citing a letter of John Bevis of August 1771. The clock was the work of John Monk, who apparently began as an apprentice clockmaker, but he may have given the trade up to engage in general mechanics—as did many clockmakers in those early days of the Industrial Revolution. See G. H. Baillie, *Watchmakers and Clockmakers of the World* (London: Methuen, 1929; rpt. of 3rd ed., London: N.A.G. Press, 1972), s.v. Monk, John.

18. I use the English translation of 1768: *A Succinct Account*, p. 25.

19. *Précis des recherches*, p. 35.

20. Ibid., p. 37; *Suite du Précis sur les montres marines de France; avec un supplément au Mémoire sur la meilleure manière de mesurer le tems en mer* (Leiden, 1774), p. 2.

21. After the death of Julien in 1759, Pierre continued to make (or to have made) watches signed with his father's name and bearing numbers in the same series. This practice gave rise to some confusion subsequently, but the matter has since been clarified by the researches of Charles Allix and Giuseppe Brusa. There are very few pocket watches signed by Pierre (the son and not the brother), and one has the impression that in his search for a marine chronometer, he left the civilian business to others. It no doubt brought him the income he needed to pursue his experiments, but this was far less than he might have earned had he given the shop his attention and some of his genius. See his *Précis des recherches*, p. 3, where he says he has spent a large part of his personal fortune on the development of an effective chronometer.

22. *"Id vero an ipsi Daemone possibile nescio."* *Précis des recherches*, p. 38. Morin was the author of an important study of longitude: *Longitudinum terrestrium et coelestium nova et hactenus optata scientia* (Paris, 1634–1639), in which he argued strongly for calculation by lunar distances.

23. Le Roy, *Suite du Précis*, pp. 1–2.

24. British horologist Anthony Randall has prepared a detailed chart of the characteristics of Berthoud's marine chronometers: "L'oeuvre chronométrique de Ferdinand Berthoud de 1760 à 1787," *Association Nationale des Collectionneurs et Amateurs d'Horlogerie Ancienne* 30 (Spring 1981): 23–35. This is a most valuable clarification. See also Conservatoire National des Arts et Métiers, *Catalogue du Musée*, sect. JB, *Horlogerie*, pp. 253–257; Gould, *The Marine Chronometer*, pp. 94–95.

25. Le Roy, *Suite du Précis*, p. 6.

26. Ferdinand Berthoud, *Traité des horloges marines* ... (Paris, 1773), p. 576; Le Roy, *Précis des recherches*, pp. 25–26.

27. Le Roy, *Suite du Précis*, pp. 40–41.

28. Le Roy, *Précis des recherches*, p. 12, citing Berthoud, *Traité des horloges*, p. 26.

29. Le Roy, *Suite du Précis*, p. 6.

30. Ibid., pp. 1–2.

31. Le Roy, *Précis des recherches*, pp. 25–26.

32. Berthoud, *Traité des horloges*, p. xviii.

33. Le Roy, *Suite du Précis*, p. 12.

34. I take for my comparison the low end of the scale for Paris watchmakers as given in the corporation (guild) tariff. Emile Levasseur, *Histoire des classes ouvrières et de l'industrie en France avant 1789,* 2nd ed., vol. 2 (Paris, 1901), p. 842. I could take the top end of the scale, which was twice as high, but then the equivalent American salary would run to $150 or more per day. On an annual basis, it should be noted that because of religious holidays, French journeymen worked at most about 250 days a year, or about what we do with two-day weekends; but they worked longer days.

35. Le Roy, *Précis des recherches*, p. 3.

36. Le Roy, *Suite du Précis*, pp. 10–11.

37. Gould, *John Harrison and His Chronometers*, p. 14.

38. Louis Berthoud's first signed pocket chronometer dates from 1786. For a lyrical appreciation of his talent, see Cecil Clutton, *Collector's Collection* (London: Antiquarian Horological Society, 1974), p. 87. Jean-Claude Sabrier is currently preparing a history of Louis Berthoud's life and work.

11. Fame Is the Spur

1. The detent escapement has generally taken one of two basic forms. In the pivoted version, the locking piece or detent swings on a spring-loaded pivot; in the spring version, the locking stone is mounted on a straight spring, which has to be snapped out of the way to release the scape wheel. Both systems have their merits. The pivoted detent is less likely to set if shaken, and hence is more suitable for pocket instruments. In the long run, however, and especially in boxed chronometers, the spring detent is more accurate, in part because it needs no lubrication. The great French *chronométriers* typically preferred the pivoted version for both marine and civilian use. The British went the other way. The Swiss in the nineteenth century made both kinds, depending on the market. (See Figure A.6.)

2. Confirmed Minutes of June 11, 1796, cited in Eric G. Forbes, "The Origin and Development of the Marine Chronometer," *Annals of Science* 22 (March 1966): 20. The invidious comparison of 1796 was actually confined to Thomas Mudge's chronometers; but in a further judgment of 1803, the board recognized that Earnshaw's chronometers had "gone better than any others

that have been submitted to trial at the Royal Observatory." Ibid., p. 21: Confirmed Minutes of March 3, 1803.

3. Rupert T. Gould, *John Harrison and His Timekeepers,* 4th ed. (Greenwich: National Maritime Museum, 1978), p. 15 (rpt. from *Mariner's Mirror* 21 [April 1935]: 115–139).

4. The above paragraphs tell the story as given in Thomas Earnshaw, *Longitude: An Appeal to the Public Stating Mr. Thomas Earnshaw's Claim to the Original Invention of the Improvements in His Timekeepers . . . and His Consequent Right to National Reward* (London: privately printed, 1808).

5. Nor did Arnold ever move to prosecute or sue Earnshaw for slander or for breach of patent, or to submit his device to the Board of Longitude for a reward, as Earnshaw did; nor did he "at the time, boast of such invention, as it might have been expected from his warm disposition, he would have done, had he been able to substantiate his claim to it." Nevil Maskelyne, *Arguments for Giving a Reward to Mr. Earnshaw for Improvements on Time-keepers, by the Astronomer-Royal; to the Commissioners of Longitude* (n.p., n.d. [signed as of December 1, 1804]), p. 7. See Forbes, "Origin and Development," p. 22.

Vaudrey Mercer, *John Arnold and Son, Chronometer Makers, 1762–1843* (London: Antiquarian Horological Society, 1972), p. 90, suggests that Arnold could not sue Earnshaw because at that point Earnshaw was not yet signing his own watches. But then why not sue Wright and the other signers of Earnshaw-made spring-detent watches? Besides, Earnshaw did set up on his own in 1790 and presumably signed his own pieces from then on. Mercer further notes that, in any case, Arnold must have seen the Earnshaw escapement as different, and hence not a violation of his patent. True enough.

6. Mercer, *John Arnold,* pp. 74–79, 87–96, and esp. pl. 83 (opp. p. 60).

7. Under the agreement with Wright, Earnshaw was free to make watches with his new escapement for other makers, on condition that he pay Wright a fee of one pound each. In this way, Wright would recover the costs of the patent. In addition, Earnshaw made a number of these watches for Wright, one of which has survived: No. 2228 of 1784. See Andrew Crisford, "Thomas Wright in the Poultry London, No. 2228," *Antiquarian Horology* 9 (June 1976): 785–788. This earliest known example of the Earnshaw escapement has only a single spring (no passing spring), as in the patent application. Note that Arnold's first version (as per his application) also used a single spring, which Earnshaw argued was proof of plagiarism. Earnshaw, *Longitude,* p. 8.

8. See Mercer, *John Arnold,* pp. 65–66.

9. *Explanation of Time-keepers Constructed by T. Earnshaw, Three of Them Having Been Tried under the Present Act of Parliament Delivered to the Board of Longitude by the Astronomer Royal on the Part of Mr. Earnshaw, March 7th, 1805* (n.p., n.d.), p. 9. These criticisms are very close to those made by the Board of Longitude on examination of Arnold's No. 176 in March of 1804 after its failure on an expedition to Australia. Among other problems, the epicycloidal teeth of the "ballance wheel" (the scape wheel is meant) were "considerably" worn. See Christie's catalogue, "Horological Books, Important Clocks, Watches and

Chronometers," sale of November 25, 1981, lot 209, pp. 76–77. For more on this question of wear and lubrication, see Vaudrey Mercer, "Thomas Earnshaw," *Antiquarian Horology* 13 (December 1981): 175–176.

Earnshaw expressed scorn not only of Arnold's design but of his skill as a watchmaker: thus his disdain for what he said was Arnold's confessed inability to make a spiral spring "go without breaking or touching." "Here then, what Arnold by his Oath declared an impossibility, I have rendered quite easy. And this is strong evidence from Mr. Arnold of my superior abilities." Earnshaw, *Longitude,* p. 8.

10. *Certificates and Circumstances Relative to the Going of Mr. Arnold's Chronometers* (London, 1791).

11. On Banks, see the short biographies in the *Dictionary of National Biography* and the *Encyclopedia Britannica,* 11th ed.; also Alan Moorehead, *The Fatal Impact: An Account of the Invasion of the South Pacific, 1767–1840* (Harmondsworth: Penguin, 1968), chs. 1–3: a fascinating portrait.

12. Rupert T. Gould, *The Marine Chronometer* (London: J. D. Potter, 1923; rpt. Holland Press, 1960), p. 123.

13. Dalrymple, *Longitude: A Full Answer to the Advertising Concerning Mr. Earnshaw's Timekeeper* (London, 1806). The title *Longitude* was clearly the inspiration for Earnshaw's book of the same name, published in 1808. On Dalrymple, see Howard T. Fry, *Alexander Dalrymple (1737–1808) and the Expansion of British Trade* (Toronto: University of Toronto Press, 1970); also some of the references in Mercer, *John Arnold,* esp. p. 136.

14. Earnshaw, *Longitude,* unpaginated.

15. Mercer, *John Arnold,* pp. 163–164.

16. On Barraud, see Cedric Jagger, *Paul Philip Barraud* (London: Antiquarian Horological Society, 1968), pp. 32–36; on Pennington, see Vaudrey Mercer, "The Penningtons and Their Balances," *Antiquarian Horology* 12 (Spring 1981): 514.

17. Gould, *The Marine Chronometer,* p. 124.

18. Vaudrey Mercer, *The Frodshams: The Story of a Family of Chronometer Makers* (London: Antiquarian Horological Society, 1981), pp. 12–13.

19. Gould, *The Marine Chronometer,* p. 125.

20. See the catalogue for the sale of June 3, 1981 (code name Eel), lot 34.

21. From a letter to me of May 22, 1981.

22. Cited in Paul Ditisheim et al., *Pierre Le Roy et la chronométrie* (Paris: Editions Tardy, 1940), p. 108. The letter is reprinted *in extenso* in *Association Nationale des Collectionneurs et Amateurs d'Horologerie Ancienne* 27 (Spring 1980): 65–67.

23. Tardy, *Dictionnaire des horlogers français* (Paris: Editions Tardy, n.d.), p. 401, presents a page of holograph manuscript by Pierre Le Roy that speaks by chance (?) to this very point: "Experience has taught me only too well that most often it is not our worthy actions or useful works that cover us with honor. It is ordinarily the reward that follows, and it happens only too often that a reward usurped honors him whom it ought to abase, while a reward withheld denigrates the man who earned it."

24. See Henri Robert (a leading French watchmaker), *Comparaison des chronomètres ou montres marines à barillet denté avec celles à fusée* (Paris, 1839), p. 13, who says the British system is to make many chronometers and then cull out the bad ones. "That is what has led people to say that the excellence of their products owed more to chance than to science."

25. See C. A. Crommelin, "The Introduction of the Marine Chronometer into Naval and Mercantile Use," *Bulletin of the National Association of Watch and Clock Collectors* 4 (October 1951): 406–407.

26. Thus John H. Parry, *Trade and Dominion: The European Overseas Empires in the Eighteenth Century* (London: Weidenfeld & Nicolson, 1971), pp. 227–228; and Gould, *The Marine Chronometer*, pp. 131–132.

27. Gould, *The Marine Chronometer*, p. 132.

28. This is the argument of Alun C. Davies, "The Life and Death of a Scientific Instrument: The Marine Chronometer, 1770–1920," *Annals of Science* 35 (1978): 509–525.

29. P. Kartaschoff, *Frequency and Time* (New York: Academic Press, 1978), pp. 2–3; and Malcolm M. Thomson, *The Beginning of the Long Dash: A History of Timekeeping in Canada* (Toronto: University of Toronto Press, 1978), p. 173. Thomson offers in his Appendix A an excellent discussion of recent advances in techniques of time measurement and their implications for the definition of the time unit. A word on the title: in Canada, "the beginning of the long dash following ten seconds of silence indicates exactly one o'clock Eastern Standard Time."

III. MAKING TIME

1. François Le Lionnais, *Le Temps* (Paris: Robert Delpire, 1959), p. 53. That amount of gasoline will take an energy-efficient car (making forty miles per gallon) less than twenty yards.

12. Clocks in the Belfry

1. On the widespread adoption of water power, see Lynn White, Jr., *Medieval Technology and Social Change* (Oxford: Oxford University Press, 1962), ch. 3. This chapter contains some pages on horological history (pp. 120–125). These have to be read with caution, for like much of the literature in the history of technology, they place too much reliance on ambiguous drawings and hypothetical connections. White's interpretations may well be justified, but I would like an extra year for every plausible alternative hypothesis. On the rapid diffusion of water power in the textile manufacture, the single most important branch of medieval industry, see E. M. Carus-Wilson, "An Industrial Revolution of the Thirteenth Century," *Economic History Review* 11 (1941): 39–60.

2. In chronological order: The account book is Ms. N. 2435, Real Patri-

monio Archivo de la Corona de Aragon, Barcelona. First reported in L. Camos i Cabruja, "Dietari de l'obra del rellotge i la campana del castell de Perpinya l'any 1356," in *Homenatge a Antoni Rubió i Lluch: Miscellania d'estudis literaris, historics i linguistics,* vol. 3 (Barcelona, 1936), pp. 423–446. (I am indebted to Gerhard Dohrn of the University of Bielefeld for a photocopy of this essay. Mr. Dohrn is currently preparing a badly needed inventory of all known clocks to about 1450.) C. F. C. Beeson, "Perpignan 1356 and the Earliest Clocks," *Antiquarian Horology* 7 (June 1970): 408–414. Idem, *Perpignan 1356: The Making of a Tower Clock and Bell for the King's Castle* (London: Antiquarian Horological Society, 1983). I am grateful to Anthony Turner of Le Menil-le-Roi (France), a specialist in medieval horology and scientific instruments, for making the galley proofs of this book available to me.

3. A word of explanation about these comparisons of money values across the centuries: they are necessary if the modern reader is to have any appreciation of sums expressed in units of currency long forgotten. These calculations pose difficult economic and ontological problems, and there is no consensus on an appropriate procedure. One possibility would be to compare gold and silver content of money then and now, on the premise that these metals are imperishable and unalterable standards of value. The fact is, though, that they are not, for precious metals are commodities that in some periods fluctuate freely in the marketplace, in others are pegged to fixed prices and rates of exchange; so that a calculation of this type would give very different results according to the date chosen for comparison. Another possibility would be to take some staple commodity, something used and consumed by everybody (say, wheat) and use that as the measure. This is not a bad idea in principle and would probably give good results for comparisons between the preindustrial eighteenth century and earlier periods. But since the Industrial Revolution, the relative prices of foods and manufactures have altered considerably; the contents of the representative market basket have changed drastically; and the influence of short-run price fluctuations is such as to make the dates of comparison crucial to the result.

In my opinion, the best standard for price comparisons, especially over very long intervals, is the cost of labor, either skilled or unskilled. The former is in a way the more useful, for it enters directly or indirectly into the production of every good and service; but the latter is sometimes more appropriate for commodities made in large part by skilled workers. Both of these measures fluctuate less than just about any other market commodity.

For the comparison between the costs of medieval clocks and 1982–83 dollars, I have used the price of skilled labor—for example, the day wage of a master mason or carpenter in the fourteenth century. Against this I have set a price of $75 a day for similar work in the United States today. This is on the low side, but it is better to provide a conservative estimate and argue *a fortiori.* If one did a similar calculation using the price of unskilled labor, then 2 sous a day, now $4 an hour (minimum wage), the result would be similar, especially if one set the length of today's workday equal to what it was then.

4. On the working and misworking of the Perpignan clock, see Jeanne

Vielliard, "Horloges et horlogers catalans à la fin du Moyen Age," *Bulletin hispanique* of the *Annales de la Faculté des Lettres de Bordeaux* 63 (July–December 1961): 163–164.

5. As cited in Pierre Dubois, *Collection archéologique du Prince Pierre Soltykoff: Horlogerie* . . . (Paris, 1858), pp. 23–24.

6. At least one chronicler credited the clock to Pierre Dailly, bishop of Cambrai and an amateur of astrology (astronomy), and dated it from 1405. Jules Houdoy, *Histoire artistique de la cathédrale de Cambrai,* Mémoires de la Société des Sciences, de l'Agriculture et des Arts de Lille, 4th ser., vol. 7 (Lille, 1880), p. 18, n. 3, citing the *Seria episcoporum* of the cathedral archives.

7. The paragraphs below are based on the discussion ibid., pp. 48ff. and the documents in annex, pp. 157ff. Extracts from the latter have been published without reference or dates in Tardy, *La pendule française, première partie: des origines au Louis XV* (Paris: Editions Tardy, 1949; rpt. 1974), p. xxii. The translation in the new English edition, *French Clocks the World Over, Part One: From Their Beginnings to the Louis XV–Louis XVI Transition Period* (Paris: Editions Tardy, 1981), pp. 24–25, is excellent, but not always faithful to the French.

8. I have trouble with *retinendo*. It might mean "adjusting," although the usual term for setting and adjusting was *temperare*. The word *retinendo* is also used elsewhere in the accounts to mean "fixing in place," as of glass. The reference here, then, may be to the task of fixing the clock in place after taking it down for repair; but I do not think so.

13. The Good Old Days That Never Were

1. One of the first to argue this was Stephen Marglin. See Marglin, "What Do Bosses Do?" in Andre Gorz, ed., *The Division of Labour, the Labour Process and Class-Struggle in Modern Capitalism* (London: Harvester, 1976), pp. 13–53.

2. On Roger de Stoke and the Norwich clock, see the references given in ch. 3, n. 1, above.

3. The Ramsay watch has been frequently reproduced in illustration. See, for example, Ernst von Bassermann-Jordan, *Ein Handbuch für Sammler und Liebhaber: Uhren,* 8th ed., rev. and ed. Hans von Bertele (Braunschweig: Klinkhardt & Biermann, 1976), fig. 101. The figures are numbered the same in the French and English editions.

4. On Cutriffin, see Marius Fallet, "Le rayonnement séculaire de l'horlogerie suisse," in Alfred Chapuis, ed., *L'horlogerie: une tradition helvétique* (Neuchâtel: Editions de la Bourgade, 1948), p. 10.

5. Several clockmakers (*Zitgloggenmeister*) had been trying to fix it for almost a decade. M. Fallet-Scheurer, *Le travail à domicile dans l'horlogerie suisse et ses industries annexes* (Bern: Imprimerie de l'Union, 1912), p. 159.

6. See Johann Georg Leutmann, *Vollständige Nachricht von den Uhren, Erste Continuation oder zweiter Theil* . . . (Halle, 1722), pt. 2, ch. 10; and William

Blakey (an English expatriate), *L'art de faire les ressorts des montres* . . . (Amsterdam, 1780).

7. Klaus Maurice, *Die deutsche Räderuhr,* vol. 1 (Munich: Beck, 1976), p. 93, n. 69; Bassermann-Jordan, *Ein Handbuch für Sammler und Liebhaber: Uhren,* p. 181: Louis Bulla, "Les montres de la Renaissance et l'horlogerie des Blésois," in Musée National des Arts et Métiers, *Les chefs-d'oeuvre de l'horlogerie* (Paris: Editions de la Revue française des Bijoutiers-Horlogers [1949]), p. 145.

8. This last may also reflect Louis Breguet's determination to establish his intellectual and technical autonomy vis-à-vis a crushing horological legacy. See E. Ferret, *Les Breguet,* 2nd ed. (Paris, n.d.).

9. This pattern of succession tended to repeat itself over generations: the widow would marry a much younger craftsman, who would then survive her and marry a much younger woman, who would then survive him and marry a much younger craftsman. Often the sequence would start with the death of a wife in childbirth, followed by a second marriage to a younger woman.

10. A brief chronology of dates of inception of clockmakers' guilds: Paris, 1544; Nuremberg, 1565; Blois, 1600; Geneva, 1601; London, 1631; Copenhagen, 1755. In Augsburg the clockmakers were part of the smiths' guild, which dated back to 1368; the first admission of a clockmaker as such occurred in 1441, but it was only in 1564 that the clockmakers were a strong enough group within the company to wrest sole authority in matters of admission of clockmakers and inspection of their work. See Eva Groiss, "Das Augsburger Uhrmacher-handwerk," in Klaus Maurice and Otto Mayr, eds., *Die Welt als Uhr: Deutsche Uhren and Automaten, 1550–1650* (Munich: Bayerischen Nationalmuseum, 1980), pp. 58, 60. In Rotterdam the clockmakers were one of six metalworking crafts in St. Eloy's guild. C. Spierdijk, *Klokken en Klokkenmakers: Zes eeuwen uurwerk, 1300–1900,* 2nd rev. ed. (Amsterdam: J. H. DeBussy, 1965), p. 234.

11. Léon Nardin and Julien Mauveaux, "Histoire des corporations d'arts et métiers des ville et comté de Montbéliard et des seigneuries en dépendant," in *Mémoires de la Société d'Emulation de Montbéliard* 38 (1910): 278.

12. Groiss, "Das Augsburger Uhrmacher-handwerk," pp. 62–63. Rent of an outside room was worth about two kreuzer a day; a bed in the master's house, probably something less, if only because there were usually more than one to a room. I have assumed that board was worth about two and a half times bed. The gulden at that time was worth 72 kreuzer. These wage figures, it should be noted, are for cabinetmakers; but there is no reason to believe that clockmakers were paid more or less.

13. On the good and bad of training apprentices, see the fascinating arguments of Jacques Duduict of Blois, responding to a suit for reimbursement of fees by the parents of a young man who died in the third year of his term. E. Develle, *Les horlogers blésois au XVIe et au XVIIe siècle* (3rd ed. 1913; rpt. Nogent-le-Roi: Librairie des Arts et Métiers, 1978), pp. 286–291.

14. On Blois, see ibid., pp. 23–24. On Friedberg, see Maximilian Bobinger, *Kunstuhrmacher in Alt-Augsburg* (Augsburg: Hans Rösler, 1969), p. 29, in the matter of Georg Roll, maverick entrepreneur. In Germany these satellite

suburbs (*faubourgs*) further contributed to commercial activity by admitting Jews to residence, and these undoubtedly played a role in opening the watch and clock trade to nonguild producers; the more so as Jews, even when admitted to residence, were excluded from guild membership.

15. Emile Levasseur, *Histoire des classes ouvrières et de l'industrie en France avant 1789,* 2nd ed., vol. 2 (Paris, 1901), pp. 409–410.

16. The above is based on the account in Bobinger, *Kunstuhrmacher,* pt. 1.

17. The above is taken from Develle, *Les horlogers blésois,* pp. 108–114, 151. In general, Develle is indignant at these breaches of guild law and morality.

18. Ibid., pp. 282–284. Also pp. 365–368, on the career of Abel Bérault.

19. On the London guild, see Samuel E. Atkins and William Henry Overall, *Some Account of the Worshipful Company of Clockmakers of the City of London* ([London]: privately printed, 1881); also G. H. Baillie, *Watches: Their History, Decoration and Mechanism* (London: Methuen, 1929), pp. 244–246, who imputes more power to the company than I do. In principle, the company did control entry; in fact, it could do no more than get interlopers to join.

On Holland, see Spierdijk, *Klokken en Klokkenmakers,* pp. 233–234. Amsterdam had no guild. Other towns incorporated clockmakers along with other metal-working trades and posed tests for admission to master's status. These were less severe than those set in Germany and France and do not seem to have constituted a difficult barrier. At least the literature shows nothing of the contradictions between rules and principle on the one hand and enforcement and practice on the other that betray the presence of barriers to entry.

14. The Ups and Downs of International Competition

1. On Tompion's pivot holes, see Godefroy, "Seconde lettre . . . sur la deffense de l'échappement à Cylindre de M. Graham . . . ," *Mercure de France* (May 1753): 147. For an example of sales by one movement maker to a variety of name "makers," see the list of Richard Wright's customers in 1713, including such leaders of the profession as Windmills, Quare & Horseman, and John Ellicott. Leonard Weiss, *Watchmaking in England 1760–1820* (London: Robert Hale, 1982), p. 51.

2. A gold quarter repeater with pulse piece, case punchmarked 1702, by Francis Gregg of London, was offered in a sale of November 2, 1981, by UTO of Zurich, lot 299. And a watch with very early pulse piece by Daniel Delander was sold by Sotheby's in London on May 26, 1978, lot 207. The watch is not numbered and probably dates from the turn of the century; the catalogue assigns a date of c. 1710, which is manifestly too late. Compare also a watch by David Lestourgeon, No. 1709, c. 1710–15 (no bell; strike on case); and also one by George Graham, No. 445 of 1714 (pulse piece) and No. 603, punchmarked 1730 (no bell).

3. On comparative sizes and advantages, see Pierre Le Roy (son of Julien), *Etrennes chronométriques* (Paris, 1760), p. 31.

4. There has been a sudden spate of interest in Mudge and his work. See Charles Allix, "Mudge Milestones: Watch Dates," *Antiquarian Horology* 12 (Summer 1981): 627–634; A. F. Cliborne, "Mudge Milestones: Addenda," ibid. 13 (December 1981): 144–145; George Daniels, "Thomas Mudge, the Complete Horologist," ibid. 13 (December 1981): 150–173.

5. Two of these are known. The first, No. 525, has survived as a movement. The second, No. 574, was long unsuspected until the first was discovered and examined. It is complete, in a case which if original anticipates the sunken hinge (as against the book hinge) of fifty years later. (The outer case is not punchmarked and, to my knowledge, has not been tested for gold content to see if it is the 22-carat gold that was compulsory in the mid-eighteenth century.) Since I have seen one other watch of this period with sunken hinge and unmarked outer case, I am prepared to think such precocity authentic; Mudge was that kind of maker. See the article by Richard Good, "Watch by Thomas Mudge, London No. 574 with Perpetual Calendar Mechanism," *Antiquarian Horology* 13 (December 1981): 178–187. The article makes no mention of the special character of the outer case of No. 574, nor does it illustrate it. But see Daniels, "Thomas Mudge," p. 165, fig. 9, which does show it.

6. This is the opinion of Winthrop Edie, consultant to Christie's in New York, who is planning an article on this subject.

7. Pierre Le Roy, *Etrennes chronométriques*, p. 30.

8. Alfred Franklin, *La vie privée d'autrefois: arts et métiers . . .* , vol. 4, *La mesure du temps* (Paris, 1888), p. 146.

9. This was a decree of May 16, 1711. Jaques Savary des Bruslons, *Dictionnaire universel du commerce . . .* , vol. 3 (Copenhagen, 1761), col. 334.

10. Ibid., cols. 334, 993.

11. M. Béliard, *Réflexions sur l'horlogerie en général et sur les horlogers du Roi en particulier* (The Hague and Paris, 1767), pp. 2–4. Béliard signs himself *valet de chambre–horloger du Roi*. I owe my copy of this brochure, which Baillie mentions but never was able to find, to Jean-Claude Sabrier.

12. Marie-Antoinette had fifty-one watches in her basket, plus fifty-two snuff boxes (here Paris was unbeatable), nine small bottles, eleven cases, thirteen pencil cases, one writing pad—all of these in gold with enamel and other decorations. Franklin, *La vie privée*, vol. 4, p. 160 and n. 3.

13. See the typescript history by Charles Constantin, "Annales de la Maison d'horlogerie Vacheron & Constantin" (Geneva, 1927–28), p. 9, regarding shipment of goods to Paris in 1813 on an approval basis. (I owe my access to this material to an introduction by François Jequier and to the kindness of Georges Ketterer, president–director general of this oldest of Switzerland's watch firms, who has a rare comprehension of and sympathy for the requirements of historical research.) Cf. George Daniels, *The Art of Breguet* (London and New York: Sotheby Parke Bernet, 1975)—the "Bible" on the subject—p. 5: "He maintained the most cordial business connections with the Swiss makers who continued to supply *ébauches* and complete commercial watches throughout his career." For more on Breguet's Swiss connections (including Geneva) and the interruptions posed by French Revolutionary pol-

itics, see Alfred Chapuis, *A.-L. Breguet pendant la Révolution française à Paris, en Angleterre et en Suisse* (Neuchâtel: Editions du Griffon, 1953), esp. pp. 26–37.

14. Phyllis Deane, *The First Industrial Revolution* (Cambridge: Cambridge University Press, 1965), p. 72.

15. This apparently perverse refusal to respond to pecuniary incentives gave rise to an otherwise strange preference on the part of economists and employers for high food prices, which might otherwise be expected to lead to higher wages. But only if the worker had to pay more to eat, would he work longer. Hence a paradoxically and painfully inverse relationship between food prices and wages: in time of scarcity, the supply of labor increased and wages fell. On all this, see David S. Landes, *The Unbound Prometheus: Technological Change and Industrial Development in Western Europe from 1750 to the Present* (Cambridge: Cambridge University Press, 1969), pp. 59–60, and the sources cited therein. Also Maurice Aymard, "Autoconsommation et marchés: Chayanov, Labrousse ou Le Roy Ladurie?" *Annales: économies, sociétés, civilisations* (forthcoming, 1983). On comparable reactions in Italy in the nineteenth century, see Carlo Cattaneo, "Le condizioni della Bassa Lombarda" (1850), in Alberto Bertolino, ed., *Scritti economici* (Florence: Le Monnier, 1956), pp. 160, 165–166. I owe this reference to Maurice Aymard.

16. This story is taken from a lecture of the late and much-missed T. S. Ashton, professor of economic history at the London School of Economics and pioneer in the study of the Industrial Revolution.

17. On all this, see the now classic article of E. P. Thompson, "Time, Work-discipline, and Industrial Capitalism," *Past and Present* 38 (December 1967): 56–97.

18. On British output, see ibid., p. 66; Weiss, *Watchmaking in England*, pp. 45–46, 49; and the too-little-known article of Chris Ellmers, "The Impact of the 1797 Tax on Clocks and Watches on the London Trade," in Joanna Bird, Hugh Chapman, and John Clark, eds., *Collectanea Londiniensia: Studies in London Archaeology and History Presented to Ralph Merrifield* (n.p.: London and Middlesex Archaeological Society, 1978), p. 388. All of these estimates are based on the assay office data published in the "Report from the Select Committee Appointed to Consider the Laws Relating to Watchmakers," in Great Britain, *Parliamentary Papers,* 1818, vol. 9, pp. 12–13. These give the numbers or weight of watch cases punchmarked each year back to 1796. The trouble is that not every such case was destined to hold a British-made movement, while on the other hand a large number of British movements went into unrecorded brass cases. The figure given in the text represents an effort to take these biases into account.

19. On European output, see Boissy d'Anglas, *Rapport et projet de décret, présentés au nom du Comité de Salut Public sur la manufacture d'horlogerie de Besançon,* 5 nivôse an III, to the Convention Nationale in Paris, p. 3, which estimates the total for Europe, including Britain, at between 350,000 and 400,000 pieces a year. (I owe this reference to Giuseppe Brusa.) Boissy does not make clear in his text whether he means to include Britain, but his printed report is based on a manuscript document (in France, Archives Nationales, F^{12} 1325B), "Observations générales sur la manufacture d'horlogerie, en

réponse à la lettre du Comité de Salut public du 22 Frimaire an 3ᵉ de la République," that gives (p. 4) these figures as comprising British watch output.

20. Adam Smith, *The Wealth of Nations* (London, 1776; New York: Modern Library, 1937), bk. 1, ch. 11: "Effects of the Progress of Improvement upon the Real Price of Manufactures," p. 243.

21. See F. A. Bailey and T. C. Barker, "The Seventeenth Century Origins of Watchmaking in South-west Lancashire," in John R. Harris, ed., *Liverpool and Merseyside: Essays in the Economic and Social History of the Port and Its Hinterland* (London: Frank Cass, 1969), p. 2; Weiss, *Watchmaking in England,* pp. 46, 64.

22. Robert W. Symonds, *Thomas Tompion: His Life and Work* (London: Batsford, 1951), p. 238. Tompion was the first maker to number his pieces serially—almost surely with a view to keeping track of the flow of components in conditions of dispersed batch production.

23. J. Aikin, *A Description of the Country from Thirty to Forty Miles Round Manchester* (London, 1795), p. 311.

24. On Britain's industrial contribution to horology: P. Le Roy, *Etrennes chronométriques,* pp. 29–30. On Huntsman and crucible steel: Landes, *The Unbound Prometheus,* p. 253. On the manufacture of watch files: T. S. Ashton, *An Eighteenth Century Industrialist: Peter Stubs of Warrington* (Manchester: Manchester University Press, 1939). On the critical dependency of continental watchmaking on British steel and files: [Alfred Pfister], *L'industrie horlogère dans le vallon de Saint-Imier* (St-Imier: Imprimerie Georges Favre, n.d. [1944]), p. 14; also France, Archives Nationales, F¹² 1325A, regarding the proposal of the English expatriate William Blakey (alias Blackey) to manufacture pinions by drawing cast steel wire. The memoir gives the date 1744, but this is surely an error. See also Blakey's *L'art de faire les ressorts de montres suivi de la manière de faire les petits ressorts de répétitions et les ressorts spiraux* (Amsterdam, 1780). In France it was an English émigré named James Jackson who introduced the manufacture of crucible steel during the Napoleonic period. See [W. F. Jackson], *James Jackson et ses fils* (Paris: privately printed, 1893).

25. See Thomas Tompion, No. 1256, a splendid half-quarter gold repeater sold by Sotheby's on December 1, 1978, as lot no. 50 of the Edward Hornby collection. This number lies outside the normal run for Tompion repeaters, and the watch is not signed under the cock. But it is much too good a piece to be a forgery: anyone who can make such a watch does not have to pretend to someone else's identity.

26. On Swiss forgeries, see T. P. Camerer Cuss, *The Camerer Cuss Book of Antique Watches,* rev. and enl. by T. A. Camerer Cuss (Woodbridge, Suffolk: Antique Collectors' Club and Baron Publishing, 1976), p. 22. He points out that watches copied from English and Dutch designs were already available in Switzerland at the beginning of the eighteenth century; and he concludes: "The ability of the Swiss to imitate and the willingness to sign a watch with whatever name was required renders it difficult to determine the qualities of both eighteenth and nineteenth century watches which were in fact produced in Switzerland."

27. Letter from J. Berthoud in Paris to Berthoud Père et Fils in Fleurier, October 12, 1789, in Henri de Surirey de Saint-Rémy, "Les événements de 1789 vus par des horlogers suisses," *Annales historiques de la Révolution française,* 34 (January–March 1962): 81–82. I owe this source to Patrice Higonnet.

28. Ellmers, "The Impact of the 1797 Tax," p. 390.

29. Ibid.; also G. H. Baillie, *Clocks and Watches: An Historical Bibliography,* vol. 1 (London: Holland Press, 1978; orig. pub. 1951), p. 329; and John Bromley, ed., *The Clockmakers' Library: The Catalogue of the Books and Manuscripts in the Library of the Worshipful Company of Clockmakers* (London: Sotheby Parke Bernet, 1977), p. 88. Note that when the British refer in this period to France and French watches, they usually mean Geneva and Geneva watches.

15. Multum in Parvo

1. A little chronology: Geneva remained an independent republic until 1798, when it was annexed by France and incorporated into the French *département* of Léman (not to be confused with the Swiss canton of Léman, formerly the Vaud, in the newly constituted Helvetic Republic). On the fall of Napoleon in 1814 the city recovered its independence; and then finally, in 1815, it was admitted to the Swiss Confederation.

2. The most astonishing record, however, was that of the Bernoullis, refugees from Antwerp who settled at Basel and produced the greatest constellation of mathematicians ever seen in a single family. See *Encyclopedia Britannica,* 11th ed., s.v. "Bernoulli."

3. On Cusin, see Anatole de Charmasse, "L'horlogerie et une famille d'horlogers à Autun et à Genève au 16ᵉ et 17ᵉ siècles," *Mémoires de la Société Eduenne* (Autun), n.s., vol. 16 (1888), pp. 175–213, which refers back to Thury's *Notice historique sur l'horlogerie suisse,* a small pamphlet written for the benefit of visitors to the World's Fair of 1878. Thury's phrasing (p. 7) is that Cusin introduced the manufacture of watches on a "lasting" basis.

4. The first French refugee watchmaker, according to the best authority, was Thomas Bayard of Vézelizes in Lorraine, who came to Geneva in 1550. On this whole question, see Antony Babel, *Les métiers dans l'ancienne Genève: histoire corporative de l'horlogerie, de l'orfèvrerie et des industries annexes,* Mémoires et documents publiés par la Société d'histoire et d'archéologie de Genève, vol. 33 (Geneva: Imprimerie A. Kündig, 1916), pp. 42–56.

5. The above is based on the account in Eugène Jaquet and Alfred Chapuis, *Technique and History of the Swiss Watch,* 2nd ed. (London: Spring Books, 1970), pp. 27–28.

6. Neuchâtel was not part of Switzerland yet. In 1707 it voted to take as prince the king of Prussia, who was related by his mother to the house of Chalon and thus heir to rights given in 1288 (that is, over four hundred years earlier) on the extinction of the line of the counts of Neuchâtel—an exceptional instance of the uses of genealogy. Frederick had two great advantages over the other claimants to the principality: he was Protestant, and so was the population of Neuchâtel; and he lived far away. Neuchâtel was essentially

independent in everything but constitution. It became a member of the Swiss Confederation in 1815, the only nonrepublican constituent, and remedied that by throwing the Prussians out in 1848.

7. Jaques Savary des Bruslons, *Dictionnaire universel du commerce,* new ed., vol. 1 (Paris, 1741), p. 236.

8. On Swiss and Genevan horological expatriates, see Babel, *Les métiers,* pp. 519–529; and idem, "L'horlogerie genevoise à Constantinople et dans le Levant du XVIe au XVIIIe siècle," *Etrennes genevoises* (1927): pp. 61–74. See also O. Kurz, *European Clocks and Watches in the Near East,* Studies of the Warburg Institute, no. 34 (London: The Warburg Institute; Leiden: Brill, 1975), pp. 54–56. For Pastor Léger, see "Genève et Constantinople, 1592–1732," *Bulletin de la Société de l'Histoire du Protestantisme Français* 10 (1861): 235.

9. On Sermand, see Eugène Jaquet, *Horlogers genevois du XVIIe siècle,* offprint from *Bulletin de l'Institut National Genevois,* vol. 2 (Geneva: Imprimerie Centrale, A. & G. Villard, 1938), pp. 31–32. On Rousseau, see Babel, *Les métiers,* pp. 527–528.

10. Jaquet, *Horlogers genevois,* p. 23.

11. On hands and fine metalwork, see Robert Pinot, *Paysans et horlogers jurassiens* (Geneva: Grounauer, 1979), pp. 204–205.

12. France, Archives Nationales, F^{12} 1325B: "Notes sur l'impôt et la contrebande de l'horlogerie," unsigned, undated. The document was prepared around 1790 in response to a memorandum from the Genevan minister to France arguing against the imposition of a higher duty on watch imports. The reporter, presumably a French official, was scornful of, even outraged by, the Genevan's arguments. The document bears the marginal note: *"Essentielles."*

The growth of the watch industry in Savoy remains to be studied. It developed earlier than has been thought and was already well established in 1749, when the workers of Cluses asked Turin for permission to establish a *maîtrise,* in effect a guild. They were turned down because the provincial authorities feared, probably with good reason, that the introduction of regulations would hurt business with Geneva. There is information on some of this in Jean Nicolas, *La Savoie au XVIIIe siècle: noblesse et bourgeoisie,* 2 vols. (Paris: Maloine, distrib. Librairie de la Nouvelle Faculté, 1978); and there is unexplored material in the Italian archives and the departmental archives of the Haute Savoie. See, for example, a memorandum of November 16, 1790, "Relation du Bureau de l'Intendance de la Province de Faucigny, sur une supplique présentée par divers particuliers de la ville de Cluses pour obtenir la division des communaux de cette paroisse, par chefs de famille," in Archives Départementales, Hte-Savoie, IV C 133, which treats of the doubling of the town population due to immigration by watch workers and the consequences for communal harmony in a time of business contraction. I owe this and other bibliographical helps to Jean Nicolas.

13. Babel, *Les métiers,* pp. 423, 487. On comparative wages, ibid., pp. 476ff. On country houses, Jaquet, *Horlogers genevois,* p. 30.

14. Antony Babel, *La fabrique genevoise* (Neuchâtel: Victor Attinger, n.d. [1938]), pp. 48–51.

15. On the early history of the Besançon manufacture, see, among others, Charles Savoye, *Notice sur l'origine de la fabrique d'horlogerie de Besançon* (Paris, 1878); Charles Sandoz, *Origine et développement de l'industrie horlogère à Besançon et en Franche Comté* (Besançon, 1893); E. Lebon, *Etudes historiques, morales et statistiques sur l'horlogerie en Franche-Comté* (Besançon, 1860); and Georges Megnin, *Naissance, développement et situation actuelle de l'industrie horlogère à Besançon* (Besançon, 1909). One of the most useful treatments has never been published: Hélène Galliot, "Le métier d'horloger en Franche-Comté, des origines à 1900" (thèse de Droit, University of Paris, 1954).

16. Babel, *La fabrique genevoise,* p. 54.

17. François Jequier, *Une entreprise horlogère du Val de Travers: Fleurier Watch Co SA* (Neuchâtel: La Baconnière, 1972), p. 50.

18. Isaac Cornuaud, *Aurons-nous du pain et du travail de reste l'hiver prochain?* (Geneva, pamphlet dated July 11, 1795), pp. 11–12.

19. Babel, *Les métiers,* p. 519.

20. Ibid., pp. 512–513.

21. Ibid., p. 514.

22. Cornuaud, *Aurons-nous du pain,* p. 10.

23. Emilie Cherbuliez, ed., *Mémoires d'Isaac Cornuaud sur Genève et la Révolution de 1770 à 1795* (Geneva: A. Jullien, 1912), p. 94.

24. This in 1761 at a meeting of watchmakers and merchants to which sundry officers were invited. Babel, *Les métiers,* pp. 510–511.

25. Alfred Chapuis, *Relations de l'horlogerie suisse avec la Chine: la montre "chinoise"* (Neuchâtel: Attinger Frères, n.d.), pp. 57, 55.

26. Traveler for Bautte, probably the leading watch firm in Geneva in the first half of the nineteenth century.

27. J.-B. Vacheron in Rome to his partner Charles François in Geneva, October 1818. Charles Constantin, "Annales de la maison d'horlogerie Vacheron & Constantin" (typescript; Geneva, 1928), pp. 22–23.

28. Ibid., p. 23.

29. Letter of J.-B. Vacheron, Geneva, April 11, 1820, to François Constantin, on the road in Italy. The firm that suffered this disaster was Térond, Allier & Bachelard of Geneva.

30. We do have an estimate of c. 1783–84 which speaks of six thousand *blantiers* (makers of *ébauches*) in the *fabrique* as a whole, including the Faucigny (Savoy) and the Pays de Gex; most of these presumably lived and worked outside Geneva; also of two thousand *cadraturiers* (makers of under-dial repeater and calendar mechanisms) working for Geneva in Switzerland proper and in Neuchâtel. But these estimates are to be found in a document designed to interest the French government in a watch venture, and hence are given to exaggeration. France, Archives Nationales, F^{12} 1325A, Mémoire, undated, by Jacques Glaesner and Louis-Baptiste Preudhomme. Note that Babel, *Les métiers,* pp. 398–399, cites the data in the Glaesner-Preudhomme memoir, but at second hand—as communicated by M. Fallet-Scheurer. I have not found the figure Babel gives, based on this material, of twenty thousand people supported by the Geneva industry, but this may be some kind of demographic

multiple of the workers employed. Besides, there is no way of knowing that the contents of carton F^{12} 1325A are the same today as seventy years ago. Compare the results of a census of the horological industry in the district of Cluses (hence within the district designated above as Savoy) in September 1794. This showed less than seven hundred workers, a figure reflecting not only a sharp contraction of production (these were hard times for the Geneva watch industry), but also perhaps a failure to include many isolated and part-time workers in the peak season of agricultural activity. Archives Nationales, F^{12} 1325B, large double-folio cardboard.

31. This is the same Glaesner-Preudhomme memorandum of 1783–84. Glaesner, a master watchmaker of Lyons with dreams of large-scale factory enterprise, was a ready reckoner whose assumptions and methods of calculation would gladden the heart of a New Economic Historian. In a later memoir, prepared again with the hope of interesting the French government in his plans, he calculated annual import of watches from Geneva and Neuchâtel at 300,000. How did he know this, since a large part of the traffic was contraband? He assumed that almost all Frenchmen owned watches. Since the population was 25 million, a conservative estimate of the stock of watches was 12 million. He then estimated the average life span of a watch at forty years. This called, then, for replacement of 300,000 units a year, to say nothing of new acquisitions. Since just about all these watches or movements came from Geneva and Neuchâtel, Q.E.D.—conservatively speaking. Archives Nationales, F^{12} 1325B "Mémoire sur la fabrication étrangère de l'horlogerie et sur les moyens de la naturaliser en France, en la perfectionnant," by Jacques Glaesner to the Committee of Public Safety, Paris, 25 nivôse an III.

32. Babel, *Les métiers*, p. 407; Edouard Chapuisat, *Le commerce et l'industrie à Genève pendant la domination française* (Geneva and Paris, 1908), p. 217. Sismondi's estimates accord with a contemporaneous comparison of the two periods by the mayor of Geneva in a letter to the prefect of the *département* of the Léman.

33. Babel, *La fabrique genevoise*, p. 72.

34. Ibid., pp. 65–67.

35. Ibid., p. 62.

16. Notwithstanding the Barrenness of the Soil

1. On conditions of everyday life and business in the Haut-Jura in the eighteenth century, see Jean-Laurent Wurflein, "Coup-d'oeil sur les moeurs de La Chaux-de-Fonds, au XVIIIe siècle," *Entrennes neuchâteloises* 3 (1864): 96–123. This is based on family papers of the period.

2. On the preconditions of successful cottage industry in a rural environment, see especially the long-forgotten anthropological report by Robert Pinot, *Paysans et horlogers jurassiens* (Geneva: Grounauer, 1979). This was originally published as "Monographie du Jura bernois," in *La Science sociale* (Paris, 1887–1889). Pinot was a follower of the French sociologist-ethnologist Le

Play, and, following the methods of the master, had spent considerable time in the field living with and observing both pastoral and industrial households. Ironically, Pinot, who began thus as a scholar in mountain cottages, finished his career as president of the Comité des Forges, the employers' association of the French iron and steel industry.

3. In addition to the sources already cited, see especially the important thesis of Suzanne Daveau, *Les régions frontalières de la montagne jurassienne: étude de géographie humaine* (Lyons: Université de Lyons; Trévoux: Imprimerie de Trévoux, J. Patissier, 1959); and François Jequier, "L'horlogerie du Jura: évolution des rapports de deux industries frontalières des origines au début du XIXe siècle," in Louis-Edouard Roulet, ed., *Frontières et contacts de civilisation* (Neuchâtel: La Baconnière, 1979), pp. 159–176. This is a volume of papers presented at a Colloque Universitaire Franco-Suisse held at Besançon and Neuchâtel in October 1977.

4. The latter outcome is not probable. In Protestant areas the upland Swiss reacted to population pressure by a wide variety of expedients, in particular by an energetic (enthusiastic?) recourse to home industry as a supplementary and principal source of income. See the classic study of Rudolf Braun, *Industrialisierung und Volksleben*, 2 vols. (Erlenbach-Zurich and Stuttgart: E. Rentsch, 1960 and 1965) on the textile manufacture in the Zurich highlands. Or listen to the remarks of the traveler-essayist Charles-Victor von Bonstetten, writing of the straw-plaiting industry in the Freiburg area: "The women plait when they stand, when they walk, when they sit; they plait in the courtroom, when they stand before the high bailiff (*Landvogt*); they must go to sleep plaiting." Cited in Albert Hauser, *Schweizerische Wirtschafts- und Sozialgeschichte* (Erlenbach-Zurich and Stuttgart: E. Rentsch, 1961), p. 153.

5. On JeanRichard and the early history of clock- and watchmaking in the Jura, see, among others, A. Bachelin, *L'horlogerie neuchâteloise* (Neuchâtel: Attinger Frères, 1888), ch. 1; and Marius Fallet, "Le rayonnement séculaire de l'horlogerie suisse: une synthèse historique," in Alfred Chapuis, ed., *L'horlogerie, une tradition helvétique* (Neuchâtel: Editions de la Bourgade, 1948), pp. 9–62. There is an abundant iconographic literature on JeanRichard, who plays in Swiss horological history something of the role that Peter Henlein (the alleged inventor of the watch) does in German. But we know a lot more about JeanRichard and even have one or two of his watches.

6. Ibid., p. 55.

7. Some writers have made the Breguets out to be French by origin, fled to Neuchâtel from persecution at home. See E. Ferret, *Les Breguet*, 2nd ed. (Paris, n.d.), p. 11. In fact they are citizens of Neuchâtel as far back as one can trace the family. Alfred Chapuis, *Histoire de la pendulerie neuchâteloise* (Paris and Neuchâtel, n.d. [1917]), p. 457.

8. When Breguet changed his name is not known. His baptismal certificate gives him as Abram-Louis. Later French records give Abraham-Louis. As a Neuchâtelois Calvinist, he certainly learned as a child the story of his namesake.

9. The present chief *chez* Piguet is the son and grandson of specialists in

complicated, custom-ordered mechanisms—the kind of artists who could make anything. The shop uses electric power and modern tools; but much of the work still has to be done by hand, and the shop would not seem unduly strange to a *cadraturier* of the eighteenth century. M. Piguet is now in his sixties, and his son, who showed a remarkably fine hand at the age of five (he could file flat even then) has chosen to devote his talents to medicine and surgery. With solid-state watches able to provide all these complications and indications for one thousandth of the cost, such work is more or less a thing of the past—in spite of an occasional *tour de force* on special order. For an example of a Piguet *chef d'œuvre*, see the fantastically complicated watch made for Patek in 1932 and illustrated in Eugène Jaquet and Alfred Chapuis, *Technique and History of the Swiss Watch*, 2nd ed. (London: Hamlyn, 1970), pl. 128. The watch is now to be found in the Time Museum in Rockford, Illinois.

10. *Les Ebauches: ouvrage publié à l'occasion du 25ᵉ anniversaire d'Ebauches S.A.* (Neuchâtel: Editions de la Baconnière, n.d. [1951]), p. 30.

11. On the early history of the mass production of *ébauches*, see especially M. Fallet-Scheurer, *Le travail à domicile dans l'horlogerie suisse et ses industries annexes* (Bern: Imprimerie de l'Union, 1912), pp. 262–265. For the Swiss version of the Japy-Jeanneret transaction, see Alfred Chapuis, *Grands artisans de la chronométrie: histoire de l'horlogerie au Locle* (Neuchâtel: Editions du Griffon, 1958), pp. 61–62; for the French version, as well as the estimate of output in 1780, see François Girod, "Un précurseur: Frédéric Japy (1749–1812)," *Association Nationale des Collectionneurs et Amateurs d'Horlogerie Ancienne*, no. 14 (May 1975): p. 37. The trouble with the story that credits Japy with the invention of these machines is that Japy could not read or write (and apparently remained illiterate long after he achieved industrial success). Ibid., p. 33, n.1. One wonders how he ran so large and complex an enterprise.

12. This is a translation of the passage in A. Pfleghart, *Die schweizerischen Uhrenindustrie: Ihre geschichtliche Entwicklung und Organisation* (Leipzig: Duncker & Humblot, 1908), pp. 38–39. Since that is itself a translation from the French, the above may diverge somewhat from the original. But the sense is there.

13. Ibid., pp. 16–17.

14. See his letter of April 26, 1675, to his brother Constantine, cited in L. Defossez, *Les savants du XVIIᵉ siècle et la mesure du temps* (Lausanne: Editions du Journal Suisse d'Horlogerie et de Bijouterie, 1946), p. 195.

15. The push of French watchmakers in the direction of thinness, even at the expense of quality, may also reflect a reaction against the extraordinary cumbersomeness of the French onion watch during the period 1675–1725. On this exchange of accuracy for style, see the article by Pierre Le Roy entitled "Lettre à Monsieur N. de L'Académie des Sciences de Bordeaux . . . ," *Mercure de France* (June 1752): 141–149; also idem, "Lettre à M. N*** de L'Académie des Sciences de Bordeaux sur la construction d'une montre . . . ," *Mémoires pour l'histoire et des beaux-arts* (Journal de Trévoux), (June 1752): 1300–14.

See also the boast of Pierre-Augustin Caron (better known later as Beaumarchais), inventor of a new escapement: "By means of the double-virgule escapement I can make watches as thin as anyone could want [*aussi plates qu'on le juge à propos*]; flatter than anyone has yet made, without sacrificing quality." *Mercure de France,* June 1755, cited by Elie-François Wartmann, "Notice historique sur les inventions et les perfectionnements faits à Genève dans le champ de l'industrie et dans celui de la médecine," in Geneva, Société des Arts, *Bulletin de la Classe d'Industrie et de Commerce* 101 (1871): 38, n. 1. This invention by Caron had no direct consequences, partly because the double virgule was a difficult escapement to make and maintain in use, partly because Caron gave up watchmaking for the life of courtier, business promoter, and playwright.

16. The best source is Adolphe Chapiro, "Jean-Antoine Lépine, 1720–1814, an 'Unknown' Maker," *Antiquarian Horology* 9 (September 1975): 443–454. If the fusee is eliminated, the mainspring drives the wheel train directly by turning the barrel in which it is housed, which is geared into the second wheel. Hence the name "going barrel."

17. Beatrix Cary Davenport, ed., *A Diary of the French Revolution by Gouverneur Morris,* vol. 1 (Boston: Houghton Mifflin, 1939), pp. xxxiii–xxxvi. I owe this delicious reference to Patrice Higonnet. Since reading it, I have come across another discussion of this exchange: James W. Gibbs, "George Washington's Watch," *Bulletin of the National Association of Watch and Clock Collectors* 15 (December 1971): 12–13. The watch is said to be on display at the Historical Society of Pennsylvania.

18. This escapement (Thomas Tyrer's patent) was invented in its first form by Jean-Baptiste Dutertre in Paris (1720s); and another Frenchman, the aforementioned Pierre Le Roy, improved it in 1750. But both these versions were experimental and the French did nothing with the duplex until they borrowed it back from the English in the early nineteenth century. Even then, they used it rarely and then usually in the two-wheel version, rather than the one-wheel variety favored by the English.

19. The earliest example of the rack lever known to us is—once again!— in the form of a design by a French maker, the Abbé Jean de Hautefeuille, in 1722. It is not clear that Hautefeuille's invention ever found expression in an actual watch.

20. Pfleghart, *Die schweizerischen Uhrenindustrie,* p. 41. The cylinder escape wheel was not easy to make by hand, which is one reason why French makers took it up so late. An important improvement was the introduction of scape wheels made of steel instead of brass, which diminished friction and lengthened the life of the cylinder. This innovation is usually attributed to Urban Jürgensen, the great Danish watchmaker and pupil of Breguet, in the first decade of the nineteenth century; in fact, it was well known and used in England from the 1770s on. David Penney has shown me at least half a dozen such watches or movements clearly anterior to Jürgensen's career.

21. Jaquet and Chapuis, *Technique and History,* p. 15, and plates 99 and 100.

22. Pfleghart, *Die schweizerischen Uhrenindustrie,* p. 50; Gaston Rub and Alfred Chapuis, *L'horlogerie au Val-de-Travers de ses origines à nos jours,* Collection

"Mon Vallon," no. 5 (Fleurier: Montandon, 1937), pp. 201–205.

23. The best source, with copious illustrations, is Alfred Chapuis, *Relations de l'horlogerie suisse avec la Chine: la montre "chinoise"* (Neuchâtel: Attinger Frères, n.d.). See also François Jequier, *Une entreprise horlogère du Val-de-Travers: Fleurier Watch Co SA* (Neuchâtel: La Baconnière, 1972), chs. 1 and 2.

24. One of the best selections of these astonishing timepieces, which generally kept very poor time, is to be found in *Collection de montres et automates Maurice et Edouard M. Sandoz*, in *Histoire de l'horlogerie*, fasc. 3 (Le Locle: Musée d'Horlogerie, 1976). See also Oswaldo Patrizzi and Fabienne-X. Sturm, *Montres de fantaisie, 1790–1850* (Geneva: Tribune Editions, 1979); and Alfred Chapuis, *Montres et émaux de Genève: Louis XIV, Louis XV, Louis XVI et Empire* (Geneva: Rolex, 1944). This last is a *catalogue raisonné* of the Wilsdorf collection.

25. On the history of the erotic watch, R. Carrera, *Les heures de l'amour* (Lausanne: Scriptar, n.d.). On the larger history of automata (but little or nothing on erotic automata), see Alfred Chapuis and Edouard Gélis, *Le monde des automates: étude historique et technique,* 2 vols. (Paris: privately printed, 1928), esp. vol. 2, ch. 17; and Alfred Chapuis and Edmond Droz, *Les Automates: figures artificielles d'hommes et d'animaux—histoire et technique* (Neuchâtel: Editions du Griffon, n.d.). Also *A Loan Exhibition of Antique Automatons . . . November 3–December 5, 1950* (New York: A La Vieille Russie, Inc., n.d.).

26. On this question of jeweling, see Henri Robert, *L'art de connaître les pendules et les montres* (Paris, 1849), pp. 216–218; and idem, *Etudes sur diverses questions d'horlogerie* (Paris, 1852), pp. 199–200.

27. The only thing comparable in Britain was the Liverpool watch, with its large stones, usually chrysoberyl, the so-called Liverpool windows. The Liverpool watches were long a success, but in the end their thickness (they used a full-plate layout) limited them to the old-fashioned side of the market, which has a way of shrinking over time.

28. Daveau, *Les régions frontalières,* p. 423.

29. Ibid., p. 426.

30. From a report of 1798–99 (year VI), cited ibid., p. 427. This would seem to be an exaggeration. As of 1830, according to Pfleghart, *Die schweizerischen Uhrenindustrie,* p. 50, there were some fifty *parties brisées* that went into the making of a watch. A generation earlier, there were surely fewer. But some of these steps and parts may have occupied more than one person.

31. This is no small matter in the watch trade. Paul Rudolph, *chronométrier* to the French navy, who worked for many years for Le Roy in Paris, tells me that it was the custom of the firm to try all new men by having them handle some watch parts, which were then sealed in an envelope and examined after a week. Rudolph's own hands (he showed me) are as dry as parchment.

32. Daveau, *Les régions frontalières,* p. 427.

33. [Samuel Frédéric (alias Samuel Friedrich) Osterwald (also Ostervald)], *Description des montagnes et des vallées qui font partie de la Principauté de Neuchâtel et Valangin,* 2nd ed. (Neuchâtel, 1766), pp. 94–95. This small book ap-

peared at first as an article in the *Journal Helvétique* of 1764.

34. John Bowring, "Report on the Commerce and Manufactures of Switzerland," in Great Britain, *Parliamentary Papers,* 1836, vol. 45, p. 36. This is Bowring's conversion of the Houriet memorandum to indirect discourse.

17. Nor Could He Compete with Us

1. These quotations are from testimony to the House of Commons on February 21–23, 1798, as cited in Chris Ellmers, "The Impact of the 1797 Tax on Clocks and Watches on the London Trade," in Joanna Bird, Hugh Chapman, and John Clark, eds., *Collectanea Londiniensia: Studies in London Archaeology and History Presented to Ralph Merrifield* (London: London and Middlesex Archaeological Society, 1978), p. 396. For Pitt's remark, made on the occasion of the introduction of income tax, see Adrian Burchall, "Pitt's Clock and Watch Tax: A Reassessment," *Antiquarian Horology* 13 (September 1982): 459. Burchall, in contradiction of the conventional wisdom on the subject, argues that the tax had little impact on the trade; that such drop as there was, was due to tight money and short supply of gold; and that the statements of clock- and watchmakers have to be discounted as interested. He notes that sales of gold watches fell proportionately more than those of silver—by about a half, as against a third, from 1796 to 1798—which, he contends, should not have been the case had the sole cause of difficulty been the new tax. This may well be true, but it does not demonstrate that the tax was not part of a conjuncture of forces pressing on demand. Indeed, it is not logical that it should not have had an adverse effect on the market for new timepieces.

2. The Clockmakers' Company appointed a committee in 1814 to look into the "illicit introduction of foreign clocks and watches." See the minute book, Clockmakers' Co. Library, MS 2728. Also documents concerning the marking and marketing of watch cases, 1810–1812, ibid., MS 3940, pp. 88–134. Most important was a Commons inquiry of 1817: "Report from the Committee on the Petitions of Watchmakers of Coventry, &c.," *Parliamentary Papers,* 1817, vol. 6 (no. 504); and "Report from the Select Committee Appointed to Consider of the Laws Relating to Watchmakers," ibid., 1818, vol. 9 (no. 135). The data on watch cases presented for assay are in the latter document, p. 12, app. 2.

3. *Parliamentary Papers,* 1817, vol. 6 (no. 504), p. 67 (testimony of Henry Clarke).

4. Ibid., pp. 63–64, testimony of Paul Philip Barraud and letter of April 21, 1817 to Viscount Sidmouth from Barraud, Robson, Arnold, Vulliamy, and Dutton on behalf of the Clockmakers' Company; and p. 30, testimony of Benjamin Pyke.

5. Ibid., pp. 41, 59. A duty of 27.5 percent *ad valorem* had been imposed in 1787 and increased in steps thereafter to 49–50 percent.

6. Thomas Schelling, "On the Ecology of Micromotives," *The Public Interest* 25 (1971): 61–98.

7. *Parliamentary Papers*, 1817, vol. 6 (no. 504), p. 66, testimony of Joseph Glenny of Clerkenwell, London. Glenny attributed such practices to Liverpool and Coventry makers and complained that this gave these unscrupulous makers an advantage over those in London.

8. Ibid., pp. 28, 31.

9. Ibid., p. 28.

10. Proclamation of July 4, 1817, cited ibid., pp. 31–32.

11. Ibid., p. 78, testimony of William Mayo.

12. Edward J. Dent, *An Abstract from Two Lectures on the Construction and Management of Chronometers, Watches and Clocks* (London, 1842), pp. 6–7.

13. On the history of the Jürgensen firm: Charles Kalish, "La dynastie des Jürgensen: trois siècles d'horlogerie en deux pays," *La Suisse horlogère et Revue internationale d'horlogerie*, 1963, no. 2, pp. 45–56; no. 3, pp. 37–48. See also Le Locle, Musée d'Horlogerie, *Les Jürgensen, une dynastie de grands horlogers*, "Histoire de l'horlogerie," fasc. 2 (Le Locle: Editions du Château des Monts, n.d. [1974]).

14. British *ébauches* were, if anything, even less standardized. On the inadequacies of "mass-produced" rough movements from Lancashire, see a talk by a Mr. Bickley on "The Present System of the English Watch Manufacture," *Horological Journal* 22 (1879–1880): 128.

15. One of the best discussions is Robert A. Howard (engineering curator at the Hagley Museum), "Interchangeable Parts Reexamined: The Private Sector of the American Arms Industry on the Eve of the Civil War," *Technology and Culture* 19, no. 4 (1978): 633–649; and idem, "Interchangeable Parts Revisited," ibid. 21, no. 3 (1980): 549–550. Howard makes much of the differences in cost constraints, quality standards, and complexity between military ordnance and civilian arms, which led to interchangeability in the military arsenals long before it was adopted in private factories. Indeed, he notes with some surprise that much gun production today is still not interchangeable, though there is a tendency to design new models with uniformity in mind.

16. The best source on Ingold's life and technical contributions is the article by R. F. and R. W. Carrington, "Pierre Frédéric Ingold and the British Watch and Clockmaking Company," *Antiquarian Horology* 10 (Spring 1978): 698–714.

17. There is a recollection of the circumstances in remarks by G. William Frodsham of Parkinson & Frodsham at a meeting of the Society of Arts in London on May 19, 1886. The remarks are given in indirect discourse: "Reference had been made to an English company which did not succeed in consequence of the opposition of Parliament; but this was, no doubt, owing to the fact that in those days there was a strong prejudice, more particularly on the part of the workmen, against the introduction of the machines designed by Mr. Ingold, so much so that at one time this gentleman went in fear of his life, and no doubt the views entertained by the watchmakers and workmen in-

fluenced members of Parliament, who, of course, had no technical views upon the subject." "The Mechanical Art of American Watchmaking," *Journal of the Society of Arts* 34 (May 21, 1886): 749. For more on the Ingold affair, see the petitions and letters in the library of the Clockmakers' Company, Guildhall, London, MS 3943, pp. 6–16.

18. See the eminently readable account in Derek Howse, *Greenwich Time and the Discovery of the Longitude* (Oxford: Oxford University Press, 1980), ch. 4: "Greenwich Time for Great Britain 1825–1880." The paragraphs that follow draw heavily on this material.

19. Ibid., chs. 5 and 6, esp. p. 153.

20. Ibid., pp. 108–109, 112.

21. See R. A. Church, "Nineteenth-Century Clock Technology in Britain, the United States, and Switzerland," *Economic History Review*, 2nd ser., 33 (1975): 618, 625.

22. Bowring, "Report on the Commerce and Manufactures of Switzerland," in *Parliamentary Papers*, 1836, vol. 45, p. 14. The same was true in 1860: "There are almost no watches made in Paris." *Journal de statistique suisse* 1 (1865): 205, citing the *Statistique de l'industrie à Paris* collected by the Paris Chambre de Commerce in 1860 and published in 1864.

23. These are the figures of gold and silver watches submitted for punchmarks at the assay office at Besançon. It should be remembered that many of these contained a non-French (Swiss) movement; that a certain number of gold and silver watches were never submitted for assay; and that a large number of watches were cased in brass and other base metals, hence not included in this count. See Viviane Isambert-Jamati, *L'industrie horlogère dans la région de Besançon* (Paris: Presses Universitaires de France, 1955), pp. 12–13.

24. Martin Huber, *Die Uhren von A. Lange und Söhne Glashütte/Sachsen* (Munich: Verlag Callwey, 1977); and Kurt Herkner, *Glashütte und seine Uhren* (Dormagen: Herkner, 1978).

25. A. Pfleghart, *Die schweizerischen Uhrenindustrie: Ihre geschichtliche Entwicklung and Organisation* (Leipzig: Duncker & Humblot, 1908), p. 52. This may have been a little boastful: twelve years later, in 1867, the director of the Neuchâtel Observatory thought 800,000 was an upper bound. But one contemporary thought this conservative and advanced the figure of a million watches a year. Laussedat, *Etude sur le développement de l'horlogerie*, reprint of an article in the *Annales du Conservatoire Impérial des Arts et Métiers* (Paris: n.p., n.d. [c. 1868]), p. 10.

26. *Les Ebauches: ouvrage publié à l'occasion du vingt-cinquième anniversaire d'Ebauches S.A.* (Neuchâtel: La Baconnière, n.d.), p. 37.

27. Switzerland, Eidgenössischen Departement des Innern, Statistisches Bureau, *Schweizerische Statistik: Eidgenössische Volkszählung vom 1. December 1870*, vol. 3, *Die Bevölkerung nach den Beschäftigungsarten* (Bern, 1876), pp. 152–153.

28. We have the following estimate of world watch output in the early 1870s (the table is in fact undated and probably underestimates Swiss production for that time:

Country	Number of watches (in thousands)	Value (in thousands of dollars)
Switzerland	1,600	17,600
France	300	3,300
Great Britain	200	3,200
United States	100	1,500
Total	2,200	25,600

Source: Edward Young, *Labor in Europe and America: A Special Report on the Rates of Wages, the Cost of Subsistence, and the Condition of the Working Classes in Great Britain, Germany, France, Belgium, and Other Countries of Europe; also in the United States and British America* (Washington, D.C., 1875), p. 608.

29. Report by Mr. Herries, Bern, August 30, 1857, in Great Britain, *Parliamentary Papers,* 1857–58, vol. 55; *Accounts and Papers* 23: Trade of Various Countries: Switzerland, p. 49.

18. Ah, But He Could!

1. On the history of Swiss chronometric competition, see Eugene Jaquet and Alfred Chapuis, *Technique and History of the Swiss Watch,* rev. ed. (London: Hamlyn, 1970), pp. 173–179, 187–195.

2. The French eventually came to the same conclusion, and their reborn industry in the Besançon area went over to lever-escapement pocket chronometers by the end of the century. On this question, see A. H. Rodanet, "Rapport sur la question: définition du chronomètre—genre d'échappement qu'il comporte," in Exposition Universelle de 1900, Congrès International de Chronométrie, 1900, *Comptes-rendus des travaux,* ed. E. Fichot and P. de Vanssay (Paris: Gauthier-Villars, 1902), pp. 34–39. But as late as 1882, Claude Saunier, dean of French horologists, was writing that a chronometer must have an *échappement libre à détente.* C. Saunier, "Origine et emploi du mot chronomètre: usage abusif moderne," *Association Nationale des Collectionneurs et Amateurs d'Horlogerie Ancienne,* no. 32 (Fall-Winter 1981): 22–23. This is a reprint of the article in the *Revue chronométrique* 12 (1882).

3. See, on the relative performance of different escapements and balance springs, the competition results as published in the *Journal suisse d'horlogerie,* passim, e.g., vol. 2 (1877–1878), p. 204. Also Elie Wartmann, "Rapport sur l'horlogerie suisse à l'Exposition Universelle de Paris, en 1867," in Geneva, Société des Arts, *Bulletin de la Classe d'Industrie et de Commerce,* no. 91 (1868), p. 72, giving results for the years 1862–1866; and an article by Paul Ditisheim, "Classification des échappements," in Congrès International de Chronométrie, 1900, *Comptes-rendus des travaux,* pp. 41–44, for the years 1862–1899.

4. Jaquet and Chapuis, *Technique and History,* p. 157; *Les Ebauches; ouvrage publié à l'occasion du vingt-cinquième anniversaire d'Ebauches S.A.* (Neuchâtel: La Baconnière, n.d. [1951]), pp. 32–34. On the history of horological schools in Switzerland, see Association des Ecoles Suisses d'Horlogerie, *Les écoles suisses d'horlogerie* (Zurich: Fritz Lindner, n.d.).

5. *Les Ebauches,* p. 34.

6. Ibid., pp. 38–41, 48–49.

7. Ibid., p. 40.

8. E. Muston, "L'horlogerie dans les montagnes du Jura: essai de statistique industrielle," *Compte-rendu de la situation et des travaux* [also listed as *Mémoires*] *de la Société d'Emulation de Montbéliard,* May 5, 1859, p. 135.

9. See the account in Elie-François Wartmann, "Notice historique sur les inventions et les perfectionnements faits à Genève dans le champ de l'industrie et dans celui de la médecine," in Geneva, Société des Arts, *Bulletin de la Classe d'Industrie et de Commerce,* no. 101 (1871), pp. 34–35.

10. On Leschot's contribution, see Charles Constantin, "1840 ... Les machines de G. Leschot," *Bulletin annuel de la Société Suisse de Chronométrie* (1940): 27–30; idem, "Annales de la Maison d'horlogerie Vacheron & Constantin" (typescript; Geneva, 1927–28), pp. 98–99, 104–107, 110–112, 118. See also Jaquet and Chapuis, *Technique and History,* p. 158 and pl. III, showing the same layout in eight different sizes, as produced by Leschot's equipment. I have seen some of Leschot's machines, which are presently in the attic of the Vacheron building in Geneva. The firm has plans to restore them to working order.

11. Great Britain, Commissioners for the Exhibition of 1851, *Exhibition of the Works of Industry of All Nations, 1851: Reports by the Juries* (London, 1852), Class Xb: "Report on Horological Instruments," p. 311.

12. Cited in a manuscript history of the Le Coultre firm by François Jequier of the University of Lausanne, soon to be published. I am grateful to Prof. Jequier for allowing me to read this study in advance of publication; to say nothing of all manner of other materials, published and unpublished.

13. Muston, "L'horlogerie dans les montagnes," p. 97. The same reliance on hand manufacture characterized the first movement factory in the Jura, that of Robert at Fontainemelon. The industrial census of 1860 showed it to have 260 employees but no power source (*point de force motrice*). Gustave Moynier, "Statistique des fabriques du canton de Neuchâtel," *Journal de statistique suisse* 1 (1865): 97. It did, however, have a prime mover, an ox gin to start with, and then a hand-turned wheel (*roue d'écureuil mue à bras*). *Les Ebauches,* p. 40. In the beginning at least, much of the work was done at home.

14. Cited in the Jequier manuscript, ch. 4.

15 *Les Ebauches,* p. 50.

16. Suzanne Daveau, *Les régions frontalières de la montagne jurassienne: étude de géographie humaine* (Lyons: Revue de géographie de Lyon; Trévoux: Imprimerie de Trévoux, J. Patissier, 1959), p. 429.

17. See Jaquet and Chapuis, *Technique and History,* pp. 164, 153.

18. See Alfred Chapuis and Eugène Jaquet, *The History of the Self-Winding Watch, 1770–1931* (Neuchâtel: Editions du Griffon, 1956), pp. 31ff.

19. On the history and work of Patek Philippe, see Horand Vogel, *Uhren von Patek Philippe* (Düsseldorf: H. Vogel, 1980); and Martin Huber and Alan Banbery, *Patek Philippe, Genève* (Zurich: Verlag Peter Ineichen, 1982).

20. Roy Church, "Nineteenth-Century Clock Technology," *Economic History Review*, 2nd ser., 28 (1975): 625, repeats in this connection the oft-heard explanation that Prest's winding was only applicable to watches with a horizontal escapement. Insofar as "horizontal escapement" is synonymous with the cylinder escapement, this is simply not true: Prest's winding would work with any escapement but a verge, and then only because the verge needed a fusee. See, for example, a lever watch with Prest winding by J. R. Arnold, Charles Frodsham, in 1851, shown in T. P. Camerer Cuss, *The Camerer Cuss Book of Antique Watches,* rev. and ed. T. A. Camerer Cuss (Woodbridge, Suffolk: Antique Collectors' Club, 1976), p. 230. The source of this piece of misinformation is probably the fact that the cylinder escapement was the only one that the British envisaged using in the early nineteenth century without fusee; and as noted above, keyless winding of fusee was difficult and costly.

21. See Cecil Clutton, "The Pocket Watch 1750 to 1850: A Review," *Antiquarian Horology* 11 (Winter 1978): 220–221.

22. On the Lancashire Watch Co., see Kenneth D. Roberts, ed., *The Lancashire Watch Company: Prescot, Lancashire, England 1889–1910* (Fitzwilliam, N.H.: Ken Roberts, 1973), with historical essay by Alan Smith, passim, esp. pp. 31 and 52. See also the catalogue of Smith & Sons Ltd., *Guide to the Purchase of a Watch,* 3rd ed. (London, n.d. [c. 1900]), p. 9.

23. Camerer Cuss, *Book of Antique Watches,* p. 290.

24. On the British and Swiss chronometric competitions, see Jaquet and Chapuis, *Technique and History,* pp. 187–198; and Charles Thomann, *Les dignitaires de l'horlogerie: la merveilleuse et tragique épopée . . . 1923–1967* (Neuchâtel: Editions du Griffon, 1981). On Paul Ditisheim, there are short biographies in Paul Chamberlain, *It's about Time* (New York: Richard Smith, 1941), pp. 463–464; and Tardy, *Dictionnaire des horlogers français* (Paris: Tardy, n.d.), s.v. Ditisheim.

25. Tony Mercer, *Mercer Chronometers: Radical Tom Mercer and the House He Founded* (Ashford, Kent: Brant Wright Associates, 1978), p. 17. A great read.

26. Ibid., p. 11. It is not clear whether these words are Tony Mercer's or a remark handed down in the family.

27. Camerer Cuss, *Book of Antique Watches,* p. 295.

28. Alan Smith in Roberts, ed., *Lancashire Watch Company,* p. 17.

29. Alun C. Davies, "The Rise and Decline of Chronometer Manufacturing in Great Britain," *Bulletin of the National Association of Watch and Clock Collectors, Inc.* 24 (December 1982): 630.

30. Church, "Nineteenth-Century Clock Technology," pp. 629–630.

31. On the changing relative importance of the two industries in the Swiss Jura, see Daveau, *Les régions frontalières,* pp. 419–420. She writes: "It is the direct competition of watchmaking that killed the other mountain industries."

32. Even in watchmaking, the British continued to innovate: thus the invention in 1892 of the *karrusel* arrangement by Bahne Bonniksen of Coventry (he had come to England from Schleswig at the age of twenty-four),

a relatively simple and reliable device for eliminating position errors; and the introduction of the self-winding wristwatch by John Harwood in the early 1920s. But the karrusel found limited application, partly because of cost, partly because ordinary watches, in spite of position errors, were accurate enough; while Harwood's wristwatch could be imitated, at lower prices, by Swiss makers. He himself bought his *ébauches* in Switzerland.

33. Bienne/Biel offers an interesting test of the cultural configurations of enterprise: it was a predominantly German-speaking city in the eighteenth century, with a French-speaking minority that grew with the watch manufacture. There were German-Swiss watchmakers in Bienne/Biel; they seem to have signed their watches as from Bienne.

34. See, in this regard, the passage from the diary of Samuel Jequier cited in François Jequier, *Une entreprise horlogère du Val de Travers: Fleurier Watch Co SA* (Neuchâtel: La Baconnière, 1972), pp. 49–50. I have already had occasion to quote this in my essay "Religion and Enterprise: The Case of the French Textile Industry," in *Enterprise and Entrepreneurs in Nineteenth- and Twentieth-Century France,* ed. Edward C. Carter II, Robert Forster, and Joseph N. Moody (Baltimore: Johns Hopkins University Press, 1976), pp. 67–68.

35. On the Girardet bookstore, see Léon Montandon, "Samuel Girardet, ses ancêtres, sa boutique, II. La boutique," *Musée neuchâtelois,* n.s., 36, no. 2 (March–April 1949): 47–58. See also the domestic scene in the cottage of a clockmaker of the eighteenth century shown in Alfred Chapuis, *Histoire de la pendulerie neuchâteloise* (Paris and Neuchâtel, n.d.), p. 157.

36. [Samuel Frédéric (or Samuel Friedrich) Osterwald (or Ostervald)], *Description des montagnes et des vallées qui font partie de la Principauté de Neuchâtel et Valangin,* 2nd ed. (Neuchâtel, 1766), pp. 95–97.

37. On the distinction between open and closed communes (villages) and their disparate potentials for industrial development, see Rudolf Braun, *Industrialisierung und Volksleben* (Erlenbach-Zurich and Stuttgart: E. Rentsch, 1960), ch. 1.

38. There were efforts to establish guild controls (*maîtrises*) in the Vallée de Joux (Bern) in the 1750s, in large part at the behest of guild masters in the towns on the north shore of Lake Geneva (Nyon, Rolle). In particular, the aim was to prevent the employment of women. The regulations were abolished in 1776. How could such a *fabrique* survive in proximity to guild-free Neuchâtel? See Daveau, *Les régions frontalières,* p. 427.

19. Not One in Fifty Thousand

1. On Polhem's clocks and techniques, see William A. Johnson, ed. and trans., *Christopher Polhem: The Father of Swedish Technology* (Hartford: Trinity College, 1963), pp. 109–115. Polhem cut his wheels in small batches by a kind of automatic broaching machine; but that was all. Tolerances were presumably generous, as they are with clocks; and it is not clear whether the wheels

needed further work. In any event, he did not make any money on his clocks, and his machine was not copied. I have not been able to read Sten Lundwall, *Stjärnsundsuren Väggurtillverkningen vid ett 1700-tals bruk* (Stjernsund's clocks: the manufacture of wall clocks in an eighteenth-century factory); or Elis Sidenbladh, "Om uhr eller uhrvärk af Chr. Polhem," *Svensk Urmakeri-Tidning* (1910).

2. As of March 1791, Blanc was *contrôleur-principal* of three artillery arsenals. His proposals to manufacture musket trigger assemblies received a most encouraging hearing from the committee of the Academy of Sciences appointed to consider his project, but nothing came of them, for reasons that remain to be studied. My own sense is that the task was far more difficult than he imagined. See the report of David Sigismond, bailli de la Tour du Pin, *Rapport fait à l'Académie Royale des Sciences, le samedi 19 mars 1791, d'un mémoire important de M. Blanc sur la fabrication des armes de guerre* (Paris, 1791). The best source on Blanc is still the long-forgotten article of W. F. Durfee, "The First Systematic Attempt at Interchangeability in Firearms," *Cassier's Magazine* 5 (April 1904): 469–477. On the link to Jefferson and Whitney, see Lionel T. C. Rolt, *A Short History of Machine Tools* (Cambridge, Mass.: MIT Press, 1965), pp. 139–141; and Jeanette Mirsky and Allan Nevins, *The World of Eli Whitney* (New York: Collier Books, 1962), ch. 5: "Jefferson and a Practical Demonstration."

3. On Maudslay, see Rolt, *A Short History of Machine Tools*, pp. 84–85; also H. W. Dickinson, "Joseph Bramah and His Inventions," *Newcomen Society Transactions* 22 (1941–42).

4. Observers of the time and historians since have been inclined to explain the popularity of this technique in the United States by the high cost of labor, particularly of skilled labor, relative to that of other factors of production. This explanation has since been challenged and has given rise to a copious literature of disagreement among economic historians. See especially W. Paul Strassmann, *Risk and Technological Innovation: American Manufacturing Methods during the Nineteenth Century* (Ithaca: Cornell University Press, 1959); H. J. Habakkuk, *American and British Technology in the Nineteenth Century* (Cambridge: Cambridge University Press, 1962); Stanley Lebergott, *Manpower in Economic Growth* (New York: McGraw-Hill, 1964); Nathan Rosenberg, ed., *The American System of Manufactures* (Edinburgh: Edinburgh University Press, 1969); idem, *Perspectives on Technology* (Cambridge: Cambridge University Press, 1976).

5. Compare the cautious judgment of Merritt Roe Smith, "Military Entrepreneurship," in Otto Mayr and Robert C. Post, eds., *Yankee Enterprise: The Rise of the American System of Manufactures* (Washington, D.C.: Smithsonian Institution, 1981), p. 68: "Although [Whitney's] efforts fell short of success, he nonetheless became a zealous advocate of the uniformity principle, popularized the concept, and persuaded many politicians to support policies aimed at standardizing the manufacture of military arms." Also on this point, see Merritt Roe Smith, "Eli Whitney and the American System of Manufacturing," in Carroll W. Pursell, ed., *Technology in America: A History of Individuals and Ideas* (Cambridge, Mass.: MIT Press, 1981), pp. 45–61.

6. Cf. Smith, "Military Entrepreneurship," p. 46: "In their minds [the U.S. Army Ordnance Department] the uniformity system meant something more than a means of production; it represented a barometer of national development, an achievement in which Americans could take pride and from which republican institutions could draw strength and vitality. In this sense, uniformity not only stood for material progress through technology but also for the maintenance of virtue in an era of developing industrialism."

7. It is interesting to observe that this (or something like this) was what a number of French projectors proposed in the 1780s and 1790s by way of breaking the Genevan monopoly. See the fascinating dossiers in France, *Archives Nationales*, F^{12} 1325A and 1325B. Nothing, however, came of any of these schemes.

One of these, by the way, came from no less an *artiste* than A.-L. Breguet, who even claimed to have already designed all the machines required. Yet he never built them, and such plans as he may have drawn have never been published. What French horology lost on the industrial side, it gained on the side of artistry and ingenuity.

8. On Maudslay's role, see Rolt, *A Short History of Machine Tools,* p. 90. On Terry and his contributions, see Carl W. Drepperd, *American Clocks and Clockmakers* (Garden City: Doubleday, 1947), ch. 6; Chris H. Bailey, *Two Hundred Years of American Clocks and Watches* (Englewood Cliffs: Prentice-Hall, 1975), pp. 103–120.

9. These paragraphs are drawn primarily from Chauncey Jerome, *History of the American Clock Business for the Past Sixty Years and Life of Chauncey Jerome, Written by Himself* (New Haven, 1860), pp. 62–64. This is something of a self-serving document.

10. Ibid., p. 64.

11. F. L. Neher, *Ein Jahrhundert Junghans: Ein Beitrag zur Technik- und Kulturgeschichte der Uhr* (Schramberg, n.d. [1961]), pp. 58–59.

12. This was the venture of the Pitkin brothers, Henry and James Flagg, of East Hartford, Connecticut, later of New York City. See Robert M. Wingate, "The Pitkin Brothers Revisited," *Bulletin of the National Association of Watch and Clock Collectors, Inc.* 24 (August 1982): 381–391, esp. 384–385.

13. Henry G. Abbott, *The Watch Factories of America Past and Present* (Chicago, 1888), p. 10. The only thing wrong with Dennison's logic in all this was his assumption that the watches he examined were made by the man who signed them. As we have seen, this was by no means certain.

14. Charles W. Moore, *Timing a Century: History of the Waltham Watch Company* (Cambridge, Mass.: Harvard University Press, 1945), p. 12.

15. Cited in Abbott, *Watch Factories,* pp. 12–13.

16. Tony Mercer, *Mercer Chronometers: Radical Tom Mercer and the House He Founded* (Ashford, Kent: Brant Wright Associates, 1978), pp. 32–33.

17. The best survey of the subject is in Charles H. Fitch, "Manufactures of Interchangeable Mechanisms, V: The Manufacture of Watches," in U.S. Department of the Interior, Census Office, *Report on the Manufactures of the United States at the Tenth Census (June 1, 1880)* (Washington, D.C.: Government

Printing Office, 1883), pp. 60–67. E. A. Marsh, *The Evolution of Automatic Machinery as Applied to the Manufacture of Watches at Waltham, Mass.*, by The American *Waltham Watch Company* (Chicago: 1896; rpt. Exeter, N.H.: Adams Brown, 1968), is a classic; but Marsh was clearly better with the lathe than with the pen. There is also much excellent material in a manuscript by Jean G. Laviolette, "L'industrie horlogère," which he kindly allowed me to see in advance of publication.

18. Moore, *Timing a Century*, p. 50.

19. This has to be a guess, because dollar watches were not normally numbered. Bailey, *Two Hundred Years,* p. 210, estimates total production of cheap watches as over 250 million by 1970. But this seems too low to me, as output of clock-type watches numbered over 105 million in the twelve years alone from 1931 to 1942—this, during a period of deep depression. United States Tariff Commission, *Watches,* "War Changes in Industry Series," report no. 20 (Washington, D.C.: Government Printing Office, 1947), p. 87.

20. See Jean G. Laviolette, "De la vente en horlogerie," *Horlogerie ancienne* 12 (1982): 55–74. The article contains some fascinating illustrations.

21. For example, Roskell, the leading maker in Liverpool, produced about a thousand watches a year from 1800 to 1900. Litherlands, another top Liverpool firm, made about 40,000 in seventy years. Letter of D. M. W. Evans in *Antiquarian Horology* 9 (March 1976): 705. Note that the so-called manufacturers of cheaper watches for the "country trade" did more—two to three thousand pieces a year around 1810. But demand also tended to be more variable in that sector of the industry and reputations ephemeral; so that it is by no means certain that such large annual volumes translated over the lifetime of the enterprise into greater aggregate sales than for "name" makers.

22. E. Muston, "L'horlogerie dans les montagnes du Jura: essai de statistique industrielle," *Compte-rendu de la situation et des travaux* [also listed as *Mémoires*] *de la Société d'Emulation de Montbéliard,* May 5, 1859, p. 92.

23. For labor productivity figures, see William A. Countryman, "Watches and Watch Cases," in United States, *Twelfth Census of the United States, Taken in the Year 1900,* Census Reports, vol. 10, pt. 4: Manufactures, p. 493. For the serial numbers of American watches, see especially George E. Townsend, *Almost Everything You Wanted to Know about American Watches and Didn't Know Who to Ask* (privately published, 1971); and idem, *American Railroad Watches* (privately published, 1977).

24. On Ingersoll, see Willis I. Milham, *Time & Timekeepers* (New York: Macmillan, 1923; rpt. 1942), p. 408.

25. On Ingraham, see Bailey, *Two Hundred Years,* pp. 217–218.

26. Edouard Favre-Perret, *Rapport présenté au Haut Conseil Fédéral sur l'industrie de l'horlogerie: Exposition de Philadelphie, 1876, Section Suisse, Groupe XXV* (Winterthur, 1877).

27. In a manuscript essay "Time Is Money" by Henry Fried, who was kind enough to allow me to read and copy his text.

28. The average price of watches exported from Switzerland to Great Britain 1867–1870 was £1.94 (equaled 48 francs). Most of these were uncased.

Roy A. Church, "Nineteenth-Century Clock Technology in Britain, the United States, and Switzerland," *Economic History Review*, 2nd ser., 28 (November 1975): 625, n. 3. Also, most of them had cylinder escapements, which cost substantially less than jeweled levers.

29. I have used the translation in the Fried manuscript. There is a somewhat less complete English version in Moore, *Timing a Century*, p. 201. The French original is to be found in *La Nationale suisse* (La Chaux-de-Fonds), November 17, 1876.

20. Who Killed Cock Robin?

1. In the meantime, Americans were able to turn out machine-made watches that were good enough for all practical purposes—good enough, for example, to keep time within thirty seconds a week, as railway companies required. For the American railroad watch and its standards, see Lawrence W. Treiman, "Railroad Watches and Time Service," *Bulletin of the National Association of Watch and Clock Collectors, Inc.* 15 (October 1972): 151–175.

2. *Rapport présenté au Conseil d'Etat par les experts horlogers délégués à l'Exposition Universelle de 1878* (Neuchâtel, 1879), p. 17.

3. Letter to me of May 16, 1981.

4. André Francillon, *Histoire de la fabrique de Longines* (Saint-Imier: privately published, 1947), p. 59.

5. *The Watchmaker and Jeweler*, trade paper of the industry, was already warning its readers of this invasion of fraudulent imitations in July 1870, when American machine manufacture had barely gotten under way. Cited in Henry Fried, "Time Is Money" (unpublished manuscript).

6. F. Scheurer, *Les crises de l'industrie horlogère dans le canton de Neuchâtel* (Neuveville: Editions Beerstecher, 1914), table g, between pp. 148 and 149.

7. The Swiss figures are to be found ibid., pp. 136–137. The U.S. figures are from the decennial censuses, as given in William A. Countryman, "Watches and Watch Cases," in *Twelfth Census of the United States, Taken in the Year 1900*, Census Reports, vol. 10, pt. 4, Manufactures, pp. 485, 489. I have converted dollars to francs at 1 dollar = 5 francs. Note that the Swiss figures include clocks as well as watches, but clocks represented a small fraction of total horological exports, which may be compensated by the omission from the figures of Swiss watches sold in Switzerland.

8. Switzerland, Département Fédéral de l'Economie Publique, *La Suisse économique et sociale*, vol. 1 (Einsiedeln: Benziger, 1927), p. 229*. These figures include finished movements.

9. Exports of these articles fell from 130,000 quintals worth 167 million francs in 1917 to 1,400 quintals worth 1.6 million francs in 1919. Ibid., p. 230*, n. 1.

10. In French, ASUAG was known as the Société Générale de l'Horlogerie Suisse S.A. This was the first time that a major organization of the Swiss watch industry took its principal designation in German, though this

gets lost in the neutral acronym; but this is customary for Federal enactments.

11. See Harvard Business School, "Note on the Watch Industries in Switzerland, Japan and the United States," prepared by Frederick Knicker-bocker (Boston: Intercollegiate Case Clearing House, 1972; rev. September 1976), pp. 7–8; Max Huber and Frédéric Baumann, "Uhrenindustrie," *Handbuch der schweizerischen Volkswirtschaft* 2 (1955): 459.

12. Circular letter of September 12, 1917, cited in Charles W. Moore, *Timing a Century: History of the Waltham Watch Company* (Cambridge, Mass.: Harvard University Press, 1945), p. 12.

13. Ibid., pp. 113, 118.

14. From 1911 to 1921, employment at Amoskeag ranged between 14,000 and 16,000 and output from a high of 236 million yards in 1912 to 143 million in 1919. Alan Sweezy, "The Amoskeag Manufacturing Company," *Quarterly Journal of Economics* 52 (May 1938): 474. (I owe this reference to William La-zonick.) This is probably still the best analysis of the decline and fall of Amoskeag, though it is curiously weak on the workers' side of the story. On the latter, see Tamara Hareven and Randolph Langenbach, *Amoskeag: Life and Work in an American Factory-City* (New York: Pantheon, 1978), passim.

15. Sweezy, "Amoskeag," p. 483. In addition to Sweezy and Moore, see Edmund L. Sanderson, *Waltham Industries: A Collection of Sketches of Early Firms and Founders* (Waltham: Waltham Historical Society, 1957), p. 98.

16. Hareven and Langenbach, *Amoskeag,* p. 301.

17. Sweezy, "Amoskeag," p. 487.

18. Ibid., pp. 503–504.

19. Hareven and Langenbach, *Amoskeag,* p. 304.

20. Vincent P. Carosso, "The Waltham Watch Co.: A Case History," *Bulletin of the Business History Society* 3 (December 1949): 176.

21. Cited in Moore, *Timing a Century,* p. 282. Lawrence M. Hughes, in a very perceptive article entitled (after the well-known children's verse) "Who Killed Waltham?" *Sales Management* ("The Magazine of Marketing"), April 15, 1950, writes: "In the long list of those who have been in on the long killing of Waltham, Dumaine stands high." And he goes on to cite *Fortune* of April 1949: "He cut wages, abolished jobs, cancelled advertising contracts; spent little on new machinery, research and new watch designs . . . He skinned through the depression on shutdowns . . . and price cutting."

Dumaine went on from Waltham to become chairman of the board of the New Haven Railroad, to which he also applied his purgative therapy, with similar results.

22. Moore, *Timing a Century,* p. 236.

23. Report by J. Freistadter, cited ibid., p. 236.

24. Letter from D. W. Leverenz, May 6, 1981.

25. D. W. Leverenz, "About Elgin" (undated typescript). I am grateful to Mr. Leverenz for his help with the story of a company that has not yet found its historian.

26. By the late 1930s imported movements constituted over 60 percent of American consumption of jeweled watches—and this, without counting a

substantial number unrecorded because smuggled. United States Tariff Commission, *Watches,* War Changes in Industry Series, report no. 20 (Washington, D.C.: Goverment Printing Office, 1947), p. 16. Only Bulova among assemblers chose to attempt watch manufacture in the United States. To do this, it sent a representative to Switzerland to copy the best available movements and smuggle out the necessary machines. These were carried piece by piece over the mountains, in a manner reminiscent of earlier watch contraband—an ironic and, some would say, just trick of nemesis.

27. On logistic curves of growth and their link to the exhaustion of the possibilities of innovation in a given technology, see the classic article of Simon Kuznets, "Retardation of Economic Growth," *Journal of Economic and Business History 1* (1929): 534–560.

28. Moore, *Timing a Century,* p. 211. On contraband and the larger problem of American-Swiss commercial relations, see Jean-Jacques Bolli, *L'aspect horloger des relations commerciales américano-suisses de 1929 à 1950* (La Chaux-de-Fonds: Editions de la Suisse Horlogère, 1956).

21. The Quartz Revolution

1. *Boston Globe,* May 1, 1981, p. 1. I owe this reference to Elly Solmitz.

2. "A Time Bomb for Watchmakers," *Business Week,* November 16, 1963, cited in Harvard Business School case study "Time Corporation" (1972), p. 6.

3. Wilhelm Hill, "Die Wettbewerbsstellung der schweizerischen Uhrenindustrie" (typescript; Betriebswirtschaftliches Institut an der Universität Basel, March 1977), pp. 5, 92–93; "Anhang," p. A III/3. I owe this manuscript to the courtesy of Frank Ruhl, participant in a seminar on the history of the Swiss watch industry that I offered in 1978 in the University of Zurich and the Eidgenössische Technische Hochschule of that city.

4. "How Timex Hit German Markets with Top Quality, Durability—and the Hard Sell," *Business Abroad,* February 5, 1968, cited in Harvard Business School, "Time Corporation," p. 9.

5. Frederick J. Britten, *Britten's Watch & Clock Maker's Handbook, Dictionary and Guide,* 16th ed., rev. and ed. Richard Good (London: Eyre Methuen, 1978), p. 262. This so-called 16th edition is really a new work of exceptional value and typographic and iconographic excellence. It is especially strong on the new timekeeping technologies and instruments. For the earlier history of the new technique, see Paul Ditisheim, *Précision des garde-temps piézo-électriques et des pendules astronomiques* . . . (Besançon, 1937).

6. On this shift of the inspection function from manufacturer to consumer, see Eric Bruton, *The History of Clocks and Watches* (London: Orbis; New York: Crescent, 1979), p. 201.

7. *New York Times,* September 18, 1976, sect. 3, pp. 1–2. The watches were from a shipment of April of that year.

8. Britten, *Britten's . . . Handbook,* p. 107. The engineering of the Bulova Accutron is a tour de force. The ratchet wheel that converts the vibrations of

the tuning fork into rotary motion is only .09 inch in diameter. Around its edge are three hundred teeth only .001 inch apart. The wheel rotates in just under a second and makes 38 million rotations a year. Over this period, if it keeps within the guaranteed variance, it deviates from exact time by no more than one part in fifty thousand. Ibid., p. 109.

9. The founders of the CEH included the Fédération Horlogère (FH), the Chamber Suisse d'Horlogerie (CSH), and ASUAG. The director, Gerard Bauer, was a man of considerable experience not only in the watch industry but in the economics and diplomacy of international trade. On the role of the CEH in horological research and development, see the chapter by André Gabus, "Introduction de l'électronique dans la montre," in Otto Hieronymi et al., "Le rôle de nouvelles technologies de production," typescript "Rapport final dans le cadre du Programme National N° 9 sur les mécanismes et évolution de l'économie suisse et leurs incidences sociales, pour le Fonds National Suisse de la Recherche Scientifique, Berne" (Geneva: Institut Battelle, July 1982). Gabus was kind enough to make a copy of this preprint available to me in response to a last-minute phone call on the day of his departure for ten months' work in Côte d'Ivoire—most timely and fortunate. I want to thank him here, as well as Messrs. Bauer and René Retornaz, director of the Fédération Horlogère, who led me to him and allowed me to use their names in making my request.

10. Charles Thomann, *Les dignitaires de l'horlogerie* (Neuchâtel: Editions du Griffon, 1981), pp. 102–114.

11. This summary discussion of the new technology is based primarily on Gabus, "Introduction de l'électronique," pp. 27, 36, 57–59.

12. On the Statek Corporation, see ibid., pp. 65–67.

13. Fédération Horlogère Suisse (FH), "Essai sur le groupe Seiko," confidential report, mimeographed (Bienne, December 1980), p. 16. The percentage given in the text is incorrect—a typographical error. I want to thank René Retornaz for making this report available to me.

14. H. Kurita et al., "Multifunctional Digital Quartz Watches," in Tenth International Congress of Chronometry (ed. Société Suisse de Chronométrie), *Proceedings* (Büren, 1979), vol. 2, pp. 301–307; K. Kawasumi and K. Ushikoshi, "Analogue Quartz Watches Consuming Low Electric Power," ibid., pp. 313–320.

15. Ibid., p. 318.

16. The watch was advertised in *The Wall Street Journal.* I am obliged to Mel Stevens of Embassy Marketing for further details of manufacture.

17. Hill, "Die Wettbewerbsstellung der schweizerischen Uhrenindustrie," pp. 80, 82.

18. Britten, *Britten's . . . Handbook,* p. 144.

19. The production data are taken from the annual reports of the Chambre Suisse d'Horlogerie and the Fédération Horlogère in Bienne. I am indebted for them to François Jequier of Lausanne. On the changing sense and content of "Swiss Made," see Hill, "Die Wettbewerbsstellung," pp. 77–78.

20. From tables provided by the Fédération Horlogère, Bienne. *Chablonnage* has been particularly important in the low price ranges, where Switzerland has been a major supplier of the competitive industries of East Asia. See Hill, "Die Wettbewerbsstellung," p. 77 and n. 6; and Hilmar Stetter, *L'exportation du Swiss Made: transfert de l'appareil de production industrielle dans le Tiers Monde* (Geneva: Editions Pierre-Marcel Favre, 1981), pp. 40–41.

21. *New York Times,* December 15, 1980, p. D-1.

22. In the meantime Piaget followed suit and in July 1981 brought out its own diamond-studded bracelet watch for the modest price of $3 million. (To be sure, the dollar had gained considerably against the franc in those months, while the price of diamonds had fallen sharply. If Vacheron had not succeeded in selling its watch in 1980, it would have been in trouble.) Ibid., August 1, 1981, p. 22.

23. Simone Oppliger, *Quand nous étions horlogers; images-portrait de mon pays* (Lausanne: Payot, 1980), pp. 72, 85, 4.

24. Ibid., p. 5 and passim.

25. Ibid., pp. 24–25.

26. Ibid., p. 100.

27. Ibid., p. 9.

28. Ibid., p. 70.

29. Ibid., p. 102. Even the giants are not exempt from market and bank pressures. It has now been announced that the two largest combines in the industry, ASUAG and SSIH (the Société Suisse de l'Industrie Horlogère) will merge and receive an injection of fresh credit. "This isn't a marriage of love," said one representative of the industry, "but a marriage of reason." *Wall Street Journal,* May 16, 1983, p. 35.

30. See Stetter, *L'exportation du Swiss Made,* pp. 37–51.

Credits for Illustrations

Plates

I. Museum für Kunst und Kulturgeschichte (St. Annen Museum), Lübeck. Photo courtesy of Klaus Maurice.

II. Bayerisches Nationalmuseum, Munich. Photo courtesy of Klaus Maurice.

III. Kunsthistorisches Museum, Vienna, Collection Schloss Ambras. Photo courtesy of Klaus Maurice.

IV. Private collection; on loan to the Bayerisches Nationalmuseum, Munich.

V. Courtesy of the Musée International d'Horlogerie, La Chaux-de-Fonds.

VI. Courtesy of the Musée International d'Horlogerie, La Chaux-de-Fonds.

VII. Courtesy of The Time Museum, Rockford, Illinois.

VIII. Courtesy of the Musée International d'Horlogerie, La Chaux-de-Fonds.

Figures

1. Courtesy of Omega Electronic Equipment (a division of SSIH Equipment SA) and the International Amateur Athletic Federation.

2. Abraham Rees, ed., *The Cyclopædia* (London, 1819–1820), pl. 1, figs. 1 and 3.

3. Courtesy of John Combridge.

4. Ernest L. Edwardes, *Weight-driven Chamber Clocks of the Middle Ages and Renaissance* (Altrincham: John Sherratt & Son, 1965), pl. 3. Reproduced with the permission of John Sherratt & Son, Altrincham, and Aldo Garzanti Editore, Milan.

5. Ernest L. Edwardes, *The Story of the Pendulum Clock* (Altrincham: John Sherratt & Son, 1977), pl. 1. Reproduced with the permission of John Sherratt & Son, Altrincham.

467

6. Joseph Needham, Wang Ling, and Derek J. de Solla Price, *Heavenly Clockwork: The Great Astronomical Clocks of Medieval China* (Cambridge: Cambridge University Press, 1960), frontispiece. Reproduced with the permission of Cambridge University Press.

7. Left: Courtesy of the Smithsonian Institution. Right: Courtesy of the Bodleian Library, Oxford; MS. Laud. Misc. 620, folio 10 verso.

8. Courtesy of Jean-Claude Sabrier.

9. Top and middle rows: Denis Diderot and Jean Le Rond d'Alembert, *Encyclopédie* (1765), vol. 8, pl. 10. Bottom row: Abraham Rees, ed., *The Cyclopædia* (London, 1819–1820), p. 218, pl. 36.

10. Courtesy of the Wuppertaler Uhrenmuseum, Wuppertal, West Germany.

11. Courtesy of the Museum of the Worshipful Company of Clockmakers, London.

12. T. P. Camerer Cuss, *The Camerer Cuss Book of Antique Watches* (Woodbridge, Suffolk: Antique Collectors' Club, 1976), p. 64. Reproduced with the permission of T. P. Camerer Cuss.

13. Photograph by Hans von Bertele; reproduced with the permission of the Staatlicher Mathematisch-Physikalischer Salon, Dresden. Drawing by David Penney; copyright 1983 by David Penney, London.

14. Courtesy of the Biblioteca Nazionale Centrale, Florence.

15. Courtesy of the Trustees of the British Museum.

16. T. P. Camerer Cuss, *The Camerer Cuss Book of Antique Watches* (Woodbridge, Suffolk: Antique Collectors' Club, 1976), p. 94. Reproduced with the permission of T. P. Camerer Cuss.

17. T. P. Camerer Cuss, *The Camerer Cuss Book of Antique Watches* (Woodbridge, Suffolk: Antique Collectors' Club, 1976), p. 99, pl. 47. Reproduced with the permission of T. P. Camerer Cuss.

18. Courtesy of the National Maritime Museum, London.

19. Ferdinand Berthoud, *Histoire de la mesure du temps par les horloges*, vol. 1 (Paris, 1802), pl. 12.

20. Courtesy of The Time Museum, Rockford, Illinois.

21. Courtesy of The Time Museum, Rockford, Illinois.

22. Reproduced with the permission of the Comptroller of Her Majesty's Stationery Office.

23. Private collection.

24. Private collection.

25. Collection of the Musée d'Art et d'Histoire, Geneva, Inv. AD282.

26. Courtesy of Jean-Claude Sabrier.

27. Collection of the Musée d'Art et d'Histoire, Geneva, Inv. MG11.

28. Collection of the Musée d'Art et d'Histoire, Geneva, Inv. E305.

29. Courtesy of Jean-Claude Sabrier.

30. Courtesy of Simone Oppliger.

31. Courtesy of Simone Oppliger.

32. Map by William Minty.

A.1–A.8. Drawings by David Penney; copyright 1983 by David Penney.

Index